Combat Robot
Tutorial

version 2.2 - August 2009

written by

Marco Antonio Meggiolaro

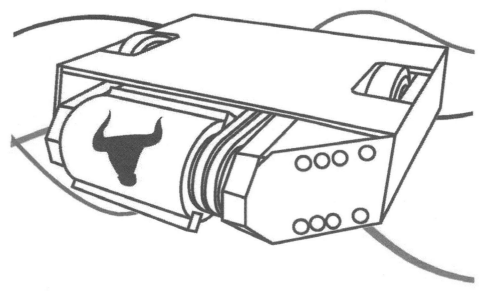

       WWW.RIOBOTZ.COM.BR

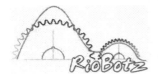

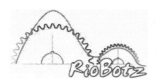

RioBotz Combat Robot Tutorial
version 2.2 – August 2009

written by
Marco Antonio Meggiolaro

Head of the RioBotz team from PUC-Rio University – www.riobotz.com.br

collaborators: *Bruno Favoreto Fernandes Soares*
Eduardo Carvalhal Lage von Buettner Ristow
Felipe Maimon

with 895 figures

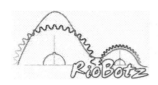 

RioBotz Combat Robot Tutorial
version 2.2 – August 2009

CONTENTS

1. INTRODUCTION	**11**
1.1. A Brief History of Robot Combat	12
1.2. Structure of the Tutorial	13
1.3. Acknowledgments	14
2. DESIGN FUNDAMENTALS	**15**
2.1. Weight Classes	15
2.2. Scale Factor	16
2.3. Combat Robot Types	20
2.3.1. Rammers	22
2.3.2. Wedges	22
2.3.3. Lifters	23
2.3.4. Launchers / Flippers	23
2.3.5. Thwackbots	24
2.3.6. Overhead Thwackbots	24
2.3.7. Spearbots	25
2.3.8. Horizontal Spinners	25
2.3.9. Sawbots	26
2.3.10. Vertical Spinners	26
2.3.11. Drumbots	27
2.3.12. Hammerbots	27
2.3.13. Clampers	28
2.3.14. Crushers	28
2.3.15. Flamethrowers	29
2.3.16. Multibots	29
2.4. Design Steps	30
2.4.1. Cost	30
2.4.2. Sponsorship	30

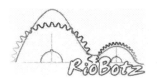

2.4.3. Designing the Robot 32

2.4.4. Calculations 34

2.4.5. Optimization 34

2.4.6. Building and Testing 38

2.5. Robot Structure 38

2.6. Robot Armor 40

2.6.1. Traditional Armor 40

2.6.2. Ablative Armor 40

2.6.3. Reactive Armor 40

2.7. Robot Drive System 41

2.7.1. Tank Treads and Legs 41

2.7.2. Wheel Types 41

2.7.3. Wheel Steering 42

2.7.4. Two-Wheel Drive 43

2.7.5. All-Wheel Drive 44

2.7.6. Omni-Directional Drive 45

2.7.7. Wheel Placement 45

2.7.8. Invertible Design 47

2.8. Robot Weapon System 48

2.9. Building Tools 48

3. MATERIALS **54**

3.1. Mechanical Properties 54

3.2. Steels and Cast Irons 57

3.3. Aluminum Alloys 62

3.4. Titanium Alloys 64

3.5. Magnesium Alloys 67

3.6. Other Metals 68

3.7. Non-Metals 70

3.8. Material Selection Principles 75

3.8.1. Stiffness Optimization 75

3.8.2. Strength and Toughness Optimization 78

3.9. Minimum Weight Design 80

3.9.1. Minimum Weight Plates 81

3.9.2. Minimum Weight Internal Mounts 83

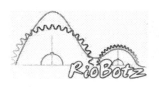 

3.9.3. Minimum Weight Protected Structural Walls   85

3.9.4. Minimum Weight Integrated Structure-Armor Walls   88

3.9.5. Minimum Weight Wedges   89

3.9.6. Minimum Weight Traditional Armor   89

3.9.7. Minimum Weight Ablative Armor   91

3.9.8. Minimum Weight Beams   93

3.9.9. Minimum Weight Shafts and Gears   95

3.9.10. Minimum Weight Spinning Bars and Eggbeaters   96

3.9.11. Minimum Weight Spinning Disks, Shells and Drums   99

3.9.12. Minimum Weight Weapon Inserts   102

3.9.13. Minimum Weight Clamper and Crusher Claws   104

3.9.14. Minimum Weight Trusses   105

3.10. Minimum Volume Design   107

3.10.1. Compact-Sized Internal Mounts   108

3.10.2. Compact-Sized Drums   109

3.10.3. Compact-Sized Shafts, Gears and Weapon Parts   112

3.11. Conclusions on Materials Selection   113

4. JOINING ELEMENTS   117

4.1. Screws   117

4.2. Shaft Mounting   121

4.3. Rivets   123

4.4. Hinges   123

4.5. Welds   124

5. MOTORS AND TRANSMISSIONS   126

5.1. Brushed DC Motors   126

5.1.1. Example: Magmotor S28-150   129

5.1.2. Typical Brushed DC Motors   130

5.1.3. Identifying Unknown Brushed DC Motors   134

5.2. Brushless DC Motors   135

5.3. Power Transmission   137

5.3.1. Gears   137

5.3.2. Belts   139

5.3.3. Chains   141

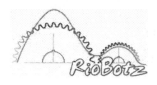 

5.3.4. Flexible Couplings	141
5.3.5. Torque Limiters	142
5.4. Weapon and Drive System Calculations	142
5.4.1. Example: Design of Touro's Drive System	142
5.4.2. Example: Design of Touro's Weapon System	145
5.4.3. Energy and Capacity Consumption of Spinning Weapons	147
5.5. Pneumatic Systems	149
5.6. Hydraulic Systems	152
5.7. Internal Combustion Engines	153
6. WEAPON DESIGN	**154**
6.1. Spinning Bar Design	154
6.2. Spinning Disk Design	156
6.3. Tooth Design	158
6.3.1. Tooth Height and Bite	158
6.3.2. Number of Teeth	160
6.4. Impact Theory	161
6.4.1. Impact Equations	161
6.4.2. Limit Cases	163
6.4.3. Impact Energy	164
6.4.4. Example: Last Rites vs. Sir Loin	165
6.5. Effective Mass	168
6.5.1. Effective Mass of Horizontal Spinners	168
6.5.2. Effective Mass of Vertical Spinners and Drumbots	169
6.5.3. Example: Drumbot Impact	170
6.5.4. Effective Mass of Hammerbots	171
6.5.5. Full Body, Shell and Ring Drumbots	172
6.5.6. Effective Mass Summary	172
6.6. Effective Spring and Damper	174
6.6.1. A Simple Spring-Damper Model	174
6.6.2. Spring and Damper Energy	175
6.6.3. Offensive Strategies	176
6.6.4. Defensive Strategies	177
6.6.5. Case Study: Vertical Spinner Stiffness and Damping	178
6.6.6. Equivalent Electric Circuit	179

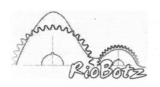

6.7. Hammerbot Design	180
6.7.1. Hammer Energy	181
6.7.2. Hammer Impact	182
6.8. Overhead Thwackbot Design	183
6.9. Thwackbot Design	185
6.9.1. Thwackbot Equations	185
6.9.2. Melty Brain Control	187
6.9.3. NavBot Control	188
6.10. Launcher Design	190
6.10.1. Three-Bar Mechanisms	191
6.10.2. Launcher Equations	193
6.10.3. Height Launcher Equations	195
6.10.4. Range Launcher Equations	196
6.10.5. Four-Bar Mechanisms	200
6.10.6. Launcher Stability	200
6.11. Lifter Design	202
6.12. Clamper Design	203
6.13. Rammer Design	204
6.14. Wedge Design	205
6.14.1. Wedge Types and Shapes	206
6.14.2. Wedge Impact	208
6.14.3. Defensive Wedges	209
6.14.4. Offensive Wedges	211
6.14.5. Example: Offensive Wedge vs. Horizontal Spinner	211
6.14.6. Angled Impacts	212
6.14.7. Wedge Design Against Vertical Spinners	213
6.15. Gyroscopic Effect	215
6.16. Summary	219
7. ELECTRONICS	**220**
7.1. Radio Transmitter and Receiver	220
7.1.1. Transmitters	220
7.1.2. Receivers	222
7.1.3. Antennas	224
7.1.4. Gyroscopes	225

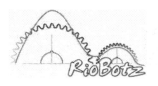

7.1.5. Battery Elimination Circuit 225

7.1.6. Servos 226

7.2. Controlling Brushed DC Motors 227

7.2.1. Bang-Bang Control 227

7.2.2. Pulse Width Modulation 228

7.2.3. H-Bridge 229

7.3. Electronic Speed Controllers 231

7.3.1. OSMC - Open Source Motor Controller 231

7.3.2. IFI Victor 232

7.3.3. Robot Power Scorpion 234

7.3.4. BaneBots 236

7.3.5. Other Brushed Motor Speed Controllers 237

7.3.6. Brushless Electronic Speed Controllers 238

7.4. Solenoids 241

7.4.1. White-Rodgers 586 SPDT 241

7.4.2. Team Whyachi TW-C1 242

7.5. Wiring 243

7.5.1. Wires 243

7.5.2. Terminals, Plugs and Connectors 244

7.6. Power Switches 245

7.7. Connection Schemes 248

7.7.1. Classic Connection Scheme 248

7.7.2. Improved Connection Scheme 250

7.7.3. Connection Scheme for Reversible Weapons 252

7.8. Developing your Own Electronics 253

7.8.1. Speed Controller Development 253

7.8.2. RC Interface Development 257

8. BATTERIES **263**

8.1. Battery Types 263

8.1.1. Sealed Lead Acid (SLA) 263

8.1.2. Nickel-Cadmium (NiCd) 264

8.1.3. Nickel-Metal Hydride (NiMH) 264

8.1.4. Alkaline 265

8.1.5. Lithium 265

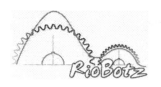

8.2. Battery Properties   268

   8.2.1. Price   268

   8.2.2. Weight   268

   8.2.3. Voltage   268

   8.2.4. Shelf Life   268

   8.2.5. Number of Recharge Cycles   268

   8.2.6. Charge Time   269

   8.2.7. Self-Discharge   269

   8.2.8. Discharge Curve   269

   8.2.9. Internal Resistance   270

   8.2.10. Capacity   270

   8.2.11. De-Rating Factor   271

   8.2.12. Discharge Rate   273

8.3. Battery Care and Tips   274

   8.3.1. Shock Mounting   274

   8.3.2. Recharging   274

   8.3.3. Battery Storage   276

   8.3.4. Assembling Your Own Pack   277

   8.3.5. Billy Moon's Rules for LiPo   278

9. COMBOT EVENTS   **279**

9.1. Before the Event   279

   9.1.1. Test and Drive Your Robot   279

   9.1.2. Prevent Common Failures   281

   9.1.3. Lose Weight   283

   9.1.4. Travel Preparations   286

9.2. During the Event   289

   9.2.1. Getting Started   289

   9.2.2. Waiting for Your Fight   291

   9.2.3. Before Your Fight   293

   9.2.4. During Your Fight   294

   9.2.5. Deciding Who Won   294

   9.2.6. After Your Fight   297

   9.2.7. Between Fights   300

9.3. After the Event   301

9.3.1. Battery Care	301
9.3.2. Inspect Your Robot	302
9.3.3. Wrap Up	303

10. RIOBOTZ BUILD REPORTS — 304

10.1. Lacrainha	304
10.2. Lacraia	305
10.3. Anubis	306
10.4. Ciclone	309
10.5. Titan	315
10.6. Touro	319
10.7. Mini-Touro	327
10.8. Tourinho	330
10.9. Puminha	336
10.10. Touro Light	339
10.11. Micro-Touro	342
10.12. Touro Jr.	343
10.13. Touro Feather	344
10.14. Pocket	348

CONCLUSIONS	349
FAQ - Frequently Asked Questions	350
Bibliography	355
Appendix A – Conversion among Brinell, Vickers and Rockwell A, B and C hardnesses	357
Appendix B – Material Data	358
Appendix C – Stress Concentration Factor Graphs	360
Appendix D – Radio Control Channels and Frequencies	367
Appendix E – Dave Calkins' Guide	368

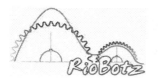 

Chapter

1

Introduction

The motivation to write this tutorial came from the great experience we've had during RoboGames 2006, in San Francisco. We were able to see how friendly competitors are, exchanging information, showing their robots in detail even for their next opponents. Several teams publish in their websites detailed build reports, with step by step information on how they've built their robots. There are also great books and tutorials showing how to build combots, however there was nothing written in Portuguese. This is why I started writing this tutorial, right after RoboGames 2006.

The tutorial was first released in August 2006, in Portuguese, as a free download both at the RioBotz website www.riobotz.com.br and at the website of the Brazilian combat robot league RoboCore, www.robocore.net. The idea was to stimulate the creation of new Brazilian combot teams, as well as to help the existing ones. It was very well received, with 1,500 downloads within the first week, 10,000 in the first 6 months, and more than 20,000 so far. A few people say that it might have helped with the increasing number of Brazilian teams that we see today.

A few builders asked me to generate an English version of this tutorial, so here it is. The tutorial was originally aimed for beginners, but its contents grew so much since the 2006 version that even veterans might find it useful. It basically includes everything that we've learned since January 2003, when RioBotz was created. We're still young compared to several great international teams, however we still hope we can contribute in some way with this text.

My biggest challenge was to try to include the maximum possible amount of information, from basic to advanced topics, in a compact way that would be easy to understand. We want to stimulate new teams to start building robots, showing that you don't need to be a rocket scientist to create a competitive combot. It is possible to do it even with little engineering background.

Feel free to distribute or print out this tutorial, I would just ask to keep it in its original form. I believe that this tutorial will help not only combat robot builders, but also anyone who wants to build robust and resistant mobile robots, to participate in any type of competition.

Excuse me if I make any mistakes in the following pages, some pieces of information include personal opinions, and therefore they can be biased. In spite of that, almost all the presented ideas have been tested in practice, in the arena, either by us or by other builders. I would love to receive your feedback in anything related with this tutorial, including comments, suggestions, corrections, anything that might improve future versions of the text, posted to the "RioBotz Combot Tutorial" topic on the RFL Forum. Thanks.

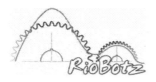

1.1. A Brief History of Robot Combat

Robot competitions have existed for a long time. They have been attracting competitors and spectators from all over the world. A very good review, along with great photos, can be found in the book "Gearheads – The Turbulent Rise of Robotic Sports" by Brad Stone [9].

I'll try to introduce the subject based on my personal experience. One of the first competitions involving robot confrontation was the Design 2.007 course (http://pergatory.mit.edu/2.007), a 2 night event that happens every year since 1970 at the Massachusetts Institute of Technology (MIT). The robots are built during one semester by undergraduate students taking the 2.007 course, Introduction to Design and Manufacturing, from the Department of Mechanical Engineering. The objective is to build a radio-controlled robot that fulfills certain tasks, such as collecting balls or transporting parts, in an arena with obstacles. Every year the task is modified to stimulate creativity.

I had the opportunity to witness as a graduate student the 1996 MIT 2.007 competition (pictured to the right). I was fascinated with the enthusiasm and mainly with the students' creativity. The best thing about these competitions is that the tasks were disputed with two robots facing each other at the same time in the arena. One wins by scoring more points collecting balls, transporting parts, it varies. At some point, you are allowed to block your  opponent. It was noticeable that this was the part that most drivers waited for and when the audience really cheered: blocking the opponent. Seeing robots confronting each other, pushing and blocking in an ingenious way the opponent was more exciting than just completing the tasks. I wish I knew back then that robot combat had already been created, 4 years earlier.

The success of Design 2.007 helped inspire the creation in 1992 of a robot competition among high school students, organized by FIRST (For Inspiration and Recognition of Science and Technology, www.usfirst.org), which is held annually. Unfortunately, it doesn't include combat robots.

In that same year, the US designer Marc Thorpe connected a vacuum cleaner to a remote control tank to help perform domestic tasks. The invention didn't work very well as a vacuum cleaner, but it caused damage, a fundamental requirement for a combat robot. At that time, he worked for Lucas Films and, inspired by the Star Wars movie, he created in 1994 the first official competition, Robot Wars. The first event was disputed in Fort Mason Center, San Francisco.

In 1997, Robot Wars was televised in the United Kingdom by BBC, starting the robot combat fever in that country. Legal disputes aside, it was such a success that Robot Wars moved to the UK. For more information on current UK combot events, check out the Fighting Robot Association (FRA) website at www.fightingrobots.co.uk.

Robotica and BotBash competitions were later created in the United States, filling the void left by Robot Wars.

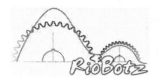

In 1999, Trey Roski and Greg Munson founded in San Francisco the BattleBots league (www.battlebots.com), creating the competition with most media exposure until today. The first event was held in Long Beach, California, in August 1999, with 70 enrolled robots. The second event was one of the most famous, held in Las Vegas in  November 1999, televised by pay-per-view. In 2000, BattleBots started to be televised by Comedy Central, quickly becoming popular, being transmitted during 5 seasons.

In 2001, the first Brazilian combat robot competition was held, based on BattleBots rules, in an arena built at the Unicamp University. In 2002, the second competition was held again at Unicamp, this time during the ENECA event (National Meeting of Control and Automation Students). Since then, Brazilian competitions have been held yearly during the ENECA, organized by the Brazilian league RoboCore (www.robocore.net), attracting an ever increasing public.

In 2002, the Robot Fighting League (www.botleague.com) was created in the US. It is the combat robot league with largest activity in the world, organizing from local events to the RFL Nationals, as well as RoboGames (www.robogames.net), which counts with several countries.

In December 2003, the RoboWars competition (www.robowars.org) had its debut in Australia.

In 2005, another Brazilian competition was created, the Winter Challenge, which is held annually in July (southern hemisphere, winter, July - you'll figure it out). The 2005 competition was held, for the first time ever, on an ice arena. At the end of 2006, the Brazilian league RoboCore became a proud member of the RFL.

1.2. Structure of the Tutorial

The tutorial is divided into 10 chapters. This chapter includes the introduction, robot combat history, and acknowledgments. Chapter 2 talks about the fundamentals of the design of several types of combots. Chapter 3 introduces the main materials used in those robots, and how to select them. Chapter 4 presents the main joining elements, such as screws and welds. Chapter 5 deals with motors used in the robot's drive and weapon systems, as well as power transmission elements, such as gears and belts. Chapter 6 deals with weapon design, and how to improve your robot's weapon system. Both chapters 5 and 6 include several equations, based on basic physics and dynamics calculations, however they are not essential to understand the text and its conclusions. Chapter 7 discusses the several electronic and electric components necessary to power the robot, while chapter 8 talks about batteries. Chapter 9 gives important tips on how to get ready to an event and how to behave before, during and after it. Chapter 10 shows build reports of all the combots from RioBotz, including the entire Touro family, exemplifying several concepts presented in the preceding chapters. I've also included, after the conclusions, a section of frequently asked questions (FAQ), a bibliography containing a few of the best books about combots, and a few appendices with useful information in a summarized form.

1.3. Acknowledgments

I would like to thank the entire RioBotz team, without whom the ideas here presented would not have left the drawing board, and for the careful revision of this tutorial. More specifically, I thank Eduardo "Dudu" Ristow for his effort as our team captain, for using the mill and lathe at the same time all night long without losing the smile; Bruno Favoreto for being able to master Solidworks even blindfolded; Felipe Maimon and Alexandre Ormiga for their effort in creating powerful and robust electronic systems; Daniel "Esguerda" Freitas and Rodrigo "Delay" Almeida for their great driving skill; Guilherme Porto for his excellent Spektrum programming lessons; Julio Guedes for his fidelity to the team since its creation; Ilana Nigri for helping us turn civilized our most frenzied pitstops; Marcio "Senador" Barros for our webpage; Gustavo "Emo" Parada for his grinding skill; and to Guilherme Franco, Thiago "Tico" Pimenta, Marcos "Pet" Marzano, Camila Borsotto, Carlos "Gotinha" Witte, Carlos "Minhoca" Nascimento, Daniel "Toioio" Lucas, Debora Almeida, Michel "Tocha" Feinstein, and Rodrigo "Cowboy" Nogueira, for all their help building combots. Thanks again to Eduardo Ristow, Felipe Maimon and Bruno Favoreto, for their contributions to this tutorial, especially in chapter 7. Thanks also to our past members, such as Felipe Scofano, Filipe "Saci" Sacchi, Claudio Duvivier, Rafael "Pardal" Moreira, Gustavo "Calouro" Lima, and several other students and alumni from the PUC-Rio University.

I cannot thank enough Profs. Mauro Schwanke, for everything he has taught us, and Mauro Speranza, for the fundamental administrative support, as well as the entire support of PUC-Rio.

Thanks to Mark Demers for his great contributions to the pneumatics section. To Mike Phillips, Matt & Wendy Maxham, Kevin Barker, Ray Billings, Hal Rucker and Carlo Bertocchini for their helpful pictures and information. Thanks to Robert "Trebor the Mad Overlord" Woodhead for the great action shots, and to the several other builders who have helped us along our way, either during a competition or through their websites or forum posts. To Dave Calkins and Simone for their warm welcome at every RoboGames, and for all their effort to promote robotic competitions. And, finally, to Paulo Lenz and Thacia Frank, whose dedication in organizing all Brazilian combot events inspired me to dedicate time to write this tutorial. Thanks to everyone!

Chapter

2

Design Fundamentals

The starting point of any combat robot design is the choice of the weight class, discussed next.

2.1. Weight Classes

The lightest combat robots ever built have about 28 grams (28.35g = 1oz), but they are so rare that there is no name yet for this weight class. Fleaweights (a.k.a. nanoweights or UK fairyweights), in general with a weight limit of 75g (or 50g depending on the event organizers), are also very rare. Fairyweights (up to 150g, known as antweights in the UK) are becoming popular, however there are still few events including them. Antweights (1lb) and beetleweights (3lb) are the most competitive among the "insect" classes (ants, beetles, fleas...). There are also autonomous ant and beetle classes.

The kilobots (1kg) events only exist in Canada, and the 15lb class is only for students between 12 and 18 years old, who participate in the competition BattleBots IQ. The Mantis (6lb) weight class has not really caught yet, there are very few robots in it. Featherweights (30lb) are becoming increasingly popular, especially in Brazil.

The 12lb and 30lb Sportsman's classes are special categories where all robots must have an active weapon, wedges of any form are forbidden, and spinners are severely restricted.

Possibly the most competitive classes are the hobbyweight (12lb), lightweight (60lb) and middleweight (120lb). Heavyweight (220lb) is the most famous class, in spite of having nowadays much fewer competitors than when BattleBots was televised.

Unfortunately, super-heavyweights (up to 340lb, or 320lb in UK events) are in decline, their apogee was also during the BattleBots era. The heaviest class is the Mechwars megaweight (390lb), exclusive to the Twin Cities Mechwars competition, however few robots exist.

There are still heavier robots, such as the MonsterBots, however events involving them are very rare due to logistic problems and high costs involved.

Back in 2006, when the first version of this tutorial was released, most Brazilian combat robots were middleweights (but not anymore, since the hobbyweight and featherweight classes started in Brazil). Because of that, several examples in this tutorial make reference to middleweights. However, the contents of this tutorial can be applied to any robot size, as it will be discussed in the next section, which deals with scale factor.

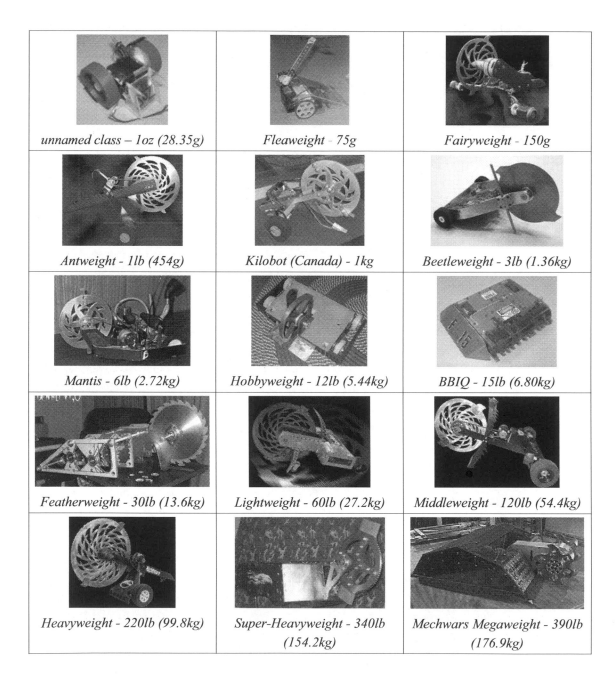

unnamed class – 1oz (28.35g)	*Fleaweight - 75g*	*Fairyweight - 150g*
Antweight - 1lb (454g)	*Kilobot (Canada) - 1kg*	*Beetleweight - 3lb (1.36kg)*
Mantis - 6lb (2.72kg)	*Hobbyweight - 12lb (5.44kg)*	*BBIQ - 15lb (6.80kg)*
Featherweight - 30lb (13.6kg)	*Lightweight - 60lb (27.2kg)*	*Middleweight - 120lb (54.4kg)*
Heavyweight - 220lb (99.8kg)	*Super-Heavyweight - 340lb (154.2kg)*	*Mechwars Megaweight - 390lb (176.9kg)*

2.2. Scale Factor

One important thing to keep in mind during the design phase of a combot is the scale factor. If you grew up in all your body dimensions, you would be twice as tall, and with eight times your weight (because your volume would be multiplied by $2^3 = 8$). However, the area of the cross section of your bones and muscles would only have been multiplied by $2^2 = 4$. Since the cross section area

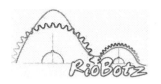 

(of a column of a building, for instance) dictates the resistance and load capacity, you would be 8 times heavier but only 4 times stronger. Conclusion: the larger the scale, the worse the force/weight ratio.

To compensate for that, your bones would have to be proportionally wider and shorter so that they wouldn't fracture or buckle. This is why rhinos and elephants have such wide and short legs. On the other hand, when reducing the scale, the inverse effect happens. An ant is about 100 times smaller than a human being, and because of that its weight is about 100^3 times smaller, however its force is only 100^2 smaller. As a result, ants can carry objects $100^3/100^2 = 100$ times heavier (relatively) than a human being would be able to. That estimate is confirmed in practice: a typical human can carry an object with half of his/her weight, while it was already proven that ants can lift loads 50 times their own weight, a factor of 100 more!

You should be wondering: what does this have to do with combots? Everything. If for instance you have designed a hobbyweight that is resistant and works well, you could take advantage of a lot of its design to build a middleweight, as long as you keep in mind the scale effect. To do that, you would need to multiply the weight by 10, which happens when you multiply all the robot dimensions by the cube root of 10, which results in a scale factor of 2.15.

The picture to the right shows a few drumbots, the middleweight *Touro*, the hobbyweight *Tourinho*, the beetleweight *Mini-Touro*, and the mock-up of a fleaweight *Pocket-Touro*. The scale factor between the 12lb *Tourinho* and the 120lb *Touro* is a little lower than 2 (which is close to the theoretical 2.15, but this suggests that *Tourinho* could still have been optimized to arrive in that 2.15 value, since both robots have similar shapes and weapons). This rule works very well in all scales, as long as the robots are similar: *Touro* is 40 times heavier than *Mini-Touro*, and the scale factor measured among them is about 3.25, very close to 3.42, the cube root of 40!

The question is: following the reasoning of the ant and the human, is it true that a middleweight such as *Touro* is, relatively, about $2.15^3/2.15^2 = 2.15$ times less strong, agile, powerful and resistant than the hobbyweight *Tourinho*? Yes and no. *Touro* will probably be relatively less strong and

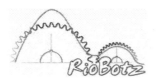 

agile. If for instance *Tourinho* used a pneumatic cylinder, which has a force that depends on the piston area, a cylinder scaled to 2.15 in *Touro* would be only 2.15^2 times stronger, while the robot would be 2.15^3 times heavier. The drive system accelerations, which depend on the ratio between the robot's traction force and mass, would be compromised as well. This is why, comparatively to their sizes, the insect robots seem to be much more agile.

However, *Touro* won't be relatively 2.15 times less powerful and resistant. In the case of a pneumatic cylinder, its energy comes from its internal volume (multiplied by the operating pressure). Therefore, a cylinder scaled to *Touro* would have 2.15^3 times more volume and energy, which is compatible with a weight increase of 2.15^3 times. The same is observed, for instance, in electrical direct current (DC) motors. In practice, the power/weight ratio of the best DC motors does not depend much on the scale factor. Otherwise, it would be worthwhile to replace a large motor with hundreds of small ones in parallel. Since power generates energy, and energy generates damage, *Touro* and *Tourinho* would have the same relative power and therefore the same relative damaging capabilities.

This conclusion is not very intuitive, especially when you consider that both *Touro* and *Tourinho* are able to fling opponents from their same weight classes up to the same 3 feet in the air. One can think that *Tourinho* would generate more destruction, because the relative throw height would be larger if compared to the robot size. But that same height is not surprising, it is verified by the expression of the potential energy $E = m \cdot g \cdot h$, where m is the robot mass, g is the acceleration of gravity, and h is the height reached in the throw. As the E/m ratios of *Touro* and *Tourinho* are approximately the same (as discussed before) and g is a constant, the height h should be the same. Although small robots are flung to a greater height with respect to their size, both energy and resistance depend on the cube of the scale factor. Therefore the destruction power (damaging capability) is relatively the same.

But why are *Touro* and *Tourinho* equally resistant, considering that the resistance of a column depends on the square of its scale and not on the cube? In fact, if *Touro* used in some way slender columns, subject to compression and buckling, it would be relatively 2.15 less resistant than *Tourinho*, following the "ant reasoning" and the dependence on the square of the scale. But the best combots are compact and robust, without slender parts. The most important loads that act in their compact structure are due to bending and torsion. But the resistances to bending and torsion depend on the cube of the scale factor (a shaft with diameter d, for instance, has bending and torsion resistances proportional to d^3), not on the square such as in buckling. Therefore, the bending and torsion resistance-to-weight ratios are still similar for both *Touro* and *Tourinho*.

The conclusion is that the scale factor can be used directly in the entire robot, without any significant loss of the power-to-weight or resistance-to-weight ratios. For instance, if you multiply by 2 the robot size, its weight is multiplied by 8. The analogy with ants would say that the diameter of a shaft in this robot would need to be multiplied by the square root of 8, about 2.83, to maintain the same resistance-to-weight ratio. That would be necessary if you were designing the column of a building, subject to buckling, but this is not the case for combots. In that case, it would be enough to multiply by 2 the shaft diameter to keep the same resistance-to-weight ratio. This is useful for two

reasons: first, this means that you can apply the same scale factor (2, in this case) to all the individual components of the robot; and second, you save weight, because the shaft with diameter multiplied by 2.83 would have twice the weight of the one multiplied by 2.

But there is another factor to consider: shafts in combat robots are usually relatively short, which are subject to high shear stresses. In addition, great impacts can generate tensile stresses or significant compression. The resistance to those traction, compression and shear stresses in a shaft with diameter d is proportional to d^2, not to d^3, taking us back to the ant analogy. As during combat we cannot predict which stresses will be more or less significant, and as the shafts are very critical components that cannot break or get bent, it is desirable to be conservative and adopt the higher factor $2^{1.5} = 2.83$ for the shaft from the previous example.

In summary, you should use the scale factor to multiply (or divide) the dimensions of all the robot components, except for the most critical ones such as shafts, where the scale factor should be raised to the power of 1.5. Don't use such larger factor in the entire robot, otherwise the robot might gain too much weight (when upsizing it) or lose strength (when downsizing it). Use the higher factor only for multiplying shaft diameters or for the dimensions of a few other critical components.

All those considerations are not just philosophical, they are verified in practice. Steel shafts used to drive the wheels of several combat robots typically have, in average, a diameter of about:

- 13mm (about 0.5") for lightweights (60lb);

- 18mm (a little less than 0.75") for middleweights (120lb);

- 25mm (about 1") for heavyweights (220lb); and

- 31mm (a little less than 1.25") for super-heavyweights (340lb).

Comparing lightweights and middleweights with similar aspect, the theoretical scale factor would be $(120lb/60lb)^{1/3} = 2^{1/3} = 1.26$, and the ratio between the shaft diameters is 18mm/13mm = 1.38, a value incredibly close to $1.26^{1.5} = 1.41$.

Between middleweights and heavyweights, the theoretical scale factor is $(220lb/120lb)^{1/3} = 1.22$, and the diameter ratio is 25mm/18mm = 1.39, very close to $1.22^{1.5} = 1.35$.

And between heavyweights and super-heavyweights, the theoretical factor is $(340lb/220lb)^{1/3} = 1.16$, and the diameter ratio is 31mm/25mm = 1.24, which also agrees extremely well with $1.16^{1.5} = 1.25$.

The bottom line is that theory, combined with common sense, is a very powerful design tool in practice. Imagine how many shafts have been broken in combats worldwide before arriving at these optimized diameters, while with a few simple calculations we've arrived at the same result.

Note however that these are average diameters, the actual values may vary depending on the steel alloy used in the shaft, number of wheels and combat robot type. The combat robot types are discussed next.

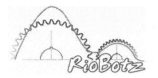

2.3. Combat Robot Types

After choosing the weight class of your robot, the next step is the choice of the robot type. There are several types of combat robots. None of them is the best. It is a rock-paper-scissors game. Or, as combot builders say, a wedge-spinner-hammer game. The wedges tend to flip over the spinners, which in turn tend to cut off hammers, which tend to puncture or damage the wedges. But they only tend to.

The truth is that a well designed robot can win against a robot of any type, independently of the trends. In the figure below there is a diagram showing such trends for several types of robots. In the figure, each robot has a tendency to win against the one it is pointing to. But a good design and a good driver can completely change this.

There are basically 16 types of combots: rammers, wedges, lifters, launchers, thwackbots, overhead thwackbots, spearbots, horizontal spinners, sawbots, vertical spinners, drumbots, hammerbots, clampers, crushers, flamethrowers and multibots, which will be described next.

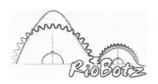

Other types of robots exist, but they can almost always be categorized into one of the 16 types above, or in a combination of them. Consider for instance the robot known as the "Swiss army knife," one with two or more weapons. The Swiss army knives in general are not very efficient, it is better to concentrate the weight on a single powerful and efficient weapon than on two or more smaller weapons. It may be a good idea when the weapons act together, at the same time against an opponent. For instance, the 2006 version of our middleweight spinner Titan (pictured to the right) used a wedge (the weapon of the wedge robots) together with its blade to lift lower opponents and hit them.

Most secondary weapons that are efficient in practice are wedges, used for instance to slow down the spinning bar of an opponent before it is safe to attack it with the main weapon.

There is also the "chameleon" robot, with weapons that can be switched during each pitstop depending on the opponent from the next fight. These robots can change their type very quickly, taking advantage of the best each type has to offer. The super-heavyweight Shovelhead (pictured to the right) has 15 different weapons that can be installed on its articulated front, one for each type of opponent.

A few accessories can make a big difference. For instance, it is not a bad idea to install some sort of bumper if you'll face a spinner. There are even specific accessories against specific robots, such as using a long stick to hold the shell spinner Megabyte by its vertical tube, as pictured to the right, to repeatedly shove it against the arena walls. However, it is not easy to design efficient weapons that can be quickly dismounted and assembled during a pitstop.

The 16 main types of robots are discussed next. Several photos below were taken from the BattleBots website, www.battlebots.com.

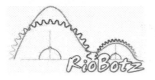

2.3.1. Rammers

Rammers are ramming robots, they damage the opponent throwing themselves against them or pushing them against the borders of the arena. They usually have 4 (or more) wheel drive, wide wheels with high traction, a sturdy drive system, robust armor, high resistance to impacts, and they don't have weapons except for their passive shields. In general, they are invertible (they can be driven upside down). They need to be capable to push at least 2 times their own weight. They are effective against robots with spinning weapons, such as spinners, drumbots and sawbots.

2.3.2. Wedges

Wedges are robots with a sloped plate shaped as a wedge. They usually have 2 or 4 wheels, with a very resistant drive system. They can be invertible or not. Despite rarely causing damage directly, they are a good tactic against spinners, making them flip over when hitting the wedge. Wedges win against their opponents by entering underneath them and dragging them around the arena, or flipping them at high speeds. Fast wedges usually reach 20 to 25km/h (12.4 to 15.5 mph). The front of the wedge should not be made out of sheet metal, because it can get easily bent and lose functionality. Use thick plates chamfered at the edge to withstand the opponents' impacts. Wedges are good against rammers and robots with spinning weapons, and they are vulnerable mainly to other lower, faster and more powerful wedges.

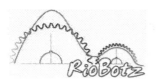

2.3.3. Lifters

Lifters are robots capable of lifting the opponent, immobilizing it or turning it upside down. They are efficient against robots that depend on traction such as rammers and wedges, or robots that have protuberating parts that can be reached by the lifting arm. They are inefficient against thwackbots and overhead thwackbots, because they are difficult to catch and they can work inverted. Lifters are vulnerable to spinners. The lifter design involves a slow linear actuator to lift the opponent, which can stop in the middle of its course. In this way, one can lift an opponent and drag it around the arena instead of just flipping it over. A few lifters use pneumatic systems, but most of them use electric motors with linear actuators. Place the batteries as far behind as possible in the robot, to act as a counterweight when lifting an opponent. The front wheels need to have high torque and high traction, because the robot weight will move forward when lifting and dragging the opponent. A few robots, such as the famous Sewer Snake, use active wedges that also work as lifters.

2.3.4. Launchers / Flippers

Launchers (or *flippers*) are lifters on steroids, being capable of flinging the opponent high into the air. The opponent not only can be flipped over, but it can also suffer great damage when hitting the ground. Therefore, launchers are good against opponents with weak chassis, or batteries and electronics without protection against impacts. Launchers need pneumatic components with large diameters actuated by high pressure air or CO_2. Eliminate all needle valves from the system, or use a big accumulator, to guarantee the high gas flow necessary to power the weapon.

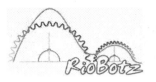

2.3.5. Thwackbots

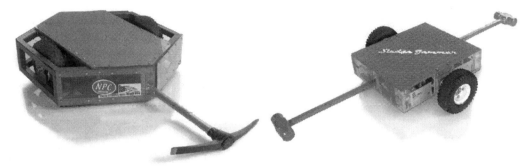

Thwackbots are usually 2-wheeled robots, invertible, which rotate all their structure in the same place at high speeds. They usually have one or more long rods with a hammer, axe, or some piercing weapon. They use the energy of their own drive motors to power the weapon, leaving more weight for their armor. The tires need to be narrow, otherwise they will suffer large friction losses when trying to turn on a dime at high speeds. The wheels, besides being narrow, cannot be too far apart. The closer they are, the faster will be the final angular speed of the robot, however the slower will be the acceleration and the harder will be to drive on a straight line if necessary. The drive motors need to have high RPM. The main problem is that most thwackbots are not capable of moving around to pursue their opponent while they are spinning. Very few thwackbots have developed successful mechanical or electronic systems with that purpose, as studied in chapter 6. Thwackbots are sometimes called full-body spinners, for obvious reasons.

2.3.6. Overhead Thwackbots

Overhead thwackbots use their weapon in an overhead movement, instead of a horizontal one such as with the thwackbots. They have 2 wheels and a long rod, which rotates when the drive motors are reversed, attacking the opponent's top. It is important that the motors have high torque, because the weapon has only a 180 degree course to acquire its maximum speed. Unlike thwackbots, the wheels should be far apart to help it move on a straight line and to increase the precision of the attack. The tires should be wide to maximize traction. The center of mass of the robot needs to be very close to the line passing through the axes of the wheels, to guarantee that it can lift the weapon to attack. They are good against rammers, wedges and lifters.

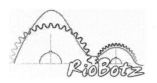

2.3.7. Spearbots

Spearbots have a long and thin penetrating weapon, usually pneumatically actuated, which tries to penetrate into the walls of the opponent's armor and damage vital internal components. The weapon needs to be resistant and sharp, reaching the largest possible speed. Some conicity in the spear tip is a must to avoid it getting stuck in the opponents. They usually have 6 wheels, to guarantee high traction, necessary so that the robot doesn't move too much backwards during the attack. They are not too efficient, except against robots with thin lateral armors or with exposed vital components. A few robots tried to implement attacks with tethered projectiles (projectiles are forbidden unless they are tethered), but they ended up converging to the spearbot design.

2.3.8. Horizontal Spinners

Horizontal spinners are the most destructive robots. They have a bar, disk, shell or ring that spins at high speeds. When the weapon spins very low near the ground, the spinner is called an undercutter. Ring or shell spinners (such as the robot Megabyte) spin their entire ring or shell-shaped armor, being capable of storing a high kinetic energy, becoming almost impossible for the opponents to reach them without being hit by their weapon. The weapon needs to spin as fast as possible, and you should be able to accelerate to a speed that can cause significant damage in less than 4 seconds. Spinners that take longer than 8 seconds to accelerate may never have a chance to damage a resistant and aggressive opponent. Spinners need to be fast to escape from their opponents while they spin up. Their greatest disadvantage is that, in general, they are not invertible, depending on luck to flip back. To compensate for that, a few robots such as The Mortician and Last Rites, called offset spinners, have moved their blade forward, making them invertible. However, by doing so the robot ends up with large dimensions, compromising its robustness, its back is vulnerable to attacks, and its center of gravity is moved too much forward, away from its wheels, decreasing traction.

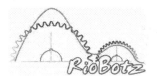

2.3.9. Sawbots

Sawbots have abrasive or toothed disks powered at high speeds by a powerful motor. They are in general combined with other designs, such as wedge-saws. The saws have little efficiency to cut through the opponent, especially if it is trying to escape. They can easily cut sheet metal and Lexan, but they can hardly cut any metal plates during a fight. Their greatest advantage is the cosmetic damage they cause, generating a shower of sparks, scratches and shallow cuts, which can impress judges and guarantee victory in a close match. Saws that rotate in such a way to lift the other robots have high risk of getting stuck on the opponent, breaking or bending. Saws that rotate downwards reduce this problem, however they increase the chance of self-flipping over.

2.3.10. Vertical Spinners

Vertical spinners are sawbots on steroids. Unlike sawbots, in general they use large diameter disks with very few teeth, or bars, spinning on a vertical plane. Damage is caused by both impacts: when the opponent is hit by the weapon and thrown into the air, and when it hits the ground. Vertical spinners need to have a wide base so that they don't tumble when turning due to the gyroscopic effect of the weapon (discussed in chapter 6). The impact force is transmitted to the ground, and not sideways such as with spinners, allowing them not to be flung to the sides due to their own impact. Their disadvantages are having their lateral and back exposed, and having a hard time making quick turns due to the gyroscopic effect. They have problems against very low wedges and tough rammers. The fights against horizontal spinners are extremely violent and fast, and they can go either way, although vertical disks with large diameter usually lose to powerful horizontal bars.

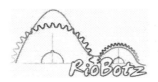

2.3.11. Drumbots

Drumbots have a spinning drum or eggbeater with teeth, in general powered by belts or chains, horizontally mounted in front of the robot. They usually rotate in such a way to launch the opponent, turning it over or causing damage from the impact with the weapon or with the ground. Drumbots are more compact versions of vertical spinners, with less moment of inertia in the weapon. This allows a shorter acceleration time for the drum, however causing less damage to the opponent. They are very stable due to their low center of gravity, they can be invertible, and they make turns more easily than vertical spinners due to the smaller gyroscopic effect (discussed in chapter 6). Wider drums allow drumbots to reach their opponent without needing a perfect alignment. The acceleration time of the drum should be at most 4 seconds. Their worst enemies are very resistant, well armored invertible robots.

2.3.12. Hammerbots

Hammerbots are robots with hammers or axes that hit their opponents' top. Usually with 4 wheels, their attack is similar to the one from overhead thwackbots, however the weapon actuation is independent of the drive system. The weapon can be fired repeatedly and quickly. It is usually pneumatically powered to deliver enough speed in its course, which has only 180 degrees. The weapon system can work as a mechanism to flip back the robot itself. They are very efficient against robots with weak top armors. Powerful hammerbots are good against rammers, wedges, thwackbots and sawbots. Their worst enemies are the spinners.

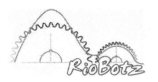

2.3.13. Clampers

Clampers are robots capable of holding and lifting an opponent, usually carrying them to the dead zone on the borders of the arena. They're usually pneumatically actuated (faster), or they use an electric system with high gear reduction (slower). Their design strategies are similar to the lifters', where the robot weight should be shifted back to avoid tipping forward when lifting the opponent. Clampers need to be fast enough to reach their opponents before they can escape from their claws. They are good against rammers, wedges and thwackbots. Hammerbots should be caught from their sides, so the clamper can avoid being repeatedly hit by the hammer while clamping them. Instead of a lifting platform, a few clampers use a *dustpan*, which is basically a wide box open at the front and top where an opponent is maneuvered into. A few dustpan designs do not include a restraining claw.

2.3.14. Crushers

Crushers are robots with hydraulic claws capable of slowly puncturing or crushing the opponents. The claws need to have long tips to penetrate efficiently, and they need to have a long course to be able to work against an opponent with large dimensions. Their main advantage is that it is almost impossible for the opponent to escape after being caught, ending the match. Crushers need to be hydraulically powered to generate enough forces to crush, which makes them very complex and heavy, leaving little weight left for the drive system. They are usually heavyweights or super-heavyweights. More sophisticated robots use a two-stage hydraulic system: the first stage is fast enough to hold the opponent before it can escape, and the second stage is slow but with enough force to puncture and crush.

2.3.15. Flamethrowers

A few competitions allow the use of flamethrowers. The *flamethrowers* are usually used together with other weapons, such as wedges. The effect is mostly visual, counting points with the judges and making the audience cheer. They are usually inefficient to disable other robots because most opponents are fireproof, except if the electronics is exposed or the wheels are flammable.

2.3.16. Multibots

Multibots are robots made out of 2 or more sub-robots, with weights that must not add up beyond the limit of the category. Most of the competitions adopt the rule that says that it is necessary to incapacitate 50% or more (in weight) of the robot to win a round. Using 2 sub-robots is therefore risky, because it is enough to have the heavier one incapacitated to lose a match. For that reason, several multibots use 3 robots of similar weights, forcing the opponent to incapacitate 2 of them to win. For instance, you can use 3 middleweights, as long as one of them drops to 100lb, to compete as a single super-heavyweight multibot (120 + 120 + 100 = 340lb). In the same way that several small weapons are less efficient than a large one, multibots have little advantage over their opponents, unless the attack (usually controlled by 2 or more drivers) is very well coordinated. In practice, it is difficult to coordinate a simultaneous attack, the opponent ends up incapacitating the multibot one by one (in general going for the smallest one in the beginning of the match). Another technique is to use, for instance, a main robot with about 90% of the weight of the category and 2 small ones with 5% each, which serve as a distraction for the opponent. In practice, the small ones are ignored and the opponent goes for the bigger one (the multibot Chiabot used 1 small robot as a distraction, but it didn't help much in practice). Another idea is to use a swarm of small autonomous robots, which would climb the opponent, get inside and destroy them from the inside out. But they are still science fiction, like the Sentries from Matrix, or Star Wars' Buzz Droids.

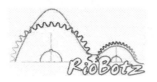

2.4. Design Steps

After choosing the weight class and type of the robot, the next concern is with its cost.

2.4.1. Cost

A middleweight robot, to be competitive internationally, has a typical cost of about US$4,000, including the radio control and spare batteries. For lightweights, about US$3,000 [10], for heavyweights US$6,000, and for super-heavyweights US$8,000. The numbers can go much higher

than that, to tens of thousands of US dollars, such as in the robot Buster (on the right, photo by Hal Rucker), a beautifully designed super-heavyweight all made in milled titanium. This doesn't mean that it is not possible to win an international competition with a much less expensive robot, everything depends on creativity. But, statistically, the above numbers are reasonable estimates. The opposite is also true, there is no guarantee that an expensive robot will win a competition.

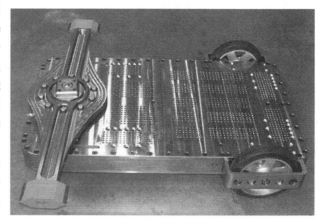

In summary, this is not a cheap sport. However, for many sponsors such numbers are low if compared with what it is usually invested in other sports. In addition, featherweights and other lighter robots can be quite inexpensive.

2.4.2. Sponsorship

A few great tips about sponsorship and several other combot subjects can be found at The Robot Marketplace (http://robotmarketplace.com/tips.html). Basically, it says that it is not an easy task to find a sponsor that will help you out if you haven't built any combot before. Probably the only exceptions are companies whose owners or directors you know well or who are your friends. Most

big companies do not bother with sponsoring robots, it's a better bet to look at smaller local shops near you that might like to help out. You might be able to get sponsorship from big companies, but it is important to meet the right people, the ones who are able to make the decisions. For instance, presenting your robot to a public relations intern won't help you a lot, he/she won't probably be as enthusiastic as you would be when presenting the proposal to their boss.

Also, you have to call and visit them in person, nobody will give you sponsorship over e-mail. Bring with you business cards with your team's logo, for a more professional look, as pictured to the right.

Prepare a presentation folder with lots of nice photos, such as the one pictured to the right. Clearly show the potential sponsors how and where their name and logo would be made visible, such as in a T-shirt layout (pictured below), on the robots, at the team website, in YouTube videos, etc. Show as well which newspapers, magazines and TV news programs have already covered the events you plan to attend. Showing videos from the fights is also a great idea, several potential sponsors have no idea of what robot combat is. They might fall in love as soon as they watch it.

Attached to the presentation folder, you should include your annual budget. Don't forget to include the cost of parts, machining time, taxes (especially if any component must be imported), marketing material (such as T-shirts with sponsor logos), event entry fees, and travel expenses. Do not cut down expenses at this stage, ask for everything you might possibly need – there's a chance you get full sponsorship for that value. If you ask for too little in the beginning, you might not be able to increase the budget later on during the same year.

But let the potential sponsors know that they don't need to provide the entire budget, that you will take partial sponsorship. You may even come up with sponsorship levels, such as bronze sponsorship for 10% of the budget, silver for 25%, gold for 50%, and platinum for 100%.

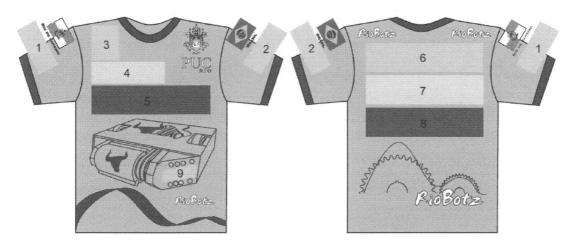

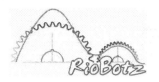

It is important to show the potential sponsors which benefits they get depending on the sponsorship level. For instance, in the T-shirt layout from the previous page, a gold sponsor would get advertising space on the areas 1, 2, 3, 6 and 9, while silver would get 4 and 7, and bronze would get 5 and 8. Needless to say, a platinum sponsor would get all areas from 1 to 9. Note that area 3 is better than areas 4 or 5, because it has a higher chance of being caught on camera during a TV interview, as pictured to the right. Usually, the silver sponsor logo in area 4 is only partially shown during an interview.

Any sponsorship help is welcome. Unless you are very well established with your sponsors, you will find it difficult to get cash from them, more often they might contribute with parts or machining time. And don't give up after getting turned down a few times, you need to put a lot of effort into it.

2.4.3. Designing the Robot

The next step is to get an estimate of the robot weight. If, after adding all the motors, wheels, structural components, weapons and batteries, the robot is way above its weight class limit, this means that it is necessary to reduce the entire scale of the robot or to use lighter components. To distribute well the robot's weight, a very useful tip is to use the 30-30-25-15 rule [10]: 30% of the robot weight should be devoted to the drive system (motors, transmissions and wheels), 30% to the weapons (weapon, motor, transmission), 25% to the structure and armor, and 15% to the batteries and electronics. Of course those numbers can vary a lot depending on the type of the robot, but they are representative average values.

When designing and sketching the robot, always have in mind the principle known as KISS: Keep It Simple, Stupid! In other words, don't complicate your design too much if not necessary, design your robot in the simplest possible way, but never simpler than that. Sketches can be made by hand, using a CAD program, or in any way that makes it quick to update and share it with all your teammates. The first sketch of our middleweight Touro was made, believe it or not, in MS Powerpoint, see the figure to the right. It is a program that the entire team had in their personal computers, either at home or in the University, unlike most CAD programs. In this way, the entire team

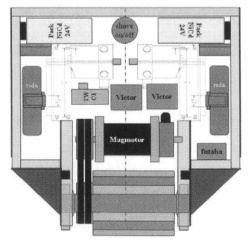

could think anytime anywhere about improvements in the robot design, using any personal computer. This technique is also known as PAD (Powerpoint Aided Design), making it easy to generate vaporbots, which are virtual robot designs that haven't been built yet. Vaporbots help a lot to stimulate creativity and to evolve your design without any building cost. The next page shows 4 vaporbots that helped generate the RoboGames 2006 version of Touro.

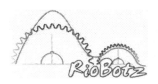

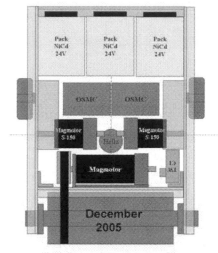

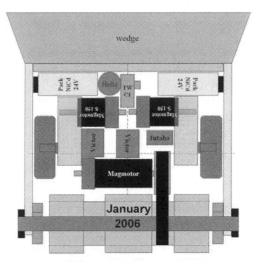

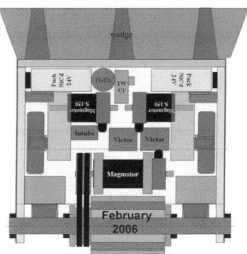

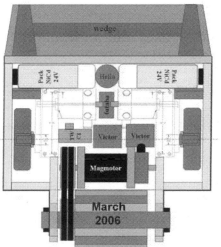

If you have access to CAD programs such as Solidworks or Rhino3D, then you can use them to create a 3D view of the robot, see the figure to the right (created using Solidworks). CAD programs are also useful to make cutting and drilling marks: just print out the layout in 1:1 scale, glue it directly onto the piece/plate with an adhesive spray (such as Spray 77), and mark the holes with a center punch.

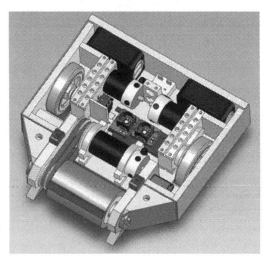

During the design phase, it is necessary to have in mind that fragile items such as electronic components should be placed well inside the robot, to be protected from cutting weapons. The robot should also be the most compact possible, so that its

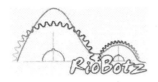 

armor can have larger thicknesses without going over the weight limit. But don't forget that too compact robots are difficult to repair during a pitstop, the parts that need to be changed may be inaccessible, so it is important to use common sense.

2.4.4. Calculations

After the first sketches, it is recommended to perform a stress analysis to calculate the resistance of each component from the robot. This subject is too vast, it is beyond the scope of this tutorial. Books about mechanics of solids and mechanical behavior of materials [8] are very useful for that. The analysis consists basically of calculating the tensile, bending, torsion and shear stresses in the structure and components, including the stress concentration factors of the eventual notches (such as holes, abrupt changes of geometry), and combining them to obtain an equivalent stress, usually known as Mises or Tresca stresses. With the equivalent stress, it is possible to design the parts against yield, rupture, plastic collapse, fatigue, etc. Finite element software, such as Abaqus, Ansys, Nastran, Adina, or several others, can be used to aid in the numerical calculation of the robot's resistance. Most of them are capable to import drawings directly from CAD programs. Their license is usually expensive, however these programs are not indispensable. With a little common sense and mechanical background it is possible to make "back of the envelope" stress analyses, which are approximate but accurate enough for design purposes. Chapter 6 will show a few examples of such dimensioning techniques.

2.4.5. Optimization

Most combat robots are born overweight. You must prepare yourself to deal with that, sooner or later. If it is too much overweight, you might need to redesign it completely. Otherwise, a few optimization techniques can be used to lose weight, improve strength, or even to do both at the same time.

One way to do that is to optimize the shape of the robot parts. This is usually done in an ad-hoc manner, using common sense, and sometimes even with the aid of finite element software to check the resulting strength, such as in the spinning disk of the middleweight Vingador (pictured to the right). The holes and voids in the disk were positioned not too close to its center, where out-of-plane bending stresses can get very high, and not too close to the outer perimeter, to avoid lowering the moment of inertia or the strength of the teeth. This process usually involves trying several hole configurations, and using finite element and CAD programs to calculate the disk strength and moment of inertia.

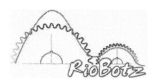

Shape optimization can also be seen in the spinning bar of the hobbyweight Fiasco, pictured to the right. Pockets were milled in the bar to relieve weight, except at the middle section, not to compromise strength, and at the ends, not to compromise its moment of inertia.

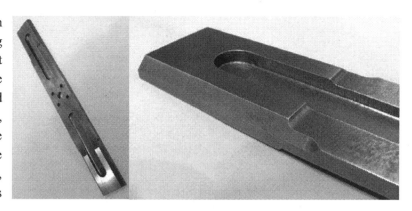

The lifter arm from BioHazard's four-bar mechanism (pictured to the right) is another example of clever shape optimization to selectively remove weight. Note that the weight saving holes near the middle pivot, where the bending moments are maximum, have smaller diameters not to compromise strength. The diameters of the holes are also directly proportional to their distance to the middle

pivot, trying to evenly distribute the stresses at the bar, because the bending moment in this system is directly proportional to the distance to the bar ends.

Such optimization tasks can also be performed automatically. Sophisticated software can perform shape and topology optimization of structural parts, to minimize their weight or maximize some property. Shape optimization software like Tosca (http://www.fe-design.de) can be run together with finite element programs to find the optimal shape of a part that will minimize its weight while achieving desired values for stiffness, strength or even moment of inertia, for instance.

For instance, suppose you need to design a single-piece bracket to be fixed inside the robot by two holes A and B, to support some vertical force F that acts at another hole C, as pictured in the next page to the left. The optimization program will require you to inform the relative positions of the holes, their diameters, their contour conditions (such as whether they allow rotations, as if attached by pins, or whether they don't, as if attached by keyed shafts), the bracket material, the direction and intensity of all the applied forces and moments, and the performance requirements. These requirements can be, for instance, the maximum allowable stress in the bracket (a strength requirement) and, at the same time, its maximum allowable deflection (a stiffness requirement), while minimizing its weight. The optimization programs usually require you to inform as well the topology of the component, which is basically the number of voids it may have. And a few programs also allow you to minimize weight together with manufacturing complexity as well, trying to achieve an optimum shape using only straight lines and circular arcs, avoiding generic curves or very small voids.

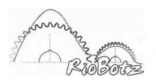

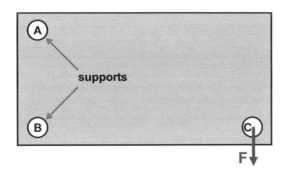

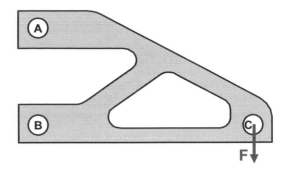

The figure above to the right shows the resulting shape for minimum weight with minimum manufacturing complexity for a version of our bracket with only one void (besides the voids from the fixed holes A, B and C). Note that this resulting shape is only optimal for specific input values,

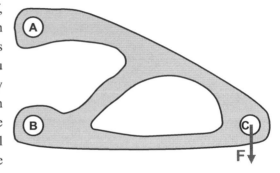

because it depends on all the given parameters. If, for instance, the maximum allowable deflection increases while the maximum allowable stress decreases, the shape will be different. Also, if you turn off the minimum manufacturing complexity requirement, you might end up with an even lighter bracket (such as the one pictured to the right), but you'll probably need a numeric control laser or waterjet cutting system to fabricate the resulting intricate part.

A few programs are also able to optimize both shape and topology, finding not only the shape but also the ideal number of voids in the component. This can be useful, for instance, to find optimal number of voids and their shapes for a spinning disk with maximum strength-to-weight and moment of inertia-to-weight ratios.

Pictured below are a few examples of bracket topologies with 1, 5 and 7 voids (not counting the voids from the holes A, B and C). The above results were obtained after choosing the 1-void topology seen below. A topology optimization program wouldn't need such user choice, it would find out by itself which topology would be the best option, and then optimize its shape.

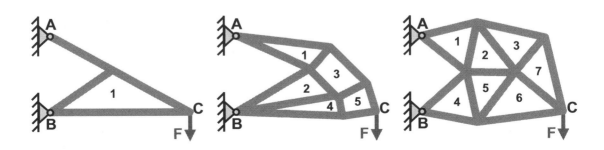

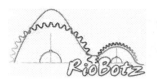

Note that the topology representations above look a lot like trussed structures, but they'll result in single-piece components, such as in the plates the form the structural frames of the hobbyweight Fiasco (pictured below to the left) and lightweight K2 (to the right). Note also that, for armor plates or other external unprotected structural elements, you'll probably want to turn off topology optimization to force a solution with zero voids. Armor plates with voids would probably be a bad idea against spearbots and flamethrowers.

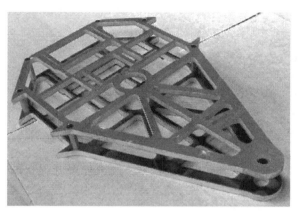

The topology and shape optimization analyses are not limited to planar problems such as parts with uniform thickness. They can also obtain the shape of optimized tri-dimensional (3D) parts, as pictured to the right. Laser or waterjet cutting won't be enough to fabricate these optimal 3D parts, you'll probably need a mill or even a CNC system.

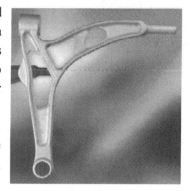

The 3D optimization process is quite similar to the planar case. You'll feed the software with an initial guess of the shape of the desired component (as pictured below to the left), along with all the required holes and contour conditions, material information, applied forces and moments, and performance requirements. The software will then optimize the topology of the component, adding voids if necessary, and finally output the optimized shape that meets the requirements with minimum weight (as pictured below to the right).

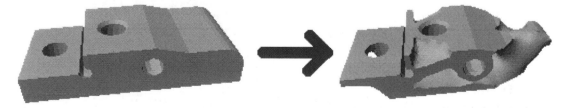

Another approach to optimize the robot is to change the material of its components. Material optimization, either to improve performance or to reduce weight, is seen in detail in chapter 3. Other weight saving techniques can also be found in chapter 9.

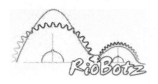

2.4.6. Building and Testing

During the robot design, building a full scale model is also very useful. We've already built several models of our robots and their components. For instance, while we were waiting for an Etek motor to arrive in Brazil during its import process, we've built a one-to-one scale model (pictured to the right) using Styrofoam, cardboard, and an old snorkel. These models guarantee that your hand will fit everywhere inside the robot, which is fundamental for quick pitstops. Unfortunately, Solidworks doesn't allow you (yet) to reach your hand inside the monitor. Always recalculate the robot weight, combots tend to easily get overweight.

At least in our experience, we've realized that the design phase usually takes most of the robot development time, perhaps about 60% or more. The other 40% would be the construction itself. In order not to waste money and material, it is a good idea to make sure that the design won't suffer huge changes before starting to build it (small changes during construction will almost always happen). Check the CAD drawings - or the cardboard prototypes - before starting to cut metal. Follow the "measure twice, cut once" rule. A lot of information on building the robot will be covered in the following chapters of this tutorial.

Finally, after finishing the robot, there's the part that everybody forgets about (including us): testing. Follow Carlo Bertocchini's law: "Finish your robot before you come to the competition." Many times the robot is finished just before the competition, leaving not enough time to test it. This is a fatal mistake, there are several things that can go wrong. With a few tests most problems can be identified and corrected. Besides, during the tests the driver is able to acquire experience in driving that specific robot, which can make all the difference during a match. This leads to one of Judge Dave Calkins' main advices: LTFD – Learn To *Freaking* Drive! Drive a lot. Hundreds of hours, not a few. Several opponents drive maybe two hours a day. This can and will make a huge difference.

2.5. Robot Structure

As for the robot structure, the three main types are: the trussed, the integrated, and the unibody. The trussed robots (such as The Mortician, pictured to the right) are built using several bars, in general welded together, resulting in a very rigid and light structure. The armor is made out of several plates, usually screwed to the bars, sometimes using rubber sandwich mounts (see chapter 4) to provide damping against impact weapons. They are the fastest type of structure to build, it is enough to use a hacksaw and welding equipment to quickly assemble the chassis. Trussed robots are also easy to work with during

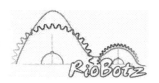

pitstops because, if one of the plates gets damaged, it is easy to unscrew it and change it for a new one. The greatest disadvantage is to depend on the welds, which are in general the weak point. Besides, the armor plates are prone to be ripped off in combat.

The integrated robots (such as our middleweight Touro, pictured to the right) receive such name because the structure and armor are integrated into a single set, using screws or welds. The same plates that work as armor are the ones where the internal components are mounted to. Sometimes there is a thinner armor layer on top of the integrated structure. Building such robots is not an easy task, however they generate very compact and resistant systems.

The unibody robots (such as our beetleweight Mini-Touro, pictured to the right) have their structure milled out of a single solid block. Through milling, it is possible to create the side walls, the bottom, and pockets to fit batteries, motors, etc. In this way, it is not necessary to weld or to use screws in the structure, except to install the components and to attach the top cover. These are the lightest and most resistant robots. However, you lose about 80% to 90% of the material from the solid block to carve its interior, not to mention the hours (or days) hogging the milling machine. The cost and material waste makes this

solution attractive only to very light robots such as the insects. Another disadvantage is that there is no way to replace a damaged part of the structure, as it is done with the armor plates from the trussed robots. If there's too much damage, it might be necessary to mill an entirely new unibody.

A unibody can also be made out of a composite frame, as in the hobbyweight VD2.0 (pictured to the right). Composite frames are basically a foam body with the shape of the unibody, covered with some fiber such as glass, carbon and/or aramid (Kevlar) fibers, which are epoxied to the surface. Carbon fibers are an excellent choice to obtain very high stiffness, while Kevlar gives high impact toughness, see chapter 3. Composite frames are not very

popular because they're expensive and difficult to manufacture, in special if the robot design requires the structure to have a high precision to mount, for instance, weapon systems.

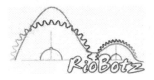 

2.6. Robot Armor

There are basically three types of armor: traditional, ablative and reactive, presented next.

2.6.1. Traditional Armor

Traditional armor plates are usually made out of very tough and hard materials that try to absorb and transmit the impact energy without getting damaged. The high hardness of the armor plate is used to break up or flatten sharp edges from the opponent weapon (which is good against very sharp horizontal spinners), while its high toughness allows the plate to withstand the blow without breaking. This sometimes can be achieved using a composite armor, which means using several layers of different materials. For instance, you can use very hard (but brittle) ceramic tiles sandwiched between two very tough (but relatively soft) stainless steel plates to use as armor.

Due to their high hardness, traditional armors need to be changed less often, and they look nicer after a match. However, traditional armors transmit a lot of the impact energy to the rest of the robot structure, as shown in chapter 6, and they usually produce sparks, which may count as trivial or cosmetic damage points depending on the judges.

2.6.2. Ablative Armor

Ablative armor plates, on the other hand, are designed to negate damage by themselves being damaged or destroyed through the process of ablation, which is the removal of material from the surface of an object by vaporization or chipping. They're also made out of tough materials, but with low hardness and low melting point to facilitate the ablation process.

Ablative armor plates are much more efficient dissipating the impact energy, which is mostly absorbed by the ablation process, transmitting much less energy to the rest of the robot. They are a good choice especially against blunt or not-so-sharp horizontal spinners. They are also good against drumbots, even the ones with sharp teeth, because most of the drum energy will be spent "eating out" chunks of the armor instead of launching the robot. Also, you won't get sparks if you're using an aluminum ablative armor, which is good even though they are only counted as trivial damage.

Thick wooden plates are also very efficient as ablative armor, however they result in a lot of visual damage that may award damage points to your opponent (even though the destruction of ablative armors should only count as cosmetic damage). Make sure that the judges know if you have an ablative armor. A disadvantage of ablative armors is that they need to be changed very often because of the ablation, and you might need to use gloves to handle your deeply scarred robot.

2.6.3. Reactive Armor

The third armor type is the reactive. Such armor reacts in some way to the impact of a weapon to prevent damage. Most of them are very effective against projectiles, which have a relatively low mass and extremely high speeds, but not much against regular combot weapons, which have much more mass but much lower speeds than projectiles.

One example is the explosive reactive armor, which is made out of sheets of high explosive sandwiched between two metal plates. During the impact, the explosive locally detonates, causing a bulge of the metal plates that locally increases the effective thickness of the armor.

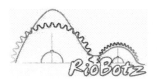 

Another example is non-explosive reactive armor, which basically consists of an inert liner, such as rubber, sandwiched between two metal plates. Against most combot weapons, this basically works as a shock-mounted armor, dissipating energy in the elastic liner. And against spearbots this armor has an additional advantage: during an angled impact, the outer metal plate will move laterally with respect to the inner plate, which may deflect or even break up any spear that eventually penetrates.

There are also studies on electric reactive armors, which would be made up of two or more conductive plates separated by air or some insulating material, creating a high-power capacitor. This could be implemented in practice using three metal layers, separated by rubber liners (acrylic tape such as VHB 4910 is also a good option due to its high dielectric breakdown strength), which would also work as a shock-mount. The middle plate is then charged by a high-voltage power source, while the other 2 plates are grounded. When the opponent's weapon penetrates the plates, it closes the circuit to discharge the capacitor, vaporizing the weapon tip or edge, or even turning it into plasma, significantly diffusing the attack. Note, however, that this system might be very difficult to implement in a combat robot, not to mention the increased battery requirements. Also, most competitions forbid the use of electric or explosive reactive armors.

2.7. Robot Drive System

The three usual types of drive systems are based on wheels, tank treads and legs, discussed next. There are also other types, based on rolling tubes (moving as a snake), rolling spheres, or air cushions (hovercrafts), but they're not very effective in combat. Flying is usually forbidden.

2.7.1. Tank Treads and Legs

Robots with tank treads are beautiful, they have excellent traction, however they waste a lot of energy when turning due to ground friction. They are also slow when turning, which allows an opponent to drive around and catch them from behind. Besides, treads can be easily knocked off by opponents with powerful weapons.

Legs have also several disadvantages. They are complex to build and control, and they usually end up not sturdy enough for combat, in special against undercutters. They tend to leave the robot with a high center of gravity, making it easy to get flipped over. Their only advantage is the weight bonus, usually 100%, allowing for instance a 240lb legged combot to compete among 120lb middleweights. But note that shufflers, which are rotational cam operated legs, are not entitled for the weight bonus. Legs are usually good options for robots in very rough and uneven terrain, which is not the case in flat floored combat robot arenas. This is why the international combat robot community has converged to the wheel solution.

2.7.2. Wheel Types

There are several types of wheels in the market. A few robots use pneumatic wheels, however they are filled internally with polyurethane foam so that they don't go flat if punctured.

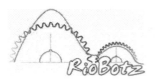

Another good solution is the use of solid wheels. To maximize traction, it is recommended that solid wheels have an external layer of rubber with hardness around 65 Shore A, at most 75 Shore A. Harder wheels tend to slide. Wheels with hardness measured in Shore D units are probably too hard.

Several robots, as well as our middleweights Touro and Titan, use the Colson Performa wheels (pictured to the right). These wheels are very inexpensive: each 6" wheel from Touro costs only US$7.25. Besides the low price, this Performa model from Colson has hardness 65 Shore A, a very good value for traction.

An interesting wheel solution was used by our team during the RoboCore Winter Challenge 2005 competition, held on an ice arena. When driving on ice, the wheel hardness is irrelevant. The important thing is the presence of sharp metal tips to generate traction. The secret of walking on ice is to know that it is not friction (very low in this case) that generates traction, but normal forces. The solution for the problem was very cheap: we've inserted several self-drilling flat head screws at angles of

about 60 degrees with respect to the wheel radius (as pictured to the right). The screw caps were also sharpened to improve traction performance. Those sharp tips generate a very small contact area with ice, generating a very high contact pressure. That high pressure makes the ice melt locally, allowing the tips to slightly sink in and lock in place. Then, when the motors spin the wheels, the "fixed" tips sunk in the ice apply normal *horizontal* forces, generating traction without sliding. The traction on ice ends up even better than the one from a regular wheel on metal. See in the picture above that we chose to use a single row of screws: our tests with 2 parallel rows generated worse traction, because with twice the number of screws to distribute the load, the pressure on the ice drops in half, and the screws sink in much less. A single well sunk screw generates much better traction than two half-sunk screws. Also notice that we've alternated the screw angles on the wheel, to guarantee that in average the traction was identical in both forward and reverse directions.

2.7.3. Wheel Steering

There are two main types of wheeled vehicles: the ones with Ackerman steering and the ones with tank (or differential) steering. Ackerman is the solution adopted by the automobile industry: a large motor is used to move the system forward or backward, and another smaller motor controls the steering of the front wheels to make turns. This is very efficient for high speeds, because it is

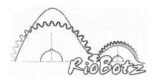

easy to drive straight, however it demands several maneuvers for the robot to spin around its own axis. Besides, the steering system is usually a weak point, needing to be very robust and, consequently, very heavy.

Tank steering receives that name for being used in war tanks. The entire left side of the robot is driven independently from the right side. To drive on a straight line, it is necessary that both sides have the same speed, which is not always easy to guarantee. Turns are accomplished when those speeds are different. The great advantage of that method is that if the speeds of both sides are equal in absolute value but have opposite senses, the robot can turn on a dime. This is perfect to always keep facing the opponent.

2.7.4. Two-Wheel Drive

There are two common options in tank steering, which are using 2 or 4 active (power driven) wheels, as pictured to the right. With 2 active wheels, it is possible to turn very fast and with less waste of energy. In addition, the robot saves weight by not needing the extra set of active wheels, shafts and bearings.

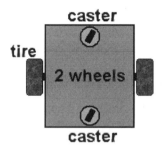

With only 2 active wheels, the robot will probably need at least another ground support, ideally 2. This is usually accomplished with skids, which are passive elements such as ball transfers or caster wheels (pictured to the right). Try to place the axis of the 2 active wheels as close as possible to the robot center of gravity, and the ball transfers or casters in the front and in the back, arranged in a cross configuration (see the previous figure for a 2-wheeled robot). In this way, you guarantee that almost the entire reaction

force from the ground will go through the 2 active wheels, where you need traction. Our middleweight Ciclone, due to lack of internal space, could not use the cross configuration. The 2 active wheels ended up in the back, as pictured to the right, supporting only half of the robot weight, compromising traction. But be careful with the cross configuration, make sure that the ball transfers or casters won't lift the active wheels off the ground, especially in an arena with uneven floor. To avoid lifting the

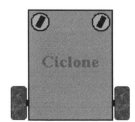

active wheels and losing traction, these robots should have their passive elements in a plane a couple of millimeters higher than the active wheels. You can also spring mount the ball transfers and casters, creating a suspension system that prevents the active wheels from being lifted off.

A disadvantage of using only 2 wheels is that it is more difficult to drive on a straight line. Several electric DC motors do not have neutral timing, which means that they spin faster in one sense, making it harder to drive on a straight line. If possible, try to set the drive DC motors in neutral timing (see chapter 5) or, if the radio control system is programmable, try to compensate the

speed differences through the trim settings. A few robots use gyroscopes to drive straight, more details can be seen in chapter 7.

If your robot continues with problems to move on a straight line, try to use rigid casters (pictured to the right) instead of the swivel ones. You will have a harder time turning, but the robot will drive straighter.

In robots with very violent weapons, the ball transfers and casters may not take the extreme forces transmitted to the ground during an impact against the opponent (these forces can easily exceed a few metric tons for middleweights). In this case, you can replace these passive elements by, for instance, button-head cap screws (pictured to the right), mounted upside down at the bottom of the robot. The round head slides very well on the arena floor. This is the technique that we use in our middleweight spinner Titan. Use hardened steel screws, the ones with high class (see chapter 4), because they are harder and do not easily wear away due to friction with the arena floor. A few robots also use wide pieces of Teflon (PTFE) for ground support with reduced friction.

2.7.5. All-Wheel Drive

Another wheel steering option is the use of 4 (or more) active wheels. Four-wheeled robots drive better on a straight line, they are good against wedges and lifters (because in general they guarantee at least 2 wheels on the ground to be able to escape if they've been lifted), and they have redundancy in case a few wheels are destroyed during a match. Experienced drivers, such as Matt Maxham from Team Plumb Crazy, are still able to drive even after 3 out of 4 wheels have been knocked off!

A few robots use 6 or 8 wheels to maximize traction as well as to increase redundancy, such as the 8-wheeled super-heavyweight New Cruelty and the famous 6-wheeled heavyweight Sewer Snake, pictured to the right. The problem with 4 or more wheels is the waste of energy while making turns, besides the additional drivetrain weight needed by the additional axes, bearings, pulleys, etc.

An interesting solution to help 6-wheeled robots to make turns more easily is to have two middle (compliant) wheels with a slightly larger diameter. In this way, the ground normal forces on the 4 outer wheels are reduced, decreasing their friction resistance while turning on a dime.

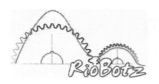 

2.7.6. Omni-Directional Drive

A very specific type of drive system is the omni-directional drive. It can be accomplished with omni-directional wheels (a.k.a. Mecanum wheels), which can move sideways without changing their direction. Those wheels have several small passive rollers, which can rotate freely, as pictured to the right.

The most common configurations are 4 parallel wheels, or 3 wheels at 120° angles. The rollers provide the wheel with traction in only the circumferential direction, rotating freely in the shaft axial direction. Coordinating the movement of the 3 (or 4) wheels, it is possible to move sideways without changing the direction of the robot. In the case of 3 wheels at 120°, the omni-directional control system is not so simple to implement, you need to program a few calculations involving sines and cosines. An off-the-shelf solution is the OMX-3 Omni-Directional Mixer (pictured to the right), a small US$45 board from Robot Logic (www.robotlogic.com), which does all these calculations automatically. This system is excellent for robot soccer competitions: the robot with the ball can move sideways to dribble an opponent without losing sight of the goal. It is possible to kick towards

the goal immediately after dribbling the opponent, without wasting time changing direction and making turns.

However, in combat robots such omni-directional capability is not necessary, because during a match you do want to be pointed towards your opponent. This is usually your goal. Moving sideways can be a good idea to dodge from an attack, but the cost-benefit is not good: the omni-directional wheels have worse traction than regular ones, they are less efficient (they waste more energy), and the rollers don't stand violent impacts.

2.7.7. Wheel Placement

Another important factor in the design of the drive system is the location of the center of mass of the robot. If it is shifted to the left of the robot, for instance, the wheels on this side will receive a larger reaction force from the ground. Because of that, they would have better traction than the right side, and the robot wouldn't move straight. Try to distribute the weight equally on both sides.

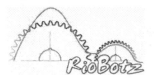 

For robots with only 2 active wheels, it would be ideal to have the robot center of mass C_1 very close to the center C of the line that joins the wheel centers, as pictured to the right. In that case, each wheel would receive about half the robot weight, guaranteeing good traction.

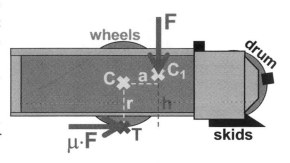

Even better would be to place C_1 slightly ahead of C, at a horizontal distance $a = \mu \cdot h$, where μ is the coefficient of friction between the tires and the ground, and h is the height of C_1, as pictured to the right.

If the combined torque of both wheels is large enough to guarantee an initial traction force of $\mu \cdot F$, the maximum value without wheel slip, where F is the robot weight, then the robot won't tilt backwards while it is accelerating if $a \geq \mu \cdot h$. It won't tilt because the forward gravity torque $F \cdot a$ with respect to the contact point T between the wheel and the ground becomes greater than or equal to the backward inertial torque $\mu \cdot F \cdot h$ (with respect to T) caused by the forward acceleration of C_1, since $F \cdot a \geq \mu \cdot F \cdot h$.

Our middleweight Touro uses this principle. Not tilting backwards is good to prevent wedges from entering underneath the skids that support Touro's drum. In addition, during the initial drivetrain acceleration, 100% of the robot weight goes to both wheels if $a = \mu \cdot h$, maximizing the initial traction force. The front skids will only feel part of the robot weight after the wheel motors drop their traction force below $\mu \cdot F$, which happens as they speed up. So, with $a = \mu \cdot h$, Touro achieves its maximum possible initial drivetrain acceleration while the front skids are barely touching the ground, until it acquires enough speed to make the front skids get some downward pressure right before they get in contact with the opponents, trying to get under them using the skids as if they were wedges. Note that, due to the almost symmetric and invertible design of Touro, the height h of its center of mass C_1 is almost equal to its wheel radius r, resulting in $a \cong \mu \cdot r$.

But it is important not to make this distance much higher than $\mu \cdot h$. As discussed before, our middleweight Ciclone, due to lack of space, had its 2 active wheels far in the back of the robot (as pictured to the right), away from the center of gravity, resulting in a large value for the distance a. Each wheel ended up just bearing about one fourth of the robot weight, the other half went to the front ground supports. With this reduced applied load on the active wheels, Ciclone had poor traction, with a lot of wheel slip. Note that magnet wheels and suction fans would be two possible solutions, although unusual, to increase the normal forces at the wheels.

The distance between the robot bottom and the arena floor is also important, this clearance needs to be large enough to avoid being trapped in debris or in uneven seams of the arena floor. If your poor featherweight will fight right after the super-heavyweight hammerbot The Judge, it will probably have to overcome arena conditions such as the one pictured to the right. But the ground clearance cannot be too

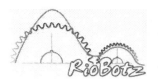

large, otherwise your robot might be vulnerable to wedges, lifters and launchers. It is also important to keep low the robot center of gravity to avoid being flipped over. In our experience, a minimum suggested clearance for any class from hobbyweight to super-heavyweight is about 1/4" (or 6mm).

2.7.8. Invertible Design

Most successful combat robots are invertible, which means that they can be driven while upside down. With the high number of wedges, drumbots and vertical spinners that we see today, not to mention launchers and lifters, it is just a matter of time until your robot gets flipped over.

If your robot is not invertible, or if its weapon has limited or no functionality when upside down, then it is a good idea to have some self-righting mechanism (SRiMech). A SRiMech is an active system that returns an inverted robot to its upright state. This mechanism can be an electric or pneumatic arm, or a passive extension on the upper surface of the robot to roll or flip it upright, such as the large white hoop on top of the featherweight Totally Offensive, pictured to the right. Launching or lifting arms, or even vertical spinning weapons, can also be used as a SRiMechs if properly designed.

The easiest way to implement an invertible design is to use wheels that are taller than the robot chassis. If the front side of your robot has a tall chassis, but not its back side, then another option is to use 2 drive wheels in the back, and skids in the front.

Another solution is to have two sets of active wheels, driven together using chains or belts, one of them for driving the robot when not flipped, and the other to be used when upside down. But note that this solution usually adds a lot of weight to your drive system.

Also, it is important to remember that your robot does not have only 2 sides. If it is box-shaped, it actually has 6 sides. But it is very simple to avoid losing a match because your robot ended up standing on its side. You only need to avoid having perfectly flat and vertical front, back and side walls.

This can be accomplished, for instance, using bolts sticking out of the walls, as pictured to the right, circled in red, in our wedge Puminha.

The bolts should not stand out too much, to avoid being easily knocked off, but enough to make sure that the robot will tilt back, as shown in the bottom picture to the right.

If the robot is invertible, without a preferential side, then you can place the screw in the mid-height of the chassis. Otherwise, you'll probably want to place it near the top of the chassis, as in both pictures to the right, to increase the chance of tilting back in the upright position. But this screw will need to stand out approximately twice as much as in the mid-height design to make sure that the robot will indeed tilt back.

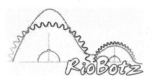 

2.8. Robot Weapon System

The wide range of weapon systems makes it difficult to give general suggestions that would apply to all of them. So, the weapon system of each robot type needs to be studied on a case by case basis. This topic is extensively covered in chapter 6, which deals with Weapon Design. You can also find several weapon design tips in sections 2.3 and 2.4. In addition, chapter 3 will show a thorough discussion on material selection for all kinds of weapon systems. Chapter 5 also deals with this subject, showing spin-up calculations for kinetic energy weapons such as spinning bars, disks and drums.

There are also very good books on the subject. For instance, Combat Robot Weapons [6] is entirely dedicated to weapon systems. Build Your Own Combat Robot [3] and Kickin' Bot [10] have chapters explaining the design details of each weapon type and the related strategies. And Building Bots [4] even presents basic physics equations. There are also a lot of weapon-related questions and answers at http://members.toast.net/joerger/AskAaron.html.

Note that there are several weapons that are usually not allowed in combat. This includes, but is not limited to, radio jamming, noise generated by an internal combustion engine (ICE), significant electro magnetic fields, high voltage electric discharges, liquids (glue, oil, water, corrosives, etc.), foams, liquefied gases (if used outside a pneumatic system), halon gas fire extinguisher (to stop an opponent's ICE), unburned flammable gases, flammable solids, explosives, un-tethered projectiles, dry chaff (powders, sand, ball bearings), entanglement weapons (nets, strings, adhesive tape), lasers above 1 milliwatt, and light, smoke or dust based weapons that impair the viewing of robots (such as the use of strobe lights to blind the opponent driver).

2.9. Building Tools

The following chapters will present the several materials and components necessary to build a combat robot. But for that it is desirable to have access to a series of tools. Below, there is a comprehensive list with everything that could be useful in the construction of a combat robot. Most of the items can be found, for instance, at McMaster-Carr (www.mcmaster.com).

The book Kickin' Bot [10] has very good sections on how to effectively use most of these tools. There's also a great 43-minute video at http://revision3.com/systm/robots, featuring RoboGames founder Dave Calkins, teaching how to use basic tools to build a combot, as well as presenting a primer on the involved components.

Of course it is not necessary to own the entire list of tools presented below to build a combot. If your robot has some special part that you're not able to build by yourself, either due to lack of experience or to restricted access to a machine shop, you can have it machined straight from its CAD drawing through, for instance, the www.emachineshop.com website.

Mechanical
- safety: safety glasses, goggles, face shield, gloves, ear muffs, first aid kit;
- wrenches: Allen wrench (L and T-handle), combination wrench, open-end wrench, socket wrench, adjustable-end wrench, monkey wrench, torque wrench;

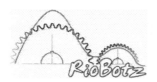 

- screwdrivers: flathead and Phillips;
- pliers: needlenose plier, cutting plier, vise-grip, slip-joint plier, retaining ring plier;
- clamping: C-clamp, bar clamp, bench vise, drill press vise;
- measuring: caliper, micrometer, steel ruler, tape measure, machinist's square, angle finder, level;
- marking: metal scriber, center punch, automatic center punch, hole transfer;
- cutting: scissors, utility knife, Swiss army knife;
- drilling: drill bit, unibit, countersink, counterbore, end mill, hole saw, reamer;
- tapping: tap, tap wrench;
- hand tools: hacksaw, file, hammer, jaw puller, keyway broach, collared bushing, telescopic mirror, telescopic magnet;
- weighing: dynamometer, digital scale;
- power tools: power drill (preferably 18V or more), jigsaw, Dremel, angle grinder, orbital sander, disc sander, circular saw;
- large power tools: lathe, bench drill, bench grinder, vertical mill, bandsaw, miter saw, belt sander, guillotine, CNC system, water jet system, plasma cutter;
- welding: oxyacetylene, MIG and TIG;
- cleanup: air compressor, air gun, vacuum cleaner (metal bits can short out the electric system).

Electrical / Electronic

- pliers: flush cutter, needle plier, crimper, wire stripper;
- soldering iron and support with sponge, desoldering tool;
- tweezers, magnifying glass, board support;
- digital multimeter, power supply, oscilloscope, battery charger;
- hot air gun, glue gun.

Fluids

- WD-40 (lubricant, it can be used to cut, drill and tap, and to clean Colson wheels);
- stick wax (to lubricate cutting discs);
- threadlocker (Loctite 242, it locks the screw in place);
- retaining compound (Loctite 601, it holds bearings);
- professional epoxy (the 24 hour version), J.B. Weld (even stronger metallic bonds);
- alcohol and acetone (metal cleanup before applying epoxy);
- layout fluid (to paint the parts and later mark holes or draw lines for cutting);
- penetrant dye (to inspect the presence of cracks);
- adhesive spray (3M Spray 77, to glue layout printouts onto plates);
- citrus-based solvent (Goo-Gone);
- solder paste and liquid electrical tape;
- wheel traction compound (Trinity Death Grip).

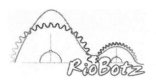

safety glasses/goggles	face shield	gloves	ear muffs
first aid kit	flathead screwdriver	Phillips screwdriver	socket wrench
open-end wrench	monkey wrench	L-handle Allen wrench	T-handle Allen wrench
needlenose plier	cutting plier	vise-grip	slip-joint plier
C-clamp	bar clamp	bench vise	drill press vise
caliper	micrometer	steel ruler	tape measure
machinist's square	angle finder	center punch	automatic center punch
metal scriber	hole transfer	drill bit	unibit

countersink	counterbore	end mill	hole saw
tap	tap wrench	hacksaw	file
telescopic mirror	telescopic magnet	torque wrench	air gun
retaining ring pliers	level	keyway broach	collared bushing
scissors	utility knife	Swiss army knife	reamer
hammer	jaw puller	dynamometer	digital scale
power drill	jigsaw	Dremel	angle grinder

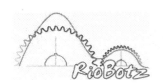

belt sander	orbital sander	disc sander	bench grinder
guillotine	miter saw	air compressor	vacuum cleaner
bench drill	vertical mill	bandsaw	miter saw
lathe	oxyacetylene welder	MIG welder	TIG welder
horizontal saw	plasma cutter	waterjet system	CNC system

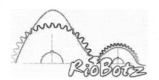 

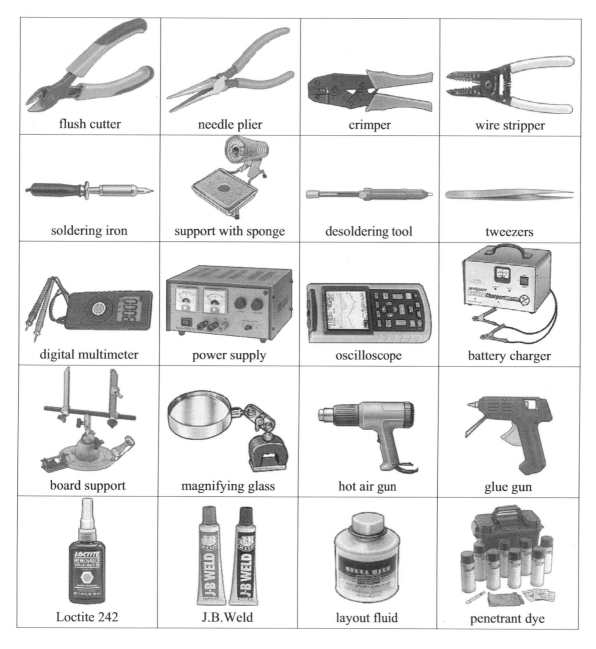

flush cutter	needle plier	crimper	wire stripper
soldering iron	support with sponge	desoldering tool	tweezers
digital multimeter	power supply	oscilloscope	battery charger
board support	magnifying glass	hot air gun	glue gun
Loctite 242	J.B.Weld	layout fluid	penetrant dye

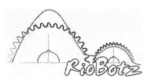

Chapter

3

Materials

The choice of structural materials is an important step to guarantee the robot's resistance without going over its weight limit. It is not a simple task to choose among the almost 100 thousand materials available, and for that it is necessary to know their mechanical properties.

3.1. Mechanical Properties

Mechanical properties quantify the several responses of a material to the loads it bears. These loads generate stresses, denominated σ, usually measured in MPa (units similar to pressure, $1MPa = 10^6Pa = 1N/mm^2$). In English units, $1MPa$ is equivalent to $0.145ksi$, where ksi stands for kilo pound-force per square inch (1 ksi = 1,000psi). For a uniform tensile stress distribution, stresses can be defined as the applied force divided by the material cross section area. These stresses also generate strains, denominated ε, which are a measure of deformation, of how much the material is elongated or contracted. The main mechanical properties can be obtained from the stress-strain curve.

The small graph to the center of the figure to the right shows the stress-strain curve of a material under small ε – in the example, smaller than 0.5% (it is the same graph as the large one, but zoomed in the region close to the origin). Note that, initially, the material has linear elastic behavior, in other words, the dependence between σ and ε can be represented using a straight line. The material stiffness is quantified by the modulus of elasticity E, or **Young modulus**, which is equal to the slope of this straight line (see figure). The larger the slope, the more rigid the material is.

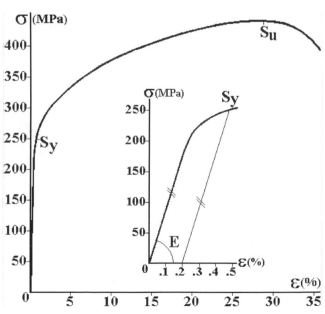

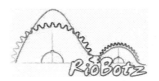

When applying increasingly larger loads, the plotted curve becomes no longer straight, becoming curved. This happens when the material begins to yield, which means it suffers permanent plastic deformations. When the stress reaches the **yield strength** S_y, the material already has 0.2% of permanent (plastic) deformation. In the previous graph, S_y is equal to 250MPa.

Looking now at the larger graph in the figure, note that the material continues yielding until the stress reaches a maximum value S_u, known as the **ultimate strength**, after which the material breaks (in the above example, S_u is about 450MPa). The **fracture strain**, ε_f, is the maximum strain that the material can tolerate before breaking. Beware that, although related, there are subtle differences between fracture strain and **ductility**. Ductility is the material capacity to plastically deform without breaking, while the fracture strain includes both elastic and plastic deformation components. So, if a material is ductile, then it has a high ε_f, but the opposite is not necessarily true: brittle tool steels can achieve a high purely elastic ε_f with almost no ductility.

The stress-strain curve is measured in slow traction tests. Therefore, S_u measures the material resistance to static loads. The resistance to dynamic loads is measured by two other properties of interest: **impact toughness** and **resilience**. Both measure the resistance of the material to impacts. But the impact toughness measures how much impact energy the material absorbs before *breaking*, while the resilience measures such energy before it *starts to yield* (plastically deform).

The impact toughness depends not only on the material strength, but also on its fracture strain. The more it can deform while resisting to high stresses, the more impact energy it can absorb. This is why it is possible to estimate the impact toughness from the area below the entire stress-strain curve (as pictured to the right). Higher values of S_u end ε_f result in a larger area under the entire curve, resulting in a higher impact toughness. The resilience can also be estimated from the area below the curve, but only in the linear elastic 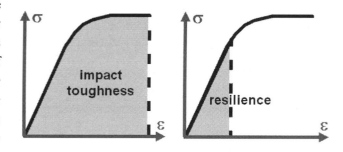 region, where the stresses are below S_y (see figure above). That area is approximated by $S_y^2/2E$.

But a tough material is not necessarily resilient, and vice-versa. For instance, the stainless steel (SS) type 304, the most used SS, tolerates large deformations but it is easily yielded. Therefore, it is very tough (because of the large ε_f), being good for armor plates that can be deformed. However, it has low resilience (because its S_y is low), and thus it should not be used in shafts (which should not get bent or distort) or in wedges (because if their edges are bent or nicked they lose functionality).

On the other hand, the steel from a drill bit, for instance, is very hard, it has a very high yield strength S_y, and thus it has a high resilience. However, its ε_f is small and therefore its impact toughness is low. This is why drill bits do not make good weapons for combat robots, because they easily break due to impacts. Titanium is an excellent choice for use in combat robots because it is very tough (good for armor) and resilient (good for wedges) at the same time, as it will be discussed further on.

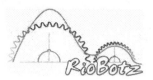 

Fracture toughness, K_{Ic} (pronounced "kay-one-see"), is the resistance of the material to the propagation of cracks. It is measured in cracked specimens, under static loads that are slowly increased until the material fractures (breaks). K_{Ic} is measured using the unusual units MPa√m or ksi√in. The higher the K_{Ic} of the material of an already cracked component, the higher the stresses it can withstand before fracturing. In most metals, it is observed that the impact toughness is very much related to the fracture toughness, even though the first is measured in dynamic and the other in static tests. More specifically, several experiments suggest that the impact toughness is directly proportional to K_{Ic}^2 / E, where E is the Young modulus. However, this is not always true for non-metals. Lexan, for instance, has a relatively high impact toughness for a polymer if cracks are not present. However, its fracture toughness is low, easily propagating cracks once they are initiated, usually around holes, which concentrate stresses.

Note that fracture toughness must be measured for very thick specimens to be called K_{Ic}. This is because fracture toughness has some thickness dependence, thinner plates can deform more easily and absorb more energy per volume than thick plates. It is found that the apparent fracture toughness of a very thin plate can reach up to twice the value of K_{Ic}. So, when searching for fracture toughness data, make sure you're getting K_{Ic} from thick specimen tests, and not a higher apparent value that cannot be compared to the K_{Ic} of other materials.

Finally, the **hardness** of a material is the resistance to penetration by other harder materials. If we press a very hard material (for instance the tip of a diamond) onto the surface of a softer one, the softer one will become dented. The larger and deeper the dent, the softer the material is. A very common hardness unit for hard metals is Rockwell C (HRc). The larger the value, the harder the material is. Another common hardness unit is Brinell (HB), measured in kg/mm^2. A conversion table between HRc and HB hardnesses can be found in Appendix A.

In general, among metals from the same family (such as among steels), the ones with higher hardness tend to have proportionally higher S_u. For instance, you can estimate within a few percent $S_u \cong 3.4 \cdot HB$ for steels, where HB is in kg/mm^2 and S_u is in MPa. This estimate is very useful in practice, because hardness tests are non-destructive and very fast to perform. For steels, this estimate is so good, with a low dispersion, that its coefficient of variation CV is less than 4%. Aerospace aluminum alloys have a relatively good correlation, $S_u \cong 3.75 \cdot HB$, with CV = 6%. There are also estimates for other alloys, but the results have higher scatter (as seen from their higher CV). For instance, aluminum alloys from the 6000 series (such as 6061) have $S_u \cong 3.75 \cdot HB$ (CV = 12%), titanium alloys have $S_u \cong 3 \cdot HB$ (CV = 16%), and magnesium alloys have $S_u \cong 4.2 \cdot HB$ (CV = 20%). These estimates are very good for quick calculations, but use them at your own risk.

Among all the properties presented above, the most important ones in combat robots, as well as in most engineering applications, are without a doubt the impact and fracture toughnesses. Robots need to tolerate impacts and cracks without breaking.

Once having presented the main mechanical properties, we can analyze the main materials used in combat robot construction, as follows.

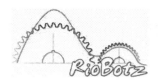

3.2. Steels and Cast Irons

Steels are metals composed basically of iron and of some other (in general few) alloy elements. Depending on the type, they can be extremely resistant, however their high density would make an all-steel robot very heavy. The density of steels does not vary much, between 7.7 and 8.0, with average 7.8 (which means 7.8 times the density of water, or 7.8kg per liter of the material).

Their stiffness also varies very little, around $E \cong 200GPa$ (notice that 1GPa = 1000MPa). This means, for instance, that to deform a piece of any steel in 0.1% it would be necessary to apply a stress of 200GPa × 0.001 = 0.2GPa = 200MPa (29ksi), the equivalent to a force of 200N for each mm^2 of cross section of the material. On the other hand, the strengths of steels can vary a lot: the best steels get to be 10 times more resistant than low strength ones, therefore it is important to know them very well.

Low strength steels are ready to be used soon after being machined. However, many steels need to go through heat treatment (HT) after machining to reach high strengths. For instance, in steels, the HT consists of heating up the material to a high temperature (typically 800 to 900°C, or 1472 to 1652°F, but it varies a lot with the steel type) and cooling it in water, oil, powder or even air (the quenching process), and later heating it up for a few hours in a not so high temperature (typically 200 to 600°C, or 392 to 1112°F, the temper process). HT can be performed in your shop with just a torch and water or oil, however specialized companies are recommended for a better result with larger reliability in the resulting mechanical properties. It may cost around US$50 to heat treat a small batch of the same material.

The following are a few of the main types of steel used in combat robots.

- 1018 steel, 1020 steel: they are mild steels, they have low carbon content, about 0.18% to 0.20% in weight respectively. They have low strength, but they are easily conformed, machined, and welded. They're usually used in shafts and in a variety of components. They are used in the robot structure due to their low cost, however their low yield strength S_y makes them easily bendable (therefore avoid using them in spinning weapon components that need to be well balanced, as pictured to the right). HT only gets to increase the strength and hardness of those low carbon materials, their interior continues with low strength.

- 1045 steel: steel with medium carbon content (0.45%), it is used when larger strength and hardness are desired. It is used in high-speed applications, gears, shafts and machine parts. It is a cheap solution for the robot shafts, however it needs HT after machining.

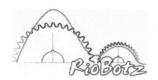 

- 1095 steel: steel alloy with high carbon content (0.95%), with hardness and strength elevated after HT. It tends to be brittle, with low impact toughness. It is typically used in springs or cutting tools that require sharp cutting edges.

- 4130 steel: steel with 0.30% carbon, with addition of chrome and molybdenum (also called chromoly) to increase strength. The low carbon content makes them a good choice for welding, allowing robots to have their structure formed by 4130 bars and tubes, which are welded together and then heat treated to reach great strengths.

- 4340 steel: steel with 0.40% of carbon, with nickel in addition to chrome and molybdenum (chromoly), with even higher strength and impact toughness after HT than 4130 or 4140 (4140 is equivalent to 4130 but with 0.40% carbon). The typical applications are for structural use, such as components of the landing gear of airplanes, gears for power transmission, shafts and other structural parts. It is an excellent material for shafts, the weapon shaft of our middleweight spinner *Ciclone* is all made out of tempered 4340 steel. To reach high impact toughness, HT the 4340 in a way to leave it with final hardness between 40 and 43 Rockwell C – much more than that and the shaft becomes brittle, breaking under a severe impact, and much less than that will allow the shaft to easily yield. Our recipe for 4340 steel is to heat it up and keep it at 850°C (1562°F) for 30 minutes, quench in oil until reaching 65°C (149°F) (important: in the case of shafts, dip it in vertically to avoid distortions), and soon afterwards temper at 480°C (896°F) for 2 hours.

- AR400 steel: high hardness steel once used in the wedge of the famous middleweight Devil's Plunger, until it was replaced with titanium. AR stands for abrasion resistant, and 400 is its Brinell hardness. AR400 has almost the same mechanical properties as 4340 steel hardened to 43 Rockwell C, which is equivalent to 400 Brinell. It is also known as Hardox AR400 steel.

- 5160 steel: steel with 0.60% of carbon, it contains chrome and manganese. Called spring steel, it has excellent impact toughness. It is usually used in heavy applications for springs, especially in the automotive area, such as leaf springs for truck suspension systems. The spinning bars of our middleweights *Ciclone* and Titan are made out of heat treated 5160 leaf springs. Be careful with the HT, the harder it gets, the lower the impact toughness – during the RoboCore Winter Challenge 2005 competition, *Ciclone*'s spinning bar was severely HT to reach a 53 Rockwell C hardness, so hard that it broke against the rammer *Panela* due to the reduced impact toughness. After that, we changed the HT without having any problems. The ideal hardness for 5160 steel in combat applications is between 44 and 46 Rockwell C, this is what we use now for *Ciclone* and Titan. Our recipe is to heat it up and keep it at 860°C (1580°F) for 30 minutes, quench in oil until reaching 65°C (149°F) (important: in the case of spinner bars, dip it in horizontally to keep any spring-back effects symmetrical, preventing unbalancing effects), and soon afterwards temper at 480°C (896°F) for 2 hours.

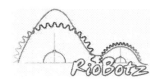

- stainless steels: they are steels with more than 12% in weight of chrome, which forms a protective film that prevents corrosion. There are 60 types of stainless steel (SS), the most used one is the SS type 304, also called 18-8 for having 18% of chrome and 8% of nickel. It has an excellent combination of impact toughness and resistance to corrosion, and it doesn't need to be HT. SS 304 is a good material for the robot armor (despite being heavy) because, besides being very tough, it increasingly hardens after suffering impacts and deformations. However, SS 304 is easily deformed, making its resilience low, therefore avoid using it in parts that significantly lose functionality if bent or distorted such as shafts. There are other SS with higher resilience, they are the martensitic SS, the most famous of them are the types 410, 420 and 440: they need to be HT, after which they reach high S_y and S_u, however their impact toughness is usually much lower than the one from 304. High end stainless steels are the precipitation hardened (PH) types, such as 17-7PH and 15-5PH, which are necessary mostly in high temperature applications.

- tool steels: tool steels can reach very high hardness values after HT. They are used to make tools and metal dies, however most of them have low impact toughness. The exceptions are the tool steels from the S series (S meaning Shock), which have a high impact toughness in addition to hardness, to be used in chisels, hammers, stamping dies, and applications with repetitive impacts. The S1 and S7 steels are the most used tool steels in combat robots, respectively in Brazil and in the US. They are mainly used in the weapon parts that get in contact with the opponent. The teeth from *Touro*'s drum are made out of S7 steel, as well as the spinning blade of the middleweight Hazard, pictured to the right. They are not too expensive, S1  steel can be found in Brazil for about US$13/kg (almost US$6/lb), while S7 steel can be found in the US at, for instance, www.mcmaster.com. Regarding HT, several robot builders adopt hardnesses varying between 52 and 60 Rockwell C (HRc) for S7 steel. Our recipe for S7 steel is to pre-heat up to 760°C (1400°F), equalize the temperature throughout the entire piece, continue heating up and keep it at 950°C (1742°F) for 30 minutes, then quench in oil (S7 can also be quenched in air, which is good to avoid warping due to the thermal shock with the oil) until 65°C (149°F), and immediately temper at a certain temperature for 2 hours. After cooling, it may be tempered again for 2 hours at the same temperature, what is

called a double temper, instead of a single one. The temper temperature depends on the desired hardness, see the graph to the right for single-tempered S7 steel: for a hardness value close to 60 HRc, use 392°F (200°C), while for values close to 52 HRc use 752°F (400°C). We've

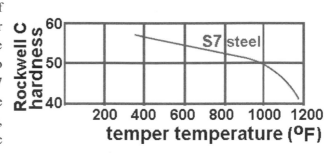

realized that there is a peak in impact toughness of the S7 steel at exactly 54 HRc, which can be achieved with a single temper with temperature around 600°F (315°C). Impact tests performed on standard unnotched Charpy specimens made out of S7 at different tempers showed that they would absorb 309 J (Joules) of impact energy before breaking when at 54 HRc. "Nothing more, nothing less than 54 Rockwell C," as experienced builder Ray Billings once told us. This is because at either 56 or 52 HRc the absorbed energy drops to less than 245 J. This energy would rise back again beyond 300 J only for lower hardnesses, below 51 HRc (324 J for 50HRc and 358 J for 40HRc). Therefore, the best cost-benefit to have both impact toughness and hardness at high levels, to prevent the component from breaking while retaining its sharpness, is to use S7 steel at exactly 54 HRc. Not 55 HRc. Not 53 HRc.

- AerMet steels: it is a class of special high strength steels with high nickel and cobalt content, patented by the Carpenter company (www.cartech.com). There are 3 types with increasing hardness but decreasing toughness: AerMet 100, AerMet 310 and AerMet 340. The most famous of them, AerMet 100, is replacing older special nickel-cobalt steels such as AF1410 and HP-9-4-30. After HT it reaches hardnesses from 53 to 55 Rockwell C, with 2.15 times higher impact toughness than S7 steel. It is probably the metal with best ultimate strength and fracture toughness combination in the world at the present time, with $S_u = 1964$MPa and $K_{Ic} = 130$MPa$\sqrt{m}$ (after HT). AirMet 100 is used in the lifting mechanism of the heavyweight BioHazard, enabling it to save weight using a compact 3/4" diameter shaft. It

was also used in the output shaft of the 24V DeWalt Hammerdrill gearbox, as pictured to the right. As one would expect, it is very expensive, more than US$55/kg (US$25/lb) in the US. To obtain its best properties, heat it up to 885°C (1625°F) for 1 hour, cool to 66°C (150°F) in 1 to 2 hours using either oil quenching or air cooling, then refrigerate to −73°C (−100°F) and hold for 1h. Then heat at 482°C (900°F) for 5 hours, never below 468°C (875°F), and finally air cool.

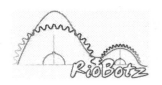 

- maraging steels: it is a class of special high strength nickel-cobalt-molybdenum steels with very low carbon content. There are four commercial types, all of them with 18% nickel: 18Ni(200), 18Ni(250), 18Ni(300) and 18Ni(350), with S_y equal to 1400, 1700, 2000 and 2450MPa, respectively, after HT. To obtain these S_y properties, heat to 900°C (1650°F) and hold for at least 1 hour, air cool to room temperature, then heat for 3 hours at 482°C (900°F), and finally air cool. The 18Ni(350) needs 12 hours (instead of 3) at 482°C. Together with the AerMet alloys, these are the best steels for high strength and high toughness applications, however they are also expensive, between US$42/kg (US$19/lb) and US$64/kg (US$29/lb) in the US, at www.onlinemetals.com. The 18Ni(200) can reach higher K_{Ic} than AerMet 100, but 24% lower S_u. The 18Ni(250) is a reasonable replacement for AerMet 100, but with 10% lower S_u and K_{Ic}. The 18Ni(350) is recommended for low impact applications because it has one third of the K_{Ic} of AerMet 100, but its S_u can almost reach the incredible mark of 2500MPa.

- K12 Dual Hardness steel: it is an armor plate with dual hardness sold by Allegheny Ludlum, with a high hardness front side to break up or flatten incoming projectiles, and a lower hardness back side that captures the projectile. The front side has a higher carbon content, reaching 58 to 64 Rockwell C hardness after heat treatment, metallurgically bonded to a lower carbon back side that reaches 48 to 54 Rockwell C, as pictured in the cross section to the right. With the hard front side facing out of your robot, you'll be able to break up or chip any sharp edges from your opponent's weapon, while the inner "softer" side will provide high toughness and prevent fractures. A careful and precise heat treatment is required to achieve optimum performance.

Soft   Hard   impact

- cast irons: they are basically steels with more than 2.5% of carbon content. The carbon excess ends up generating graphite inside the microstructure, which is very brittle (anyone who's used a pencil knows that). In combat robots they are used in bearing housings and in a few gears. Be careful with this material, it has a low impact toughness – the 2004 version of our spinner *Ciclone* used cast iron flanged housings (pictured to the right) to hold the bearings of its weapon shaft, however one of them cracked from its own impact against other robots. Luckily, it still resisted until the end of the competition, despite the cracks. Since 2005 we've stopped using cast iron housings, and started to embed the weapon shaft bearings into the aluminum plates of the robot structure. We haven't had cracking problems with bearing housings ever since.

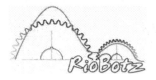 

3.3. Aluminum Alloys

Aluminum is a very light metal, it has about 1/3 of the density of steels, about 2.8, which makes it very attractive for the robot structure. Its stiffness is also around 1/3 of the one of steels, with a Young modulus $E \cong 70GPa$. Many types of aluminum exist, usually denominated by a 4 digit number. The aluminum alloys from the 1000, 3000 and 5000 series (for instance the aluminum 1050, used in electric equipment, the 3003, used in kitchen utensils, and the 5052, resistant to sea corrosion) are low strength and should not be used in combat. Cast aluminum is even less resistant, and it should be avoided – the wheels of the 2004 version of our middleweight *Ciclone* were made out of cast aluminum and rubber, luckily they didn't break but they most likely would if hit by another spinner.

A few aluminum alloys from the 6000 series (such as the 6061-T6 and 6351-T6) have medium strength, becoming a reasonable choice for the robot structure. The alloys from the 2000 and 7000 series (such as the 2024-T3 and the 7075-T6) are called aerospace or aircraft aluminum due to their extensive use in aircrafts. With high S_y and S_u, they are naturally the most expensive. The 7000 series alloys usually have higher S_y and S_u than the 2000 series, but sometimes this comes along with a lower fracture toughness.

Aluminum alloys already come heat treated from factory, which saves us time and money when building a combot. Be careful with the denominations with the letter T, the number after it indicates which heat treatment was used: for instance, the aluminum 6061-T6 has much higher strength than 6061-T4, which suffered a different HT. The main types of aluminum alloys are discussed next.

- 6063-T5 aluminum: it is the aluminum alloy used in almost all the architectural extrusions in the market, because it has high corrosion resistance, and it is relatively cheap. However, it has low strength, therefore avoid using it in the robot external structure. It can be used in the internal structural parts, to stiffen the robot or to support batteries. Because all aluminum alloys have roughly the same stiffness (due to their Young modulus always close to 70GPa), the 6063-T5 is as effective as any other more expensive aluminum alloy to stiffen the structure, its problem is just its low strength. Note that stiffness and strength are two different things: for instance, glass is much more rigid than Lexan (polycarbonate), however Lexan has a much higher ultimate tensile strength than glass. Several internal parts from our middleweight *Touro* are made out of 6063-T5 extrusions. Because it is difficult to find in

    Brazil C-channels or I-beams made out of structural aluminum such as 6061-T6, the side walls of our middleweight *Ciclone* ended up using 6063-T5 extrusions, as pictured to the right – but, to make up for that, they were reinforced with an outer layer of grade 5 titanium sheet. Depending on the quantity, 6063-T5 (or 6063-T52) costs between US$6 and US$13 per kg (between US$2.7 and US$5.9 per lb). A few stores only sell extrusions in 6 meter lengths.

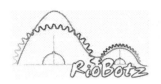

- 6061-T6 aluminum: it is the most common structural aluminum alloy, used in several applications such as bicycle frames, structures, naval and truck components. It has medium

strength, about twice the S_u of 6063-T5, and it can also be welded. Compared to aerospace alloys, 6061-T6 has lower S_y and S_u strengths, with a similar impact toughness. Its greatest advantage is its good weldability, much better than in most aerospace aluminum alloys. All the famous robots from Team Plumb Crazy are made out of 6061-T6 extrusions, as pictured to the right, as well as our hobbyweights *Tourinho* and *Puminha*. Extrusions can be found in the US, for instance, at Online Metals (www.onlinemetals.com).

- 5083-H131, 5086-H116, 5086-H32 aluminum: despite their low yield strength, common to the marine alloys of the 5000 series, they have such a high impact and fracture toughness that they are used as armor plates in light weight military vehicles. They are good material candidates for very thick armor plates.

- 2024-T3, 7050-T7451, 7075-T6, 7075-T73, 7475-T7351 aluminum: high strength aerospace alloys, with about 3 times the strength of 6063-T5. They are useful for structures that demand high strength-to-weight ratio, usually to manufacture truck wheels, fuselage of airplanes, screws, orthopedical belts, and rivets. They are the best commercially available aluminum alloys, in Brazil the 7050-T7451 and 7075-T6 cost around US$17/kg (almost US$8/lb). Considering that a middleweight with its entire structure made out of aluminum would need around 15kg (33lb) of this material, about US$250 would be enough to build it using the best aerospace alloys available, a good investment with a relatively low cost.

- 2324-T39 Type II, 2524-T3, 7039-T64, 7055-T74, 7055-T7751, 7085-T7651, 7150-T77, 7175-T736, 7178-T6 aluminum: high end aluminum alloys with improved mechanical properties over the traditional aerospace alloys. They are not readily available commercially.

- Alusion – very light aluminum foam (pictured to the right), available in several densities. Despite its low strength, it can be used as thick ablative armor plates mounted on top of the robot structure.

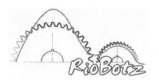

3.4. Titanium Alloys

Titanium is one of the best materials for combat robots. With little more than half the density of steels (between 4.4 and 4.6), it reaches strengths 2.5 higher than 1020 steel. Or up to four times higher in a few military grade titanium alloys, making their strength-to-weight ratio so attractive that they're used in 42% of the F-22 fighter aircraft. Its Young modulus is $E \cong 110GPa$, about half the one of steels. They are non-magnetic, non-toxic, and extremely resistant to corrosion, even in the presence of biological fluids, which explains their use in prosthetics and medical implants. Titanium generates beautiful white sparks when it is ground. Care should be taken with titanium chips from machining, they are flammable. We've carefully made several mini-bonfires with titanium chips in our lab, they generate a very intense white light.

Titanium alloys are difficult to cut and drill. The secret to drill them is to use low spindle speeds in the drill and a lot of pressure on the part (always use a bench drill with them, never a manual one). And, most importantly, do not let the piece get hot, therefore use plenty of fluid. If there is heat build-up, titanium forms a thin oxide layer that is harder than the drill bit, and then several bits will be worn-out in the process. Use special cobalt drill bits to drill titanium, they will last longer. Practice is also important.

A curiosity about titanium (as well as niobium) is that its surface can be colored without paints or pigments, just using Coke (or Pepsi) in a technique called electrolysis or anodizing. The figure to the right shows an artistic painting made on a titanium plate. Note the range of colors that it is possible to obtain.

To color it, you need a piece of stainless steel (SS) with equal or larger area than the one of the titanium to be colored, a SS screw, a titanium screw, Coke, and a DC power source (of at least about 30V). The scheme is pictured to the right. Polish well the titanium surface and clean it with alcohol or acetone – do not leave any fingerprints. Place the titanium part (which will be the anode) and the SS one (the cathode) submerged in Coke (the electrolyte, which can also be replaced with Trisodium Phosphate Na_3PO_4), very close together but without

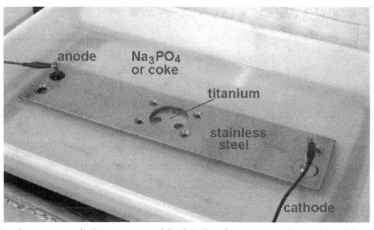

making contact. Make sure the titanium screw is in contact with the titanium part to be colored but not with the SS plate (we used a rubber grommet to guarantee this, as shown in the picture), and the

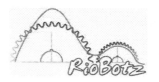

SS screw only touches the SS piece. Connect the positive of the DC power source to the titanium screw and the negative to the SS one, without letting the wire contacts touch the electrolyte. Apply a certain DC voltage between 15 and 75V for a few seconds and it is done, the titanium part is colored!

A few of the colors that can be obtained are pictured to the right. The titanium color obtained by the electrolysis process depends on the applied voltage. The higher the voltage, the thicker will be the titanium oxide layer that is formed on the plate (anode), changing its color. This color change happens because the oxide layer causes diffraction of the light waves. The colors are gold (applying 15V), bronze (20V), purple (25V), blue-purple (30V), light blue (35V), white bluish (40 to 45V), white greenish (50V), light green (55V), yellow-greenish (60 to 65V), greenish gold (70V) and copper (75V). There are other colors up to 125V, but they are opaque, not very brilliant.

Coke works well, but it is not the best electrolyte. We've discovered that Diet Coke is a little better because it doesn't have sugar, which accumulates on the contacts. But the best option would be to use Trisodium Phosphate (Na_3PO_4, known as TSP), diluted at about 100 grams for each liter of distilled water (about 13oz/gallon). Besides being transparent (which allows you to see the colors as you increase the voltage), TSP is a detergent that helps to keep the titanium surface clean during

the electrolysis, resulting in a more uniform color. In the picture to the right you can see Titan's side walls, the top two plates before the process and the bottom one after being colored using TSP and 30V. Note the masking that we've used on the top plate, written TiTAN, made out of waterproof adhesive contact paper. The mask protects the region during electrolysis, leaving afterwards letters with the original color of the titanium (as it can be seen in "RioBotz" written on the bottom plate).

Commercially pure titanium, the most common of which is grade 2 titanium, has lower strength and higher density than aerospace aluminum, therefore it should not be used in combat robots. Use only high strength alloys such as grade 5 titanium, known as Ti-6Al-4V. Ti-6Al-4V has twice the strength of the best aerospace aluminum alloys and much higher impact toughness, with only 60% higher density. However, when welding grade 5 titanium, it is a good idea to use grade 2 as a filler

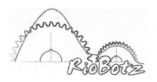

material. This is because welds are prone to cracking due to thermally induced residual stresses, and grade 2 titanium filler, despite its lower strength, has a higher ductility that prevents such cracks and improves the overall impact and fracture toughness.

Ti-6Al-4V is also known as Ti-6-4, for having 6% aluminum and 4% vanadium in weight, mixed with 90% titanium. It is the most used high strength titanium alloy, combining excellent mechanical strengths and corrosion resistance with weldability. It is extensively used in the aerospace industry in a variety of applications in turbines and structural components up to 400°C (752°F). It can be heat treated (STA – Solution Treated and Aged), however the increase in ultimate strength is small, with the drawback of a 43% lower K_{Ic}. In practice, most combat robots use Ti-6-4 in the annealed condition, without further heat treating. It is usually available in the mill annealed condition. It can also be found in other two annealed conditions: recrystallization anneal (8% higher K_{Ic}) or beta anneal (33% higher K_{Ic} but much lower S_u). Unfortunately, titanium grade 5 is expensive, about US$55/kg to US$80/kg (US$25/lb to US$36/lb). Notorious resellers are Titanium Joe (www.titaniumjoe.com), President Titanium (www.presidenttitanium.com), and Tico Titanium (www.ticotitanium.com).

Ti-6-4 has an unbelievable impact and fracture toughness for its weight, we've used it in all side walls and bottom plate of our spinner Titan, as armor plates covering the aluminum walls of *Touro* and *Ciclone*, and in the wedges of Titan and *Puminha*. If you need an even higher fracture toughness, you could use the more expensive Ti-6Al-4V ELI (Extra Low Interstitial), which presents lower impurity limits than regular Ti-6Al-4V, especially oxygen and iron. The lower oxygen content increases the fracture toughness in 22% over mill annealed Ti-6Al-4V, however it lowers in about 10% the yield and ultimate strengths.

The graph to the right shows a comparison among steels, aluminum alloys and Ti-6Al-4V titanium used in combat robots, through their stress-strain curves. The curves stop at the strain where the material breaks. Remember that the higher the curve gets, the larger the S_u strength to static loads until rupture. The farther the curve gets to the

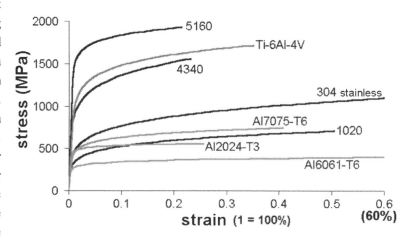

right, the higher the material can be plastically deformed before breaking, in other words, the higher their ductility and their ε_f. Note that the 7075 and 2024 aluminum alloys behave in a similar way to the 1020 steel (except for their lower impact toughness), however with only 1/3 of the weight. The stainless steel 304 has the largest area under the curve, resulting in a very high impact toughness, however it begins to yield under relatively low stresses. Note from the areas under the curves that Ti-6Al-4V has similar impact toughness to 5160 steel, but with almost half the weight.

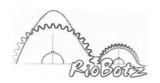

3.5. Magnesium Alloys

Magnesium is the third most used structural metal, after steels and aluminum alloys. The magnesium alloys ZK60A-T5 and AZ31B-H24 are excellent for the robot structure, because they have strength similar to 6061-T6 aluminum however with only 65% of its weight: the density of the magnesium alloys is only about 1.8, instead of 2.8 from aluminum. Their Young modulus is relatively low, $E \cong 45GPa$, however their low density allows

the use of very thick plates, resulting in very high stiffness-to-weight ratios. The impact toughness of the best magnesium alloys is similar to the one of high strength aluminum alloys.

The largest drawback of magnesium alloys is their extremely poor corrosion resistance: magnesium is in the highest anodic position on the galvanic series. Also, when tapping magnesium, choose coarse instead of fine threads to avoid stripping.

The ZK60A-T5 (US$62/kg or US$28/lb for small quantities) is the commercially available magnesium alloy with highest fracture toughness, however it is difficult to find large plates of that material. The alloy AZ31B-H24 (US$42/kg or US$19/lb for small quantities) is a little less resistant, but it is easier to find. The heavyweight lifter BioHazard has used these magnesium alloys to stay under the weight limit.

There are other magnesium alloys, such as Elektron WE43-T5 and Elektron 675-T5, however all of them have lower fracture toughness than ZK60A-T5 and AZ31B-H24. The new experimental alloy Elektron 675-T5, which was in its final stages of development in 2008, has the highest ultimate strength of all Mg alloys, $S_u = 410MPa$.

Note that there are often misconceptions regarding the flammability of magnesium and its alloys. It may ignite when in a finely divided state such as powders, shavings from magnesium fire starters (pictured to the right), ribbon or machined chips, exposed to temperatures in excess of 445°C (833°F, the lowest Solidus temperature of all Mg alloys). However, in solid form, magnesium is very difficult to ignite. It has a high thermal conduction, quickly dissipating any localized heat. Also, most alloys self extinguish in the event of ignition,

because of the oxide skin that forms over any molten alloy (in special in the presence of yttrium, such as in the two mentioned Elektron alloys). In practice, magnesium alloy ignition only happens due to sustained major fires (such as major fuel fires following an accident), similarly to aluminum ignition. The US Army is starting to use thick magnesium alloy plates as armor in its light weight vehicles, without any problems even during severe ballistic tests.

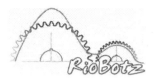 

3.6. Other Metals

A few other metals than can have structural application are:

- copper alloys: copper is an excellent electric conductor, and the bronze alloys (a copper alloy with, usually, tin) it generates are great for statues – but not for the structure of combat robots. Besides having lower strength than most steels, copper alloys are heavier, with density around 9.0. Bronze bushings (pictured to the right, showing the regular and flanged types), on the other hand, are a good option to be used as plain sleeve bearings in shafts for the wheels and weapons. The SAE 660 bronze,  a.k.a. alloy 932, is hard, strong and nonporous, offering excellent resistance to shock loads and wear, it is the best option for sleeve bearings under high impact loads. Another option is the SAE 841 bronze, a.k.a. Oilite, a porous sintered material impregnated with roughly 18 percent SAE 30 oil – it is cheaper and it provides less friction than SAE 660, but it has lower strength and impact toughness. Brass (a copper alloy with zinc) also has low strength, but it is an excellent material for shim stock (pictured to the right), to be inserted in between parts to avoid slacks.

- nickel superalloys: they are a little heavier than steels, and they only present advantages if used at very high temperatures. The best superalloys can retain their high strength even at up to 80% of their melting temperatures. They can easily work at temperatures between 700 and 1000°C (1292 to 1832°F), which is great for components inside jet engines, but useless for combat, unless the competition is held in Venus or Mercury.

- beryllium alloys: theoretically, they are by far the best metals in the world to make a light and rigid structure. A few alloys such as the S-200 have Young modulus E = 303GPa (more than 4 times the stiffness of aluminum alloys) with density lower than 1.9. They find applications in nuclear reactors, inertial guidance instruments, computer parts, aircraft, and satellite structures. They would be a marvel in combat except for three problems: they are relatively brittle; beryllium must be processed using powder metallurgy technology, which is costly; and beryllium powder and dust, which can be released during the machining process, as well as in the wear and tear during combat, is highly toxic and cancerous. Because of that, competitions usually forbid their use. Do not use it, berylliosis disease can kill you. A curious fact is that beryllium and its salts taste sweet. Early researchers (who are not among us anymore) used to taste beryllium for sweetness to verify its presence. This is why it used to be called glucinium, the Greek word for sweet. Do not taste it, just take my word that it is sweet.

- tungsten alloys: very high density alloys, their application in combat robots is mainly for counterweights of spinning weapons. Tungsten, meaning "heavy stone" in Swedish, has in

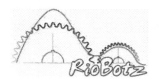 

its pure form an amazing 19.35 density, with the highest melting point at atmospheric pressure among metals (3,422°C or 6,192°F), losing only to diamond's 3,547°C (or 6,416°F). It has also the lowest coefficient of thermal expansion of any pure metal. In its raw state, it is brittle and hard to work. However, when alloyed with 3% to 10% of nickel, copper and/or iron, it becomes relatively tough, extremely machinable, and it reaches ultimate strengths S_u between 758MPa and 848MPa. Their machining properties are similar to gray cast iron These tough high density alloys are known as ASTM-B-777-07 or Densalloy, with 4 different classes: class 1 (or HD17, with 90% tungsten, and a density of 17), class 2 (HD17.5, 92.5% tungsten, 17.5 density), class 3 (HD18, 95% tungsten, 18 density) and class 4 (HD18, 97% tungsten, 18.5 density). They can be found, for instance, at www.mi-techmetals.com, www.marketech-tungsten.com or www.hogenindustries.com, with a typical price between US$50 and US$100 per pound for special orders in small quantities. Small inexpensive tungsten weights can be found at www.maximum-velocity.com.

- other high density alloys: besides tungsten, there are several other high density alloys, however they're all extremely expensive. They usually have low strength, low toughness, or they are too dangerous to use in combat. The most famous high density materials are tantalum (with density 16.65 and reasonable mechanical properties with S_u greater than 450MPa), depleted uranium (density 18.95, reasonable S_u between 615MPa and 740MPa, toxic, used in tank armor and armor-piercing projectiles), gold (density 19.32, ductile but with low strength due to S_u = 120MPa), rhenium (density 21.04, good mechanical properties, S_u = 1070MPa), platinum (density 21.45, S_u = 143MPa, low strength), iridium (density 22.4, S_u = 1000MPa, brittle), and osmium (the material with highest density, 22.6, about twice the density of pure lead, with S_u = 1000MPa, very brittle and toxic).

- metallic glasses: they are amorphous metals, which have a disordered atomic-scale structure similar to common glasses, in contrast to most metals, which are crystalline with a highly ordered arrangement of atoms. They can be produced from the liquid state by a cooling process so fast that the atoms don't have time to organize themselves as crystals. They usually contain several different elements, often a dozen or more, causing a "confusion effect" where the several different sized atoms cannot coordinate themselves into crystals. Also, they don't have a melting point. Instead, they become increasingly malleable as the temperature increases, just like most plastics, making them good candidates for injection molding. Liquidmetal is a company that sells glassy metals such as Vitreloy, an alloy with mostly zirconium and titanium that reaches S_y = 1,723MPa, nearly twice the strength of Ti-6Al-4V. But since the atoms are "locked in" in their amorphous arrangements, most currently available glassy metals cannot plastically deform at room temperature, resulting in low impact strength, which limits their use in combat. They also have a coefficient of restitution close to 1, meaning almost perfectly elastic impacts. Versions with improved impact toughness could be created in the future by embedding ductile crystalline metal fibers into the metallic glass, forming a metal matrix composite (as discussed later). In 2004 the first iron-based metallic glass was created, called "glassy steel," with very high strength.

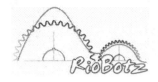

3.7. Non-Metals

Several non-metals need to be mentioned in combat robot design. The main ones are:

- polycarbonate: also known as Lexan, it is a polymeric thermoplastic (which softens and melts when heated, instead of burning), transparent to light waves and radio-control signals. It has high impact toughness, and it is very light, with density 1.2. It is used in combat robot armor, it absorbs a lot of energy as it is deformed during an impact. In spite of that, fewer and fewer robots have been using this material, because of its disadvantages: it has very low Young modulus (E = 2.2GPa, about 1% of the stiffness of steels, making the robot structure very flexible even for high thicknesses), it easily cracks (the cracks usually appear starting from the holes, and they propagate without absorbing much of the impact energy), and it is easily cut (becoming vulnerable to sawbots). To avoid cracking, chamfer all holes to remove sharp corners and edges, and provide the Lexan support with some damping, for instance using a thin layer of rubber or neoprene. Avoid tapping Lexan, if you must do it then guarantee that the hole is tapped very deeply with several threads, or else they might break. Never use threadlockers such as Loctite 242 in Lexan, because besides not locking, it causes a chemical reaction that makes it brittle. Acetone should also be avoided.

  Very thin sheets of Lexan make great drilling templates for top and bottom covers of the robot. This classic technique is very simple: once all the robot walls are finished and assembled, firmly attach the Lexan sheet on top of it, as if it were a robot cover. Since Lexan is transparent, it is easy to mark with a center punch the centers of the holes to be drilled, which must align with the already finished holes from the walls. If the Lexan sheet is very thin, it will bend as a cone into the holes from the walls as it is pressed by the center punch, improving the centering precision. After marking all hole centers, the Lexan sheet is ready to be fixed on top of the actual cover plates to be drilled.

- acrylic: good to build fish tanks, but do not use it in combat, because it has the same density as Lexan but with 20 to 35 times less impact toughness.

- PETG: it is a modified type of PET (polyethylene terephthalate) with an impact toughness in between the values for acrylic and Lexan. It is a cheap substitute for Lexan, but with worse properties. We've tried it in combat, and decided that it would be better used to make a nice transparent trophy shelf.

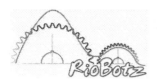

- Teflon (PTFE, politetrafluorethilene): very low friction, it can be used as a sliding bearing for moderate loads, or as a skid under the robot to slide in the arena. Its main problem is its high cost.

- UHMW: Ultra High Molecular Weight polyethylene is a high density polyethylene that also has

very low friction. Known as the "poor man's Teflon," it doesn't slide as well as Teflon, but it is cheap and it has higher strength. Shell spinners, such as Megabyte, use internal spacers made out of UHMW (circled in red in the picture to the right) between the shell and the inner robot structure, guaranteeing that the shell won't hit the internal metal parts of the robot even if it is bent, allowing it to slide with relatively low friction in case it makes contact. The high toughness of UHMW makes it a good choice even for structural parts, such as the motor mounts of the hobbyweight Fiasco, as pictured to the right.

- nylon, delrin (acetal): they are thermoplastic polymers with high strength, low density and relatively high toughness. They are good for internal spacers in the robots, and even as motor mounts, similarly to UHMW.

- rubber, neoprene, hook-and-loop (velcro): excellent materials to dampen the robot's critical internal components, such as receiver, electronics and batteries. High-strength mushroom-head hook-and-loop (pictured to the right) is also excellent to hold light components.

- epoxy: excellent adhesive, good to glue fiberglass, Kevlar and carbon fiber onto metals. Clean the metal part with alcohol or acetone before applying it, to maximize holding strength. Always use professional epoxy, which cures in 24 hours, not the hobby grade.

- phenolic laminate: it is an industrial laminate, very hard and dense, made by applying heat and pressure in cellulose layers impregnated with phenolic synthetic resins, agglomerating them as a solid and compact mass. Also known as celeron, it is an excellent electric insulator. We mount all the electronics of our robots on such laminates, which are then shock-mounted to the robot structure using vibration-damping mounts (see chapter 4) or mushroom-head

hook-and-loop, resulting in electrical insulation as well. The regular phenolic laminates are relatively brittle, but a high strength version called garolite (available at www.mcmaster.com) has already been used even in the structure of antweights and beetleweights. The top cover of our beetleweight *Mini-Touro* was made out of garolite, however it was replaced with a titanium cover with same weight. Although thinner, the titanium top cover has a higher impact strength than the garolite version, which is important when facing offset horizontal spinners that know how to skillfully pop a wheelie to deliver an overhead attack with their weapon. The first prototype of our hobbyweight *Tourinho* was made out of garolite (a green variety for the side walls and a black one for the top and bottom covers, as pictured to the right), transparent to radio signals and very resistant. However, we ended up changing it to aluminum for two reasons: the threads tapped in garolite, or in any other phenolic laminate, are brittle and easily break, and the better impact toughness of aluminum made up for its increased density (aluminum has density 2.8, and garolite 1.8).

- wood: it has low impact toughness if compared to metals. It should not be used in the structure, unless your robot is very skillfully driven, such as the wooden lightweight The Brown Note, which got the silver medal at Robogames 2008 after losing to the vertical spinner K2 (pictured to the right). A few builders have mounted wooden bumpers in front of their robot when facing spinners, to work as ablative armor: while a shell spinner chews up the wooden bumper of its opponent little by little, it loses kinetic energy and slows down, becoming vulnerable.

- ceramics: they are very brittle under traction, but under compression they are the most resistant materials in the world, so much that they are used underneath the armor plates of war tanks: the ceramic breaks up the projectiles, while their fragments are stopped by an inner steel layer. Ceramics are also extremely resistant to abrasion. The famous lifter BioHazard used 4" square 0.06" thick alumina tiles (Al_2O_3, which forms sapphires when in pure form) glued under its bottom to protect it against circular saws that emerged from the BattleBots arena floor.

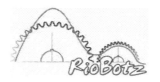

- fiberglass: known as GFRP (glass fiber reinforced polymer), it is made out of very thin glass fibers held together by a polymeric adhesive (known as the polymer matrix) such as an epoxy resin. It is very used in boats. It has potential use in the robot structure for being rigid and light, however its impact toughness is low if compared with the one of most metals.

- Kevlar: known as KFRP (Kevlar fiber reinforced polymer), it is a yellow fabric (pictured to the right) made out of aramid fibers, a type of nylon, 5 times more resistant than steel fibers of same weight. Used in bulletproof vests, it has extraordinary impact toughness. *Touro* uses a Kevlar layer covered with professional epoxy (the polymer matrix) sandwiched between the aerospace aluminum walls of the structure and the external Ti-6Al-4V plates of the armor, to  increase its impact toughness. The fabric is very difficult to cut, it is recommended to use special shears, found at www.mcmaster.com. Kevlar fabric is not expensive, we've used less than US$12 in *Touro* – more specifically, we've used the aramid fabric KK475, which costs about US$60/m^2 (less than US$6/ft^2) in Brazil.

- carbon fiber: known as CFRP (carbon fiber reinforced polymer), and available in several colors (as pictured to the right), it is very expensive but extremely rigid and light, and because of that it has been used in racing cars and in the fuselage of the new Boeing 787 and AirBus A350 (pictured in the next page). It is excellent to mount the robot's internal parts due to its high stiffness. But it is a myth that carbon fiber has high impact toughness. It surely has a high strength under static loads, but it does not take severe impacts. The undercutter Utterly Offensive is a good example of that, its carbon fiber baseplate (pictured to the right) self-destructed when it was scraped by its own spinning blade. The plate was later switched to titanium. Carbon fiber is not a good armor material, unless it is combined with Kevlar to achieve high impact toughness. Surely you could get away without Kevlar, using a very thick carbon fiber armor plate, but probably the added weight would have been better employed using, for instance, a titanium armor.

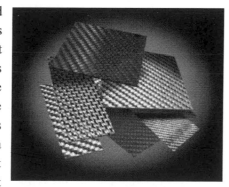

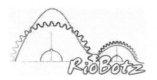

- other polymer matrix composites: there are several other composites that use a polymer matrix (such as epoxy or polyester) besides plain GFRP, KFRP and CFRP. For instance, you can tailor lay-ups of aramid and carbon fibers, cured (bonded) together with a polymer matrix, to achieve optimum impact toughness (due to Kevlar) and stiffness (due to carbon). It is possible to generate complex unibody structures by combining several parts into a single cured assembly, reducing or even eliminating the need for fasteners, saving weight and assembly time. This unibody can be joined together in three ways: cocuring, cobonding, or adhesive bonding. In cocuring, the uncured composite fabric plies are cured and bonded together at the same time using the same polymer matrix. In cobonding, an already cured part, usually a stiffener, is bonded to an uncured one, usually a skin, at the same time the skin is cured. In adhesive bonding, cured composites or metals are bonded to other cured composites, honeycomb cores, foam cores or metallic pieces. The pictures to the right show two very rigid sandwich panels with respectively a polypropylene honeycomb core and a polymethacrylimide foam core, sandwiched by CFRP sheets (available at The Robot MarketPlace). Besides increasing the panel bending stiffness, the foam core also works as a shock mount, increasing the impact strength, becoming a good option for the robot structure and even armor. An even higher stiffness-to-weight ratio can be obtained if the core is made out of balsa wood, as in the DragonPlate pictured to the right, however its impact toughness is relatively low.

- metal matrix and ceramic matrix composites: instead of having their fibers embedded and held together in a polymer matrix, these composites use either a metal or a ceramic matrix. The fibers (or even tiles in a few cases), which can also be made out of metal or ceramic, tend to increase the ultimate strength and stiffness of the matrix material. However, most ceramic matrix composites have low impact strength, which limits their use in combat, not to mention their very high cost. On the other hand, when part of a multi-layer composite armor plate, such as the Chobham armor, ceramic tiles embedded in a metal matrix can be very effective to shatter kinetic energy weapons.

3.8. Material Selection Principles

After presenting the main materials used (or not) in combat robots, the question is: which material should I use? "The most resistant" is not the correct answer. The most resistant materials per volume are steels, but a robot entirely made out of steel would be very heavy.

For instance, a 4mm (0.16") thick steel plate weighs as much as an 11mm (0.43") thick aluminum one. Which one is better, the 4mm steel or the 11mm aluminum?

The answer is not so simple. It depends on the function that the material will have, as it will be seen next.

3.8.1. Stiffness Optimization

Classic solid mechanics calculations (summarized in the table to the right) show that a beam under traction, working as a trussed element (such as in the structure of trussed robots), has the largest possible stiffness with minimum mass if the material has the largest possible ratio between the Young modulus E and the density ρ. Steels have in average E = 200GPa and ρ = 7.8, therefore $E/\rho \cong 26$. Aluminum (Al) alloys usually have E = 72GPa and ρ = 2.8, thus $E/\rho \cong 26$. Titanium (Ti) alloys have E = 110GPa and ρ = 4.6, resulting in $E/\rho \cong 24$. And, in magnesium (Mg) alloys, E = 45GPa and ρ = 1.8, resulting in $E/\rho \cong 25$. In summary, there is almost no difference in choosing among

structural element type or function	min. mass for max. stiffness	strength
truss Young modulus E specific mass ρ elastic strength S	$\Delta L = \dfrac{FL}{EA}$ $m = \rho LA = \dfrac{FL^2}{\Delta L}\dfrac{\rho}{E} \therefore$ maximize $\dfrac{E}{\rho}$	$\sigma = \dfrac{F}{A}$ $m = \rho LA = FL\dfrac{\rho}{\sigma} \therefore$ maximize $\dfrac{S}{\rho}$
beam $b = \alpha a$ $A = \alpha a^2$ $I = \alpha a^4/12$	$y = \dfrac{4FL^3}{E\alpha a^4}$ $m = \rho L\alpha a^2 = 2\sqrt{\dfrac{\alpha FL^5}{y}}\dfrac{\rho}{\sqrt{E}}$ maximize $\dfrac{\sqrt{E}}{\rho}$	$\sigma = \dfrac{6FL}{\alpha a^3}$ $m = \rho L\alpha a^2 = L\alpha\left[\dfrac{6FL}{\alpha}\right]^{\frac{2}{3}}\dfrac{\rho}{\sigma^{2/3}}$ maximize $\dfrac{S^{\frac{2}{3}}}{\rho}$
plate load: pressure p $b = \alpha c$	$y = \dfrac{5pc^4}{32Ee^3}$ $m = \rho e\alpha c^2 = \alpha c^3\left[\dfrac{5pc}{32y}\right]^{\frac{1}{3}}\dfrac{\rho}{\sqrt[3]{E}}$ maximize $\dfrac{\sqrt[3]{E}}{\rho}$	$\sigma = \dfrac{3pc^2}{4e^2}$ $m = \rho e\alpha c^2 = \dfrac{\alpha c^3}{2}\sqrt{3p}\dfrac{\rho}{\sqrt{\sigma}}$ maximize $\dfrac{\sqrt{S}}{\rho}$

$S = S_y$ if material is ductile or $S = S_u$ if fragile

steels, aluminum, titanium or magnesium alloys for a trussed element if the only requirement is to have a high stiffness-to-weight ratio, their E/ρ ratio is very similar, between 24 and 26.

However, for a plate under bending, which would be the case of most of the robot structural parts, such as side walls and top/bottom covers, stiffness is maximized with minimum weight if the material has the largest possible $E^{1/3}/\rho$ ratio. In this case, magnesium alloys are much better, with

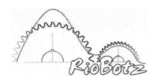 

ratio 2.0, against 0.8 for steels, 1.0 for titanium alloys and 1.5 for aluminum alloys. The results are summarized in the table to the right.

material	E/ρ	$E^{1/2}/\rho$	$E^{1/3}/\rho$
Steels	26	1.8	0.8
Al alloys	26	3.0	1.5
Ti alloys	24	2.3	1.0
Mg alloys	25	3.7	2.0
Lexan	2	1.3	1.1
Delrin	2	1.3	1.0
UHMW	0.7	0.9	0.9
wood	3 - 19	2 - 5.1	1.8 - 3.4
GFRP	8.6 - 16	2.2 - 3	1.4 - 1.7
CFRP	44 - 96	5.3 - 7.9	2.6 - 3.4
Be alloys	164	9.4	3.6

As seen in the table, beryllium (Be) alloys would result in extremely light and rigid structures, however its use is usually prohibited in combat due to health issues.

Among the allowed materials, carbon fiber (CFRP), Kevlar (KFRP) and fiberglass (GFRP) are the best choices for stiff and light beams and plates, however there are still the problems with the low impact toughness of carbon fiber and fiberglass, and the challenge in making an entire structure out of Kevlar fabric. In addition, their properties are not the same in all directions, they vary considerably. This is also true for woods, their stiffness and toughness perpendicular to their fibers are almost 10 times lower than parallel to them.

Lexan (polycarbonate) or delrin (acetal) would be awful as a trussed element under traction, their E/ρ is only 2. UHMW is even worse in that sense. Aluminum (Al) and magnesium (Mg) alloys have excellent stiffness with minimum weight, much better than Lexan, delrin, UHMW, steels, and even titanium alloys, for use in beams under bending, maximizing $E^{1/2}/\rho$, and for use in plates under bending, maximizing $E^{1/3}/\rho$.

Note from the weight optimization equations that the beam element in the previous figure assumes that both its width b and thickness a can be changed, only their aspect ratio $\alpha = b/a$ is assumed fixed. This could be true for internal structural components and for shafts, however all the robot's walls and covers cannot change their length c and width b without changing the robot design, only their thickness is a free design parameter.

Therefore, the plate element in the previous figure, which only allows its thickness e to change to optimize weight, without modifying b and c, is more appropriate for most structural parts. In summary, except for trussed elements, which are optimized by E/ρ, most of the robot's structural parts have their stiffness optimized by $E^{1/3}/\rho$ (as with plates), while shafts depend on $E^{1/2}/\rho$ (as with beams).

Since E and ρ do not vary much within the same type of material, it is possible to generate a large diagram comparing the applicability of each one. We've generated a graph in logarithmic (log-log) scale for several types of materials, whose ρ are represented in the horizontal axis and E in the vertical one. Using the log-log scale, we obtain guidelines that show materials with same E/ρ, $E^{1/2}/\rho$ and $E^{1/3}/\rho$ ratios, as explained below.

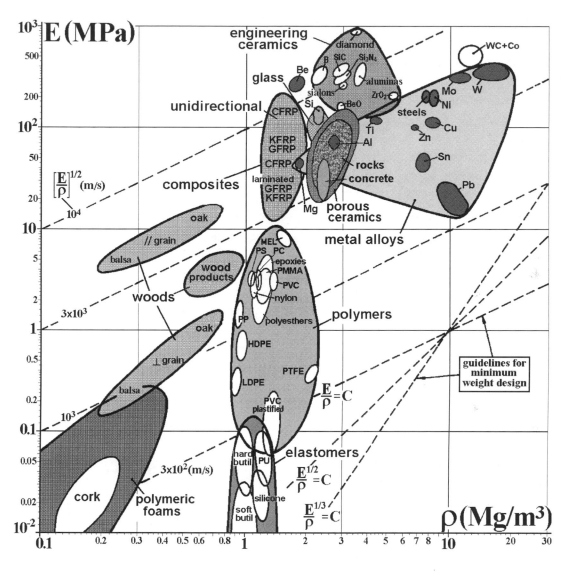

To choose materials to be used in light and stiff truss elements, consider the dashed guideline associated with constant E/ρ (labeled $E/\rho = C$). All the materials in the same straight line are equivalent, in other words, trusses of same weight made out of these materials would have the same stiffness. Now, draw parallel lines to this guideline. The higher the parallel line, the better will be the material. For instance, in the lowest guideline with constant E/ρ, we can see that plastified PVC is equivalent to cork. Going up to the next parallel E/ρ guideline, we reach the polyesters. The next parallel E/ρ guideline is a little below copper (Cu) alloys. A little above, note that, as expected, all steels, titanium (Ti), aluminum (Al) and magnesium (Mg) alloys are aligned, due to their $E/\rho \cong 25$, as calculated before. The highest line with constant E/ρ in the figure goes through unidirectional carbon fibers (CFRP). A little further above we can see the infamous beryllium alloys (Be). This means that, to make a light and stiff truss, beryllium alloys would be better than CFRP, which is

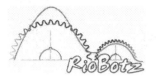 

much better than copper alloys, which in turn is much better than polyesters, which are much better than corks (which have very low stiffness because they are foams).

To choose materials for light and stiff beams under bending, the procedure is similar, except that we'll use lines parallel to the guideline for constant $E^{1/2}/\rho$. And for plates under bending, use the guideline for constant $E^{1/3}/\rho$.

Note that steels, aluminum (Al), titanium (Ti) and magnesium (Mg) alloys are practically on the same straight line parallel to the guidcline for trussed elements (constant E/ρ), so they are similar for such application, as we've already verified. However, when drawing parallel lines to the guideline for plates under bending (constant $E^{1/3}/\rho$), Mg is above Al, which is above Ti, and all of them are above steels. Therefore, it is not efficient to use steel to obtain light and rigid plates, as we had verified, it is much better to use Mg alloys.

An interesting result is that balsa wood can be the best material to make light and stiff plates (at least in a direction parallel to its grains). It is even better than titanium alloys or carbon fiber, you can easily check this from lines parallel to the constant $E^{1/3}/\rho$ guideline in the figure above. Anyone who's worked with model airplanes knows this very well. The internal structure of our fairyweight wedge Pocket is made out of balsa wood. Commercial airplanes would be much stiffer and lighter if they were made out of balsa wood, however aluminum alloys are used instead because of their higher impact toughness and weather resistance. Making a combat robot entirely out of balsa wood, including its external structure and armor, would be suicide. It would be extremely rigid, but it would break at the first impact. Thus, we must take into account other properties, not only stiffness.

3.8.2. Strength and Toughness Optimization

The yield and ultimate strengths S_y and S_u are also very important, and they need to be considered. High S_y is important for parts that should not have permanent deformations, such as shafts. And, naturally, high S_u is also important to avoid rupture and to increase the fatigue life. As the strength (denominated by the letter S) varies a lot within the same alloy family, it isn't possible to generalize conclusions to all steels, aluminum alloys, etc, as we did for stiffness. It is necessary to study each particular material separately. The best materials for trusses, beams and plates are, respectively, the ones with highest S/ρ, $S^{2/3}/\rho$ and $S^{1/2}/\rho$, as shown before in the solid mechanics calculation table.

The results for yield strength ($S \equiv S_y$) are in the table to the right, for several representative

material	S_y/ρ	$S_y^{2/3}/\rho$	$S_y^{1/2}/\rho$
UHMW	24	8	5.0
Delrin	44	11	5.6
Lexan	50	13	6.4
1020 steel	33	5	2.1
304 stainless	34	5	2.1
4340 (43HRc)	171	16	4.7
S7 (54HRc)	194	17	5.0
AerMet 100	215	18	5.2
18Ni(350)	303	22	6.1
Al 6063-T5	54	10	4.5
Al 6061-T6	102	16	6.2
Al 2024-T3	124	18	6.7
Al 7075-T6	169	22	7.8
Ti-6Al-4V	208	21	6.9
AZ31B-H24	84	16	6.9
ZK60A-T5	109	19	7.7
Be S-200	228	30	11

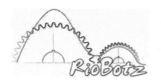 

materials. If we disregard CFRP, KFRP and GFRP, which would be the best options but they still have the problems mentioned before, it is noticeable that a trussed element under traction (such as in a trussed robot) has the largest yield strength with lowest weight if made out of 18Ni(350) steel (ratio $S_y/\rho = 303$, see the table), followed by the S-200 beryllium alloy ($S_y/\rho = 228$), AerMet 100 steel ($S_y/\rho = 215$) and Ti-6Al-4V ($S_y/\rho = 208$). If beams under bending are considered, the ranking is completely changed, the best choice against yielding would be S-200 beryllium ($S_y^{2/3}/\rho = 30$), followed by 7075-T6 aluminum and 18Ni(350) steel ($S_y^{2/3}/\rho = 22$). Plates under bending also have maximum yield strength with minimum weight if made out of S-200 beryllium ($S_y^{1/2}/\rho = 11$), followed by 7075-T6 aluminum ($S_y^{1/2}/\rho = 7.8$), however the third best option changes to ZK60A-T5 magnesium ($S_y^{1/2}/\rho = 7.7$). Note that, for plates under bending, most steels would be poor options, even worse than UHMW.

A similar table can be generated for $S \equiv S_u$, to evaluate the best options for minimum weight if ultimate strength is considered. The trends are similar to the ones obtained from the S_y analyses, with the dominance of beryllium alloys, with high strength steels playing an important role in trussed elements (such as 18Ni(350) and its $S_u/\rho = 305$), with high strength Al, Ti and Mg alloys showing similar performance in beams ($S_u^{2/3}/\rho$ between 22 and 25), and with ZK60A-T5 magnesium as a good option for plates ($S_u^{1/2}/\rho = 9.6$).

A similar reasoning can be applied to optimize fracture toughness with minimum weight. The effect of a crack in a structure is directly proportional to the applied stresses. Therefore, the same analysis can be performed considering $S \equiv K_{Ic}$ in the previous equations. The results are shown to the right. It can be seen that 304 stainless steel is the best material to achieve tough trussed elements ($K_{Ic}/\rho = 27$). It is also great for tough beams, second only to high strength magnesium alloys, which have $K_{Ic}^{2/3}/\rho$ between 5.2 and 5.7. These magnesium alloys are also the best option for tough plates

material	S_u/ρ	$S_u^{2/3}/\rho$	$S_u^{1/2}/\rho$
UHMW	43	13	6.8
Delrin	54	13	6.2
Lexan	54	13	6.7
1020 steel	56	7	2.7
304 stainless	77	9	3.1
4340 (43HRc)	184	16	4.8
S7 (54HRc)	251	20	5.7
AerMet 100	251	20	5.6
18Ni(350)	305	23	6.1
Al 6063-T5	69	12	5.1
Al 6061-T6	115	17	6.5
Al 2024-T3	174	22	7.9
Al 7075-T6	196	24	8.4
Ti-6Al-4V	224	22	7.1
AZ31B-H24	143	23	9.0
ZK60A-T5	169	25	9.6
Be S-200	415	45	15

material	K_{Ic}/ρ	$K_{Ic}^{2/3}/\rho$	$K_{Ic}^{1/2}/\rho$
UHMW	1.7	1.5	1.4
Delrin	2.1	1.5	1.2
Lexan	1.8	1.4	1.2
1020 steel	17	3.3	1.4
304 stainless	27	4.5	1.8
4340 (43HRc)	11	2.5	1.2
S7 (54HRc)	7.7	2.0	1.0
AerMet 100	17	3.3	1.5
18Ni(350)	5.2	1.5	0.8
Al 6063-T5	9.3	3.2	1.9
Al 6061-T6	10	3.3	1.9
Al 2024-T3	13	4.0	2.2
Al 7075-T6	8.7	3.0	1.8
Ti-6Al-4V	16	3.9	1.9
AZ31B-H24	16	5.2	3.0
ZK60A-T5	19	5.7	3.2
Be S-200	6.6	2.9	1.9

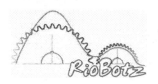

($K_{Ic}^{1/2}/\rho$ between 3.0 and 3.2), followed by 2024-T3 aluminum.

The material selection principles presented above allow us to choose a material to optimize a single mechanical property. For instance, the K_{Ic} calculations showed that 304 stainless steels and high strength magnesium alloys result in the toughest shafts with minimum weight, because shafts can be modeled as beams. But a 304 steel shaft would not be a good idea because of its low yield strength, allowing the shaft to easily get bent. And a magnesium shaft, despite being light, would need to have a very large diameter to achieve the desired toughness, which might not fit in the robot or require very heavy large diameter bearings and mounts. So, other considerations need to be introduced to decide which material is the best option for each part of the robot. This will be done next.

3.9. Minimum Weight Design

Minimum weight design has the goal to find the best dimensions and materials to optimize the performance of a component while minimizing its weight. It assumes that the dimensions of the component can be changed without interfering significantly with the robot design. If a component is performing as expected, then the idea is to reduce its weight by changing its materials and dimensions without losing functionality. Alternatively, if a component is failing in combat, then the idea is to improve its mechanical properties through material and dimension changes, while adding as little weight as possible. In this last case, if the redesign is wisely performed, it may be even possible to achieve the improved functionality and lose weight at the same time.

The following analyses will focus on typical structural materials that have potential use in combat. Note that beryllium alloys and composites won't be included in the following sections. Even though they would be, in theory, the best choices for minimum weight design, they have limitations in their use as structural elements.

Beryllium alloys would be a great option to maximize stiffness and strength of trusses, beams or plates, however they would not be the best choice in the presence of impacts, due to their low K_{Ic}. In addition, they are usually not allowed in combat due to health issues, as discussed before.

And composites, such as CFRP, despite their outstanding mechanical properties, also have several issues regarding their use. Composites are difficult to fabricate (in special high precision parts), they have poor mechanical properties perpendicular to the direction of the fibers, they may delaminate, and they lose toughness if drilled. Not to mention their high cost, which usually limits their application to insect weight classes. If these issues are addressed, then CFRP is the best option for light structures with high stiffness and strength. This is true not only for trusses, but also for beams (such as the CFRP spinning bar pictured to the right) and plates (such

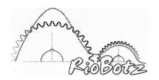 

as the CFRP structure of the same robot in the picture). If impact toughness is also necessary, then CFRP must be combined with, for instance, Kevlar. But unless you have experience with composites and a high budget, stick with the traditional structural materials: metal alloys.

The following analyses are also limited to general-purpose structural materials. Application-specific materials such as bronze, copper, PTFE (Teflon) and neoprene are not studied below. They are important in the robot to minimize friction (oil-impregnated bronze bearings, PTFE slide surfaces), to shock-mount parts (neoprene sandwich mounts), to lower electrical resistance (copper wires), and in several other tasks as described in the previous sections, but their applications are too specific for them to be compared with structural materials.

Finally, note that a few choices might seem subjective, but they are always backed up by measured properties. Note also that a few of the studied materials may be very difficult to find, such as the 2324-T39 Type II aluminum alloy used in the Boeing 777 plane, however they were included anyway for comparison purposes. Other alloys may also be unavailable in plates or bars, which might limit their applicability. For instance, the K12 Dual Hardness steel is only available in plates up to 1/2" thick, making it almost impossible to use it in shafts. And due to its dual hardness property, it will only be considered for plates that work as armor elements, its originally intended purpose.

3.9.1. Minimum Weight Plates

As seen above, magnesium (Mg) and aluminum (Al) alloys are excellent materials to increase the stiffness of structural plates that must have their weight minimized. So, if you need to lose weight, it is in general a good idea to replace steel plates with high strength Mg or Al alloy versions. But if in this case you simply change the material without increasing the plate thickness, it is easy to see that you will lower the robot stiffness and strength. To calculate the increased thickness to avoid that, we'll need to use the equations shown in section 3.8 for bending stiffness of plates. It is easy to show that the scale factor for the thickness to keep constant the plate stiffness is $(E_{old}/E_{new})^{1/3}$, where E_{old} and E_{new} are the Young modulii of the old and new materials.

So, to replace a steel plate without compromising its bending stiffness, you'll need, for instance, an aluminum one that is $(E_{steel}/E_{Al})^{1/3} \cong (205GPa/72GPa)^{1/3} \cong 1.42$ times thicker. This new thicker plate will still be lighter than the original one, because of the low density of Al alloys, which is 2.8 in average, instead of the steel average 7.8. The new plate will then have $1.42 \cdot 2.8 / 7.8 \cong 51\%$ of the weight of the original one, but with the same bending stiffness. This is a smart diet!

Similar calculations for constant bending stiffness can show that a steel plate can be switched to $(205/110)^{1/3} \cong 1.23$ times thicker titanium (Ti), weighing $1.23 \cdot 4.43 / 7.8 \cong 70\%$ of the original weight. So, if stiffness is your major concern, then switching to Al will save more weight than switching to Ti. Actually, the best choice would be Mg alloys: steel plates can be replaced with $(205/45)^{1/3} \cong 1.66$ times thicker Mg, weighing only $1.66 \cdot 1.8 / 7.8 \cong 38\%$ of the original weight, without changing their stiffness.

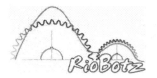 

But other material properties besides E are also relevant, depending on the functionality of the component. The table below shows important mechanical properties of several relevant structural materials, such as S_u (measured in MPa), S_y (in MPa), K_{Ic} (in MPa$\sqrt{m}$), HB (hardness, using the Brinell scale), as well as E (in GPa) and the relative density ρ. Note that the hardnesses of the 3 polymers in the table are measured in Shore D, which would translate into very low Brinell values.

If we want to compare the performance of the listed materials as structural plates, then section 3.8 showed that we must calculate their $E^{1/3}/\rho$ ratio to evaluate stiffness, their $S_y^{1/2}/\rho$ for yield strength, $S_u^{1/2}/\rho$ for ultimate strength, and $K_{Ic}^{1/2}/\rho$ for fracture toughness. Note that hardness is a local property, it only depends on the material, not on the dimensions of the component, therefore it can be directly compared without the need to consider any ratio with the density.

	material	physical and mechanical properties						min. weight plate				
		ρ	E	S_u	S_y	K_{Ic}	HB	E*	S_u*	S_y*	K_{Ic}*	HB'
Mg alloys	AZ31B-H24	1.78	44.8	255	150	28	77	100	86	76	93	11
	ZK60A-T5	1.83	44.8	310	200	34	70	97	93	86	100	10
	Elektron WE43-T5	1.84	44	250	180	15.9	95	96	83	81	68	14
	Elektron 675-T5	1.95	44	410	310	16	114	91	100	100	64	17
Aluminum alloys	Al 6063-T5	2.7	68.9	186	145	25	60	76	49	49	58	9
	Al 6061-T6	2.7	68.9	310	276	27	95	76	63	68	60	14
	Al 2024-T3	2.78	73.1	483	345	32	120	75	76	74	64	18
	Al 2324-T39 Type II	2.77	72.4	475	370	48	118	75	76	77	78	18
	Al 5086-H32, H116	2.66	71	290	207	49	78	78	62	60	83	12
	Al 7050-T7451	2.83	71.7	524	469	31.5	140	74	78	85	62	21
	Al 7055-T74	2.86	71.7	524	469	39.6	140	73	77	84	69	21
	Al 7055-T7751	2.86	71.7	638	614	27.5	171	73	85	96	58	26
	Al 7075-T6	2.81	71.7	551	475	25	150	74	80	86	56	22
	Al 7075-T73	2.81	72	505	435	29.7	135	74	77	82	61	20
	Al 7175-T736	2.8	72	550	485	34	145	74	81	87	65	22
	Al 7475-T7351	2.81	71.7	496	421	45	135	74	76	81	75	20
Ti	Ti-6Al-4V (36HRc)	4.43	110	992	923	72	336	54	68	76	60	50
	Ti-6Al-4V ELI	4.43	110	896	827	88	326	54	65	72	66	49
Steels	1020 steel	7.87	203	441	262	130	108	37	26	23	45	16
	304 stainless	8.03	193	621	276	220	153	36	30	23	58	23
	4340 (43HRc)	7.85	205	1448	1344	88	402	38	47	52	38	60
	4340 (39HRc)	7.85	205	1310	1207	121	361	38	44	49	44	54
	4340 (34HRc)	7.85	205	1172	1069	148	320	38	42	46	49	48
	S7 (54HRc)	7.83	207	1965	1520	55	544	38	55	55	30	81
	AerMet 100 (53HRc)	7.89	194	1965	1724	118	530	37	54	58	43	79
	AerMet 310 (55HRc)	7.89	194	2170	1900	71	560	37	57	61	34	84
	AerMet 340 (57HRc)	7.89	194	2380	2070	37	596	37	60	64	24	89
	HP-9-4-30 (51HRc)	7.75	200	1585	1280	126	495	38	49	51	45	74
	18Ni(200) (46HRc)	8	183	1502	1399	142	426	36	47	52	47	64
	18Ni(250) (51HRc)	8	190	1723	1702	121	491	36	50	57	43	73
	18Ni(300) (54HRc)	8	190	2067	1998	80	544	36	55	62	35	81
	18Ni(350) (61HRc)	8.08	200	2467	2446	42	670	36	59	68	25	100
	K12 Dual Hardness	7.86	205	1785	1626	72	670	38	52	57	34	100
Polym.	Delrin	1.41	3.1	76	62	3	86SD	52	60	62	39	1
	Lexan	1.20	2.35	65	60	2.2	83SD	56	65	71	39	1
	UHMW-PE	0.93	0.689	40	22	1.6	66SD	48	65	56	43	1

To make things easier, we've normalized hardness and all the above ratios using the best materials from the table, resulting in a system of grade points between 0 and 100. The normalized hardness is called here HB', while the grades for minimum weight plates are represented by the property followed by the * symbol, namely E^*, S_y^*, S_u^* and K_{Ic}^*, shown in the table. For instance, the best material from the table for a stiff plate is the Mg alloy AZ31B-H24, therefore its stiffness grade for plates is $E^* = 100$. Aluminum alloys have E^* between 73 and 76, Ti alloys between 52 and 54, and steels between 36 and 38. With these low grades for minimum weight plate design, steels would certainly flunk a "Stiffness 101" course!

These grades are also proportional to the weight savings you'll get. For instance, a 4340 steel plate can be replaced with a 1.42 times thicker 7075-T6 aluminum one that is $E_{4340}^* / E_{7075-T6}^* = 38 / 74 = 51\%$ lighter, as calculated before, without losing stiffness.

The best material for light weight plates with high ultimate and yield strengths is the experimental Mg alloy Elektron 675-T5, therefore its $S_u^* = 100$ and also $S_y^* = 100$. Unfortunately, there is no material that can optimize all properties at the same time. For instance, this very same Mg alloy has only HB' = 17, a very low score for hardness. Regarding hardness, the best materials in the table are the 18Ni(350) maraging and the K12 Dual Hardness steel alloys, hardened to 61 Rockwell C, equivalent to a 670 Brinell hardness, resulting in HB' = 100. Finally, the best material for light weight plates that must sustain impacts and avoid fracture in the presence of cracks is the Mg alloy ZK60A-T5, with $K_{Ic}^* = 100$.

To decide which material to choose from the table for a light weight plate, we must know as well which of the above properties are more important. This depends a lot on the functionality of the plate in the robot. Except for shafts, bars and trusses, most of the robot's structural parts can be modeled as plates for minimum weight design. This is because these parts usually have two fixed dimensions, width and length, obtained from the robot geometry, while their thickness and material can be changed. This is true for most internal mounts, structural walls, top and bottom covers, wedges, shields and armor plates. We'll study these plate-like structural members next.

3.9.2. Minimum Weight Internal Mounts

The most desired property of internal mounts is stiffness. All the impacts they suffer are indirectly transmitted, being relatively damped by the chassis, so K_{Ic} is not that important. Usually, internal mounts that have sufficiently high stiffness are made out of plates that are thick enough to satisfy S_u and S_y requirements.

Therefore, if only the stiffness grades E^* are considered, then all Mg alloys are by far the best choice, with grades between 91 and 100, followed by all Al alloys, grading between 73 and 76. Even the low strength 6063-T5 aluminum is a good choice if only stiffness is concerned. But forget about steel internal mounts, their low E^* between 36 and 38 will end up adding unnecessary weight to your robot.

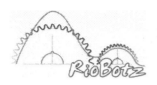

But if S_u and S_y are also critical, besides stiffness, then we must include them in the selection process. High S_u may be important for high stress mounts of heavy weapon motors, such as the 1/2" thick Etek and Magmotor mounts (pictured to the right, which can be found for instance at the Robot Marketplace).

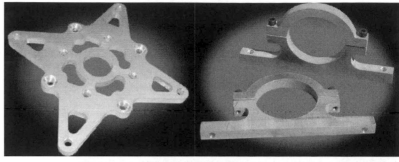

High S_y is also important for internal mounts that support wheel shafts, such as the drivetrain pillow blocks pictured to the right (sold at www.teamdelta.com), which must preserve a relatively accurate alignment without getting permanently bent.

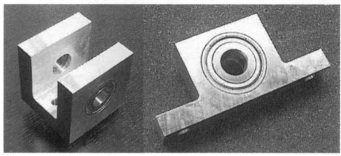

If we (arbitrarily) choose to maximize the average between the grades E^*, S_u^* and S_y^*, maximizing a certain grading parameter $X^* = (E^* + S_u^* + S_y^*) / 3$, then the best choices are the high strength Mg alloys Elektron 675-T5, ZK60A-T5, AZ31B-H24, Elektron WE43-T5, all of them with $X^* > 85$. The next choices are high-strength Al alloys from the 7000 series, such as 7055-T7751, 7175-T736, 7075-T6, 7050-T7451 and 7055-T74, in that order, all with $X^* > 77$. Steels and even Ti alloys usually result in unnecessarily heavy internal mounts, due to their lower E^* and X^*.

Lexan, delrin and UHMW mounts have much lower E^*, between 48 and 56, but they can make good internal mounts for very small parts in insect weight robots, because their higher resulting thickness will allow them to have threaded holes.

For instance, the Lexan motor mount pictured to the right only weighs 4 grams, while its 1/2" thickness allows the use of a threaded hole for the 4-40 screw. If the mount was made out of aluminum, it would need to be $(E_{Al}/E_{Lexan})^{1/3} \cong (72GPa/2.35GPa)^{1/3} \cong 3.13$ times thinner to have same stiffness (or $\rho_{Al}/\rho_{Lexan} \cong 2.8/1.2 \cong 2.33$ times thinner to have the same 4 gram weight). The lower 0.16" thickness for same stiffness would make it impractical to use threaded holes to hold

the 0.11" diameter 4-40 screw without compromising strength.

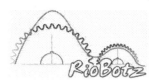

3.9.3. Minimum Weight Protected Structural Walls

Stiffness is very important for structural walls of robots with active weapons. Large structural deformations can make, for instance, a drum touch the floor when hitting an opponent, a spinning bar hit your own robot during a sloped impact, or even cause mechanism jamming due to severe misalignments.

This is why, unless you're building a passive rammer or wedge (as pictured to the right), making your entire structure and walls out of plastic is a bad idea. For instance, a UHMW plate, despite its good impact toughness, would need to be $(70GPa/0.7GPa)^{1/3} \cong 4.64$ times thicker than an aluminum one to have the same bending stiffness, and it would end up $4.64 \cdot 0.93 / 2.8 \cong 1.54$ times heavier, instead of lighter. A Lexan structure is also a bad idea, the plates would need to be $(70GPa/2.7GPa)^{1/3} \cong 3$ times thicker than aluminum ones, they would end up $3 \cdot 1.2 / 2.8 \cong 1.29$ times heavier. The thicker plates would also require longer screws to be mounted, adding even more weight. Not to mention that Lexan has cracking problems around holes, and plastics in general are easily cut by sawbots. Trust in aluminum! And in magnesium alloys, if available.

If a structural wall is not exposed, such that the opponent cannot hit it directly, or if there's some shock-mounted armor plate over it, then this wall is considered to be protected. Protected walls, besides high stiffness, should have high S_u^* to support static loads, and high S_y^* to avoid getting permanently bent. Since they are protected, K_{Ic}^* is not that important because they'll only suffer indirect impacts. Therefore, these walls basically behave as high stress internal mounts, being optimized by high strength Mg alloys and aluminum alloys from the 7000 series, as discussed in the previous sub-section. The pictures below show 7050-T7451 inner aluminum walls from our middleweight Touro, which are protected by titanium and Kevlar layers, shown in detail on the right.

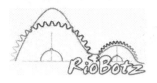

Note that most bottom plates (as pictured to the right) can be modeled as protected structural walls, therefore high strength Mg and Al alloys are usually a good option for them.

But, if you have an invertible robot, these bottom plates could get exposed while the robot is upside down, so it is also a good idea to check the optimized materials for integrated structure-armor walls, in the next sub-section. This is especially useful against vertical spinners with large bars or disks, which could flip your robot upside down with one blow, mount on top of it, and then hit again with the blade on the now exposed bottom plate. It is also useful against vertical spinners with small diameter disks, such as K2, which can lift your robot and hit its bottom plate during the same attack, as pictured to the right.

We can conclude as well that the best materials for other internal structural components such as gearbox blocks (pictured to the right, from the TWM 3M gearbox) are also high strength Al and Mg alloys, as long as the gearbox is well protected inside the robot, of course.

If you're changing the material of an existing protected wall, you'll need to find its new thickness depending on what property you want to keep constant. If it is stiffness, then we've seen that the scale factor for the thickness in plates is $(E_{old}/E_{new})^{1/3}$. It is also easy to show that the scale factor for the thickness to keep constant the bending strength of a plate is $(S_{old}/S_{new})^{1/2}$, where S_{old} and S_{new} are the strengths (either the yield S_y or the ultimate S_u) of the old and new materials.

Because both strengths vary a lot within the same alloy family, it isn't possible to generalize conclusions to all steels, aluminum alloys, etc, as we did in stiffness optimization calculations. It is necessary to study each material separately, evaluating its particular yield or ultimate strength, and its actual density (although the densities do not vary much within the same alloy family).

It is easy to see that all steels, even high strength steels such as a S7 steel tempered to 54 Rockwell C, are not a good choice for a light weight high strength structure. For instance, to replace a tempered S7 steel plate that works under bending without compromising its ultimate strength, you could use a 7075-T6 aluminum plate $(S_{S7}/S_{7075-T6})^{1/2} = (1965MPa/551MPa)^{1/2} \cong 1.89$ times thicker, weighing $1.89 \cdot 2.81 / 7.83 \cong 68\%$ of the original weight. An almost equivalent choice would be to replace the S7 steel with Ti-6Al-4V, which would need to be $(1965MPa/992MPa)^{1/2} \cong 1.41$ times thicker, with $1.41 \cdot 4.43 / 7.83 \cong 80\%$ of the original weight. Both 7075-T6 aluminum and Ti-6Al-

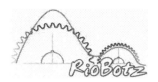

4V would have better ultimate strength-to-weight ratios than tempered S7, so both would be good options to replace any steel alloy in this case, with the high strength aluminum resulting in better values. This can be readily verified from the ultimate strength grade S_u^* of S7 steel (equal to 55), Ti-6Al-4V (equal to 68) and 7075-T6 aluminum (graded 80). Magnesium alloys would be even better for plates, since ZK60A-T5 has $S_u^* = 93$ and Elektron 675-T5 excels at $S_u^* = 100$. The use of S7 steel would only be wise if that part needed to remain sharp, which is not the case for structural walls.

But what if you want to increase some property by a factor of n, instead of keeping it constant? Well, the first step is to optimize the material, finding the new material and thickness that keep constant the desired property. After the material has been optimized, you'll need to increase its thickness to improve the property by the n factor. From the analysis of plates under bending, it is easy to show that the scale factor for the thickness is $n^{1/3}$ for improved stiffness, and $n^{1/2}$ for improved yield strength, ultimate strength, or fracture toughness.

For instance, if you want to double the bending stiffness of a 1/4" thick 4340 steel plate, the first step is to switch it to a better material, such as the Mg alloy AZ31B-H24, which has $E^* = 100$, instead of $E^* = 38$ from that steel. The same stiffness of the original plate would be obtained using a Mg alloy plate that was $(E_{4340}/E_{AZ31B-H24})^{1/3} = (205GPa/44.8GPa)^{1/3} \cong 1.66$ times thicker (0.415" thick). Now, to improve the stiffness by a factor of n = 2 with the same Mg alloy, just multiply the thickness by $n^{1/3} = 2^{1/3} \cong 1.26$, resulting in a 0.523" thick plate. This new plate, even with twice the stiffness of the original 4340 plate, would only have $1.26 \cdot 1.66 \cdot 1.78 / 7.85 \cong 47\%$ of the original weight of the steel version. Clearly, since there is no commercially available 0.523" thick plate, you'll probably have to choose between a lighter 1/2" or a stiffer 5/8" or 9/16" thick plate, or mill it down to the desired thickness. The numbers are even more interesting if you don't need to lose weight: a 1.1" thick Mg alloy plate would have the same weight as the original steel version, but its bending stiffness would be 18.75 times higher!

Note that, if the material is already optimized, then you'll need to add thickness and weight to the plate to improve its mechanical properties. If you can't afford the extra weight, then you'll have to start optimizing the entire geometry, not only the thickness. A simple way to do that is by getting a thicker plate and milling pockets in it, until the final piece has the same weight of the original thinner one. We had to mill very deep pockets in the inner walls of our hobbyweight Touro Jr not to go over its weight limit, as seen on the right. The idea is to make the plate work as an I-beam, with thick outer sections (where the bending stress is maximum) and thinner mid-sections. But to calculate the new stiffness, strengths and toughnesses, you'll probably need the aid of computer software such as Finite Element or even CAD programs.

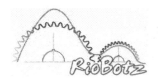

3.9.4. Minimum Weight Integrated Structure-Armor Walls

If the external structural walls of your robot are unprotected, working as well as armor (such as in our lightweight Touro Light, pictured to the right), then fracture toughness K_{Ic} plays a major role. To be used in the robot structure, the material must have high E^* and S_y^*, as discussed before. But now K_{Ic}^* is more relevant than S_u^*, because while also working as armor the plate will mostly suffer dynamic loads (impacts) instead of static ones.

Top cover plates are also included in this category, because they must have high K_{Ic}^* to survive the attacks of vertical spinners, hammerbots and overhead thwacks. If they also act as structural elements, helping for instance to support the drive system, then stiffness E^* and yield strength S_y^* grades are also important, so the following analysis also applies to them.

Since K_{Ic}^*, E^* and S_y^* are the most relevant properties for structure-armor walls, we'll (arbitrarily) choose the average of their grades, $X^* = (K_{Ic}^* + E^* + S_y^*) / 3$, to evaluate the best materials. We'll also choose only the materials with $K_{Ic}^* > 50$, $E^* > 70$ and $S_y^* > 70$, to avoid any distortions that the average X^* might carry. For instance, the 5086-H32 aluminum alloy has a relatively good $X^* = 73.5$, however it has an undesirably low $S_y^* = 60$.

It is found that the Mg alloys continue in the lead, with the best alloy from the studied table being ZK60A-T5, followed by AZ31B-H24, Elektron 675-T5 and Elektron WE43-T5. Aerospace aluminum alloys follow in this ranking: 2324-T39 Type II, followed by 7475-T7351, 7175-T736, 7055-T7751 or T74 and 7050-T7451, all of them with $X^* > 73$.

Interestingly, 2024-T3, which is not one of the best options for protected walls, is a good option for structure-armor plates, with $X^* = 71$, despite its lower S_y^*. It is almost as good in this application as 7075-T6, which has $X^* = 72$. This is because 2000 series Al alloys have lower S_y than 7000 series, but they usually make it up with better K_{Ic}, which is crucial for armor elements.

Forget once again about steels. Although steels have high K_{Ic}, their high density results in plates with relatively low K_{Ic}-to-weight ratios, with K_{Ic}^* between 24 and 58 for the ones in the studied table. These low ratios, together with their very low E^*, makes them very unattractive, due to their $35 < X^* < 47$.

Titanium Ti-6Al-4V, despite its relatively good $S_y^* = 76$ and $K_{Ic}^* = 60$ (or 66 for the tougher ELI version), is not one of the best choices for an integrated structure-armor because of its relatively low stiffness-to-weight ratio, which results in $E^* = 54$ and thus $X^* = 63.4$. This lower stiffness could be good enough to make a very tough all-titanium rammer or wedge, but it would be a poor choice for an integrated structure-armor robot with active weapons. In fact, even medium

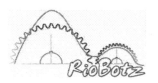

strength 6061-T6 is a better choice than titanium, despite its lower $S_y^* = 68$: it combines the same $K_{Ic}^* = 60$ of Ti-6Al-4V with a much better $E^* = 76$, resulting in a higher $X^* = 68.2$. In addition, 6061-T6 is much cheaper than high strength Ti and Al alloys, it is easier to machine, it has good weldability, and it is readily available in extrusion forms. No wonder Matt & Wendy Maxham love this material!

3.9.5. Minimum Weight Wedges

Wedges and integrated structure-armor plates behave in a similar manner. They must have high K_{Ic}^* to withstand impacts. They must have high S_y^* to avoid getting permanently bent, because bent wedges won't remain flush to the ground to scoop the opponents. And they must have high E^* to become very stiff and launch the opponents. A very flexible wedge can be effective as an armor element, as a defensive element damping the impact, however it won't be able to effectively use the opponent robot's kinetic energy against it, as an offensive element. Stiff wedges will transmit much higher reaction forces from the arena floor to the opponent.

So, good material choices for wedges must have, similarly to structure-armor plates, a high average grade $X^* = (K_{Ic}^* + E^* + S_y^*) / 3$. The main difference here is that wedges must keep their edges sharp to stay effective, so high hardness is also important. We'll choose materials with hardness higher than 32 Rockwell C, which translates to grades HB' > 45. Higher hardness materials will rule out Mg and Al alloys, so we'll need to relax a little the restrictions on the minimum values of the other properties, to be able to find any match. We only choose then materials with $K_{Ic}^* > 25$, $E^* > 35$ and $S_y^* > 40$, as well as HB' > 45, that maximize X^*.

The result is that Ti-6Al-4V ELI and Ti-6Al-4V (pictured to the right) are by far the best choices from the table, with X^* equal to 64 and 63, respectively. High strength steels would be the next choice, the best one being AerMet 100, followed by 18Ni(250), HP-9-4-30, 18Ni(200), 18Ni(300), tempered 4340, AR400 and tempered 5160, all of them with $X^* > 42$.

3.9.6. Minimum Weight Traditional Armor

If a structural plate only works as armor, such as a rammer shield (pictured to the right), then there are basically three mechanical properties of interest: fracture toughness, impact toughness, and hardness.

The armor needs to withstand impacts, as well as tolerate the cracks that will eventually be formed after receiving some serious hits, therefore K_{Ic} is a major concern. Armor plates usually do not carry any static loads besides their own weight, they're just "hanging

out" waiting to be hit, so S_u is not nearly as important as K_{Ic}. Also, they do not lose functionality after yielding, as long as their permanent deformations do not interfere with the inner structure, drivetrain or weapon system, so S_y is not that important either. And the armor plates are already mounted either over stiff internal structural walls or over shock mounts (as pictured to the right, with shock mounts made out of curled steel cables), therefore they don't need too much stiffness themselves.

Top cover plates can be included in this category if they do not act as structural elements, if they are just plates used to protect the robot interior, without having to support any internal components such as the drivetrain system. So, stiffness and yield strength are not that important, as long as the robot has a few internal supports that will prevent the top cover from bending too much into the robot and smash some critical component. Therefore, non-structural top cover plates that are well supported can be modeled as traditional armor elements, so the following analysis also applies to them.

Note that we're assuming here that the fracture toughness K_{Ic} can also be a measure of impact toughness. This is true for most metals, unless they have notch sensitivity problems, explained later. In addition, the value of K_{Ic} depends on the loading rate (the impact speed). All K_{Ic} measurements included in the previous table were made under very slow tests, therefore they are also called static K_{Ic}. If the tests had been performed at very high load rates, to evaluate the effect of the impact speed, the resulting dynamic K_{Ic} values would probably be much lower. For instance, one of the Ti-

6Al-4V armor plates from our middleweight Touro shattered almost like glass due to a high speed impact from the bar spinner The Mortician, during RoboGames 2006. Very little plastic deformation can be seen in the remaining plate (pictured to the right), and the removed portion shattered into tiny pieces. At lower impact speeds, the very ductile Ti-6Al-4V would certainly absorb more energy, meaning that its static (low speed) $K_{Ic} = 72 MPa\sqrt{m}$ is probably much higher than its dynamic (high speed) K_{Ic}.

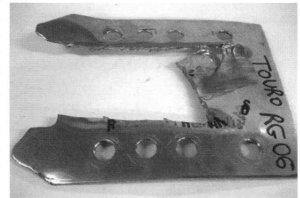

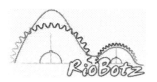 

However, if we assume that the ratio between dynamic and static K_{Ic} is similar for all materials (which is not completely true), then we can still compare them directly only using the available static K_{Ic} values.

As explained in chapter 2, traditional armors are usually made out of tough and hard materials that try to absorb and transmit the impact energy without getting damaged. They are a good option against very sharp horizontal spinners, since the high hardness will help chipping or blunting the edge of the opponent's weapon. So, in addition to a high grade K_{Ic}^*, it is desirable that the armor plate has a high hardness grade HB'.

Thus, we'll choose materials that maximize the average $X^* = (K_{Ic}^* + HB') / 2$, while guaranteeing some minimum toughness and hardness requirements $K_{Ic}^* > 30$ and HB' > 45. We find out that the best materials are, in that order, K12 Dual Hardness, AerMet 100, HP-9-4-30, AerMet 310, 18Ni(250), 18Ni(300), Ti-6Al-4V ELI, 18Ni(200) and Ti-6Al-4V, all of them having $X^* > 50$. Tempered 4340 steel is not a bad option, it has $X^* > 48$. Tempered 5160 steel could also be used, but it's not one of the best choices.

If you have trouble finding or heat treating AerMet, HP-9-4-30 or maraging steel alloys, then your best bet is to go for Ti-6Al-4V armor plates. Note that the above calculations assume that Ti-6Al-4V is in its commercially available annealed condition, because when heat treated to improve its ultimate strength it can end up with K_{Ic} lower than 45MPa√m, instead of 72MPa√m. If available, Ti-6Al-4V ELI (Extra Low Interstitial) is even better for armor plates, since $K_{Ic} = 88$MPa√m, despite its 10% lower yield and ultimate strengths. Titanium alloys with higher K_{Ic} are usually not commercially available, such as Ti-6Al-2Sn-4Zr-2Mo and Ti-11.5Mo-6Zr-4.5Sn.

If you want a low budget traditional armor, then 304 stainless steel is a reasonable choice. Its K_{Ic}^* is 58, similar to annealed Ti-6Al-4V, but it is much cheaper. However, due to its low E^* and S_y^*, it will only be a good choice if backed up by a stiff structure. Touro has a few 304 steel spare armor plates, used when we're out of Ti-6Al-4V.

Forget about other steels, because their K_{Ic}^* is usually below 50. With the same weight of the steel armor you can get a 77% thicker Ti-6Al-4V plate that will be much tougher. Even the high strength steels S7, 18Ni(350) and AerMet 340 should be avoided, because of their medium-low toughness grades $K_{Ic}^* \leq 30$. It is true that S7 tempered to 54 Rockwell C is one of the tool steels with highest toughnesses, with $K_{Ic} = 55$MPa√m, however this is a low value compared to most steels: for instance, the low strength 1020 steel can reach $K_{Ic} = 130$MPa√m. Beware, S7 steel is not a panacea! It may only be a good option when sharpness is also required.

3.9.7. Minimum Weight Ablative Armor

As explained in chapter 2, ablative armors are designed to negate damage by themselves being damaged or destroyed through the process of ablation, which is the removal of material from the surface of an object by vaporization or chipping. They're also made out of tough materials, but with low hardness and low melting point to facilitate the ablation. Most of the impact energy is absorbed by the ablation process, by breaking apart when hit by an opponent, transmitting much less energy to the rest of the robot.

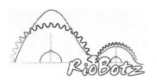

To find out if a material will result in an ablative armor, you must consider its hardness and melting point. If you want a traditional armor, we've seen that a good option is Ti-6Al-4V, which has a 1660°C (3020°F) melting point and 36 Rockwell C hardness, equivalent to 336 Brinell. But if you want an ablative armor, we'll see that the best options are high toughness Mg and Al alloys, which have low Brinell hardnesses between 60 and 171, and relatively low melting points, close to 660°C (1220°F). Low hardness helps ablation by making the armor material deform during an impact, while low melting point helps ablation by allowing the armor material to locally melt or even vaporize during high energy impacts.

Ablative armors are a very good choice against blunt or not-so-sharp weapons. But, against very sharp horizontal spinners, very hard traditional armors are a better option, as discussed before. This is because ablative armor materials such as Mg and Al alloys have low hardnesses, not being able to blunt or chip the edge of the opponent's blade during combat. An opponent with a sharp spinning blade during the entire combat is not a pleasant thought. In special because Mg and Al alloys, used in the ablative armor, tend to be easily cut by sharp tools at high speeds, drastically reducing their K_{Ic}, S_u and S_y, which had been measured under static conditions. It's like cutting butter with a hot knife. Thus, traditional armor materials such as Ti-6Al-4V are better choices against very sharp spinners, their higher hardness and melting point prevent such weakening.

In addition to their ablative properties, high toughness Mg and Al alloys also have better fracture toughness grades K_{Ic}* than traditional armor materials. For instance, let's compare Ti-6Al-4V with 5086-H32 aluminum, used in armor plates of light weight military vehicles. In average, 5086-H32 has K_{Ic} = 49MPa√m, while Ti-6Al-4V has K_{Ic} = 72MPa√m. Section 3.8 showed that strength and toughness analyses have similar equations because the effect of a crack in a structure is directly proportional to the applied stresses. Therefore, for plates under bending, it is not a surprise that the scale factor between old and new plate thicknesses for a minimum weight design with constant fracture toughness is $(K_{Ic,old}/K_{Ic,new})^{1/2}$. Because almost all armor plates don't have restrictions about having their thickness increased, since they are outside the robot, they can be analyzed using a minimum weight design.

So, a Ti-6Al-4V plate would have the same toughness as a 5086-H32 aluminum plate that was $(72/49)^{1/2} \cong 1.2$ times thicker. This aluminum replacement would only have $1.2 \cdot 2.66 / 4.43 \cong 72\%$ of the weight of the Ti-6Al-4V version. In summary, we can conclude that, for weight optimization, high toughness titanium alloys (such as Ti-6Al-4V) are not as good as high toughness aluminum alloys when it comes to fracture toughness.

This conclusion might seem weird, but it is true. Surely a titanium plate would be a better armor than an aluminum one with same thickness, but remember that this is a weight (and not a volume) optimization problem. An aluminum plate with the same weight as a titanium one is about 1.6 times thicker, this is what makes the difference. This is easily seen from the fracture toughness grades K_{Ic}* = 60 for Ti-6Al-4V and K_{Ic}* = 83 for 5086-H32 aluminum.

To find out the best ablative armor materials from the table, we'll choose the ones with low hardness HB' < 30 and rank them by their K_{Ic}*. For instance, depending on the impact orientation, 7475-T7351 can have K_{Ic} from 36 to 55MPa√m, resulting in average in K_{Ic}* = 75. The 7000 series aluminum alloys have K_{Ic}* between 55 and 75.

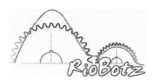

But the ultimate ablative armor would be made out of, believe it or not, the Mg alloys ZK60A-T5 (K_{Ic}* = 100) and AZ31B-H24 (K_{Ic}* = 93). And their very low hardness, between 70 and 77 Brinell (HB' between 10 and 11), helps even more in the ablation process. In summary, the best ablative armor materials are, in that order, the Mg alloys ZK60A-T5 and AZ31B-H24, followed by the Al alloys 5086-H32, 2324-T39 Type II, 7475-T7351 and 7055-T74, all of them with K_{Ic}* > 69.

Touro Feather can testify the effectiveness of aluminum ablative armors. At Robogames 2008 it withstood several powerful blows from the featherweight undercutter Relic on its 3/4" thick 7050 aluminum front armor plates (pictured to the right). These scarred plates worked as an ablative armor, slowing down Relic's weapon and allowing Touro Feather to go for the knockout.

Avoid using polymer armor plates, such as Lexan or UHMW, they have K_{Ic}* < 45. Wood might be a good option for ablative armor, although its K_{Ic} varies by a factor of 10 depending on the impact direction. In average, wood would grade K_{Ic}* = 77, a very high score. But hope that your opponent

doesn't have a flamethrower, and that the opponent robot doesn't hit you in the brittle transverse direction of the wood fibers. And be careful with wood splinters when handling your robot, it will have plenty of them.

3.9.8. Minimum Weight Beams

The previous components were all modeled as plates, because they have by design 2 fixed dimensions (width and length), while only thickness can be changed. However, there are other components (such as shafts) that might only have one fixed dimension, their length, while their diameter can be varied. If a beam basically works under bending and/or torsion, then we've seen in section 3.8 that its mechanical properties are optimized for minimum weight using materials with high $E^{1/2}/\rho$ ratio for stiffness, high $S_y^{2/3}/\rho$ for yield strength, high $S_u^{2/3}/\rho$ for ultimate strength, and high $K_{Ic}^{2/3}/\rho$ for fracture toughness.

Similarly to what we did for plates, we've normalized all the above ratios using the best materials from the table, resulting in a system of grade points between 0 and 100 for beams. The normalized hardness is still HB', because it is a local property, while the grades for minimum weight beams are represented by the property followed by the ** symbol, namely E**, S_y**, S_u**

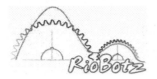

and K_{Ic}**, shown in the table below. Note that the beam grades can differ a lot from the plate grades. For instance, the best material from the table for a beam with high yield strength is not a Mg alloy, as it was for plates, but the 7055-T7751 Al alloy, grading S_y** = 100. Note that steels are poor choices for high stiffness elements not only for plates, but also for beams, with E** < 50.

In the table there are also properties followed by the *** and ' symbols, they will be used respectively in minimum weight truss design and in minimum volume design, discussed later.

	material	min. weight beam				min. weight truss				minimum volume				
		E**	S_u**	S_y**	K_{Ic}**	E***	S_u***	S_y***	K_{Ic}***	E'	S_u'	S_y'	K_{Ic}'	HB'
Mg alloys	AZ31B-H24	100	80	63	90	94	47	28	57	22	10	6	13	11
	ZK60A-T5	97	88	74	100	92	55	36	68	22	13	8	15	10
	Elektron WE43-T5	96	76	69	60	90	45	32	32	21	10	7	7	14
	Elektron 675-T5	90	100	93	57	85	69	53	30	21	17	13	7	17
Aluminum alloys	Al 6063-T5	82	43	40	55	96	23	18	34	33	8	6	11	9
	Al 6061-T6	82	60	62	58	96	38	34	37	33	13	11	12	14
	Al 2024-T3	82	78	70	63	99	57	41	42	35	20	14	15	18
	Al 2324-T39 Type II	82	78	74	83	98	56	44	63	35	19	15	22	18
	Al 5086-H32, H116	84	58	52	88	100	36	26	67	34	12	8	22	12
	Al 7050-T7451	80	81	84	61	95	61	55	41	35	21	19	14	21
	Al 7055-T74	79	80	84	71	94	60	54	51	35	21	19	18	21
	Al 7055-T7751	79	92	100	56	94	73	71	35	35	26	25	13	26
	Al 7075-T6	80	85	86	53	96	64	56	32	35	22	19	11	22
	Al 7075-T73	80	80	81	60	96	59	51	39	35	20	18	14	20
	Al 7175-T736	81	85	87	65	96	64	57	44	35	22	20	15	22
	Al 7475-T7351	80	79	79	79	96	58	49	58	35	20	17	20	20
Ti	Ti-6Al-4V (36HRc)	63	79	85	68	93	73	69	59	53	40	38	33	50
	Ti-6Al-4V ELI	63	74	79	78	93	66	62	73	53	36	34	40	49
Steels	1020 steel	48	26	21	57	97	18	11	60	98	18	11	59	16
	304 stainless	46	32	21	79	90	25	11	100	93	25	11	100	23
	4340 (43HRc)	49	58	61	44	98	60	57	41	99	59	55	40	60
	4340 (39HRc)	49	54	57	54	98	55	51	56	99	53	49	55	54
	4340 (34HRc)	49	50	53	62	98	49	45	69	99	48	44	67	48
	S7 (54HRc)	49	71	67	32	99	82	64	26	100	80	62	25	81
	AerMet 100 (53HRc)	47	70	72	53	92	82	72	55	94	80	70	54	79
	AerMet 310 (55HRc)	47	75	77	38	92	90	80	33	94	88	78	32	84
	AerMet 340 (57HRc)	47	80	81	25	92	99	87	17	94	96	85	17	89
	HP-9-4-30 (51HRc)	49	62	60	57	97	67	55	59	97	64	52	57	74
	18Ni(200) (46HRc)	45	58	62	59	86	61	58	65	88	61	57	65	64
	18Ni(250) (51HRc)	46	63	71	53	89	71	70	55	92	70	70	55	73
	18Ni(300) (54HRc)	46	72	79	40	89	85	83	37	92	84	82	36	81
	18Ni(350) (61HRc)	47	80	89	26	93	100	100	19	97	100	100	19	100
	K12 Dual Hardness	48	66	70	38	98	74	68	33	99	72	66	33	100
Polym.	Delrin	33	45	44	26	8	18	15	8	1	3	3	1	1
	Lexan	34	48	51	25	7	18	17	7	1	3	2	1	1
	UHMW-PE	24	44	33	26	3	14	8	6	0	2	1	1	1

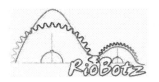

To decide which material to choose from the table for a light weight beam, we must know which of the E**, S$_y$**, S$_u$**, K$_{Ic}$** and HB' properties are more important, which depends on its functionality in the robot. We'll study these beam-like structural members next.

3.9.9. Minimum Weight Shafts and Gears

Shafts are, basically, beams with circular cross-section under bending and torsion. They must have high K$_{Ic}$** grade to withstand impacts, high stiffness grade E** to prevent vibration and gear misalignment, and certainly high S$_y$** grade not to get bent. We'll look then at the average grade X** = (K$_{Ic}$** + E** + S$_y$**) / 3 to evaluate the fitness of each material.

It is easy to see that high strength Mg and Al alloys would be the best choices, with X** between 72 and 90. This is true for most beams, but not for shafts. It is true that Mg and Al shafts would have the best minimum weight properties, however they would need to have a much larger diameter than an equivalent one made out of steel or titanium, which would imply in larger bearings, larger gears and pulleys, and larger gearboxes, increasing the robot weight.

So, to avoid shafts with very large diameters, we'll limit our choices to denser materials such as steels and Ti alloys. We can easily do that by adding another restriction, which is HB' > 45. This medium-high hardness requirement will not only filter out polymers, Mg and Al alloys, but it will also help in the mounting process, since low hardness shafts would easily become dented, making it difficult to mount, for instance, a tight tolerance roller bearing. Also, higher hardness shafts will prevent wear due to, for instance, bronze bearing friction.

Also, to avoid distortions in the X** average, we'll also limit the choices to materials with K$_{Ic}$** > 25, S$_y$** > 40 and E** > 40. It is found that the best shaft materials from the studied table are Ti-6Al-4V ELI (with X** = 73) and Ti-6Al-4V (X** = 72, see the shaft pictured on the right). The next choices are, in that order, AerMet 100, 18Ni(250), 18Ni(200), HP-9-4-30, 18Ni(300), tempered 4340, AR400, and tempered 5160, all of them with X** between 50 and 58. Avoid using S7 steel shafts, they only grade X* = 49, and they're more expensive and with lower K$_{Ic}$** than 4340 steel.

Therefore, hardened steel shafts are better than Mg or Al ones. Note that high strength titanium is only better than high strength steels if the weight saved by the shaft is not gained by the slightly larger bearings, gearboxes, etc.

Note that minimum weight gears also fall in this very same category, if their thickness is a free parameter that can be changed. They also need high K$_{Ic}$**, E** and S$_y$**, for the same reasons. And a higher hardness grade HB' > 45 will prevent wear on the gear teeth. So, if the volume of the gear is not important, the best choice for minimum weight would be high strength Ti alloys. If you need more compact gears and with less tooth wear, go for the same high hardness steels chosen for shafts (such as in the hardened steel gear pictured above, from the TWM 3M gearbox).

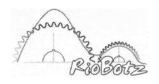

3.9.10. Minimum Weight Spinning Bars and Eggbeaters

The material choice for the bars (blades) of horizontal and vertical spinners is very critical. Spinning bars can be modeled as beams or plates, depending on your design requirements. If their

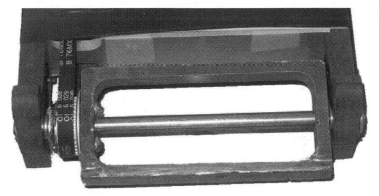

width and thickness are not restrained, which is usually true, then they can be analyzed as beams for minimum weight design. Eggbeaters (pictured to the right) also fall in this category, because they are basically two vertical spinning beams connected by two horizontal beams.

Note that minimum weight design does not mean that your spinning bar will have a low inertia, it only means that it will have high ratios between mechanical properties and weight. After optimizing your material, you'll be able to increase the bar thickness until reaching the desired weight or moment of inertia, and still have optimized mechanical properties.

All spinning bars need to have high K_{Ic}** grades, since they'll have to withstand their own inflicted impact, high E** grades to avoid hitting themselves due to excessive deflection at a sloped impact, and high S_y** not to get warped and therefore unbalanced.

There are several different material choices depending whether the bar itself needs to be sharp, blunt, or if it will have inserts made out of a different material at its tips, as discussed next.

Sharp one-piece spinning bars and eggbeaters

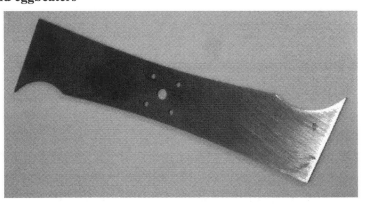

Sharp one-piece spinning bars (such as the one from the antweight Collision, to the right) and eggbeaters must have high hardness, to retain their sharpness. If we only select materials with HB' > 75, then we only have steels to choose from. The E** of steels is basically constant (it only varies between 45 and 49 for them). And all steels with very high hardness also have high yield strength, so we don't have to worry too much with S_y**, they would break before the yield deformations were high enough to cause unbalancing. We then decide to maximize the remaining grade, K_{Ic}**.

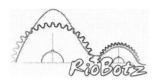 

The result is that the best materials are AerMet 100 (K_{Ic}** = 53), 18Ni(300) (K_{Ic}** = 40) and AerMet 310 (K_{Ic}** = 38).

Other good choices are S7 steel at 54HRc (as in Hazard's sharp bar pictured to the right), 18Ni(350) and AerMet 340, but only if the bar or eggbeater is not heavily notched, because of their lower K_{Ic}** < 33.

If the bar or eggbeater is notched, it would probably be safer to use 5160 steel

tempered to 46 HRc instead of S7, you will have a lower hardness (HB' = 65) but higher impact toughness. Note that these results also apply to sharp one-piece spears.

Blunt one-piece spinning bars and eggbeaters

If you want a minimum weight blunt one-piece spinning bar or eggbeater, then your hardness requirements can be somewhat relaxed. You still need medium-high hardness to be able to inflict damage, but it does not need to be too high since the weapon doesn't need to be sharp. This reasoning would also apply to one-piece hammers for hammer, thwack or overhead thwack bots.

We'll then choose materials with medium-high HB' > 45, with grade K_{Ic}** > 30, and order them by the average between these two grades, X** = (K_{Ic}** + HB') / 2, since both are important.

The result is that the best materials are AerMet 100, HP-9-4-30, 18Ni(250), Ti-6Al-4V ELI, 18Ni(200), 18Ni(300) and AerMet 310, all of them grading X** > 60. Ti-6Al-4V is the next choice (X** = 59). And 4340 steel is not too bad, as long as it is tempered to lower hardnesses between 34 and 39HRc to improve toughness (despite losing strength), instead of the usual 40 to 43HRc. But note that 4340 steel is usually sold in bar form, you might need to look for AR400 or 5160 steel plates.

Spinning bars and eggbeaters with inserts

If you want to improve the impact and fracture toughness of your spinning bar or eggbeater, then it is a good idea to use two different materials: a softer one for the bar itself, and a very hard one for inserts to be attached at its tips.

Now we have to worry again with E** and S_y**, since we won't be limiting our search to high hardness materials. So, to minimize the weight of spinning bars and eggbeaters with inserts, we'll choose materials with K_{Ic}** > 50, E** > 60 and S_y** > 60, and order them by the average grade X** = (K_{Ic}** + E** + S_y**) / 3.

The best choices are, in that order, ZK60A-T5, AZ31B-H24, Elektron 675-T5, 2324-T39 Type II, 7475-T7351, 7055-T7751 or T74, 7175-T736 and 7050-T7451, all with X** > 75. Other good

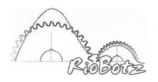

options are Elektron WE43-T5, 7075-T73 or T6, Ti-6Al-4V ELI, Ti-6Al-4V and 2024-T3, all with $X^{**} > 70$. Avoid using steels ($X^{**} < 58$) or polymers ($X^{**} < 40$).

The use of Mg and Al alloys has also the advantage of increasing the moment of inertia of the bar for a given weight. This is because you'll have a lighter material close to the spin axis, leaving more weight for heavier steel inserts at the tips, which will contribute much more to the moment of inertia due to their higher distance. Remember that the moment of inertia is proportional to the square of the distance of a certain mass to the spin axis. The aluminum spinning bars pictured to the right, from the middleweight The Mortician, not only have heavy steel inserts at their tips, but they also have pockets milled close to the spin axis or through-holes to optimize the weight distribution.

Note that the above optimum materials also apply to hammer handles (as pictured to the right), thwack or overhead thwack handles, and spears that have inserts.

Lifter and launcher arms would also be lighter if made out of these materials, since such arms are basically beams under bending stresses. But if using Mg or Al alloys in the lifter/launcher arm, make sure you'll have high hardness inserts at its tip to help wedge and scoop the opponent.

Low maintenance spinning bars and eggbeaters with inserts

As seen above, Mg and Al alloys make great spinning bars that have inserts. But, unless your inserts are large enough to shield and prevent any direct hits on such low hardness bar, you'll realize that the bar itself will need to be changed very often, due to ablation.

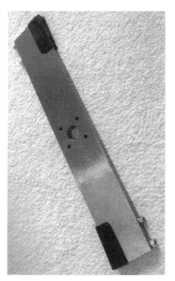

A low maintenance bar would need to have a higher hardness grade, for instance $HB' > 45$, in addition to the high average grade $X^{**} = (K_{Ic}^{**} + E^{**} + S_y^{**}) / 3$.

But these are exactly the requirements we used for minimum weight shafts and gears, so the optimum materials are the same: high strength Ti alloys are the best choice, followed by AerMet 100, 18Ni(250), 18Ni(200), HP-9-4-30, 18Ni(300), tempered 4340 (or AR400), and tempered 5160. Once again, avoid using S7 steel in the bar, leave it for the inserts. The spinning bar from the middleweight Terminal Velocity (pictured to the right) is an excellent example of optimum material choice: Ti-6Al-4V with S7 inserts.

Note that these materials also apply to low maintenance hammer, thwack or overhead thwack handles, to low maintenance spears that have inserts, and to low maintenance lifter and launcher arms, such as in the titanium hammer handle from the middleweight Deadblow (pictured to the right).

Spinning bars with minimum width restrictions

In all the above spinning bar analyses, it was assumed that both the bar width and thickness could be changed. But the bar must have a minimum width of, for instance, twice the diameter of its center hole, where the weapon shaft goes through. Smaller widths will probably compromise strength, as it will be shown in chapter 6. So, if the above calculations for minimum weight beam design result in a bar width that is smaller than, for instance, twice the center hole diameter, then the design strategy must be changed.

If this happens, then the bar width must be kept constant at this minimum value, while only the bar thickness can be changed in the design process. Thus, the problem is now a minimum weight plate design, instead of a minimum weight beam design. We'll have to use the plate grades K_{Ic}^*, E^* and S_y^* instead of K_{Ic}^{**}, E^{**} and S_y^{**}. The material choice for this bar with fixed width will then be the same as for a minimum weight spinning disk, as shown next.

3.9.11. Minimum Weight Spinning Disks, Shells and Drums

Spinning disks and shells usually have their diameter defined by the robot design, while only their thickness is a variable design parameter. The same is valid for drums, their external diameter and width are defined by the design of the robot structure, only leaving the drum thickness as a variable value. So, spinning disks, shells and drums are modeled as plates. Spinning bars with fixed width also fall in this category, as explained above.

Sharp one-piece spinning disks, shells and drums

One-piece spinning disks, shells and drums (pictured below) are usually a good option to maximize tooth strength and to minimize the number of parts.

The analysis is very similar to the one for sharp one-piece spinning bars, except that plate grades are used, instead of beam grades.

The disk, shell or drum also needs to retain its sharpness, so we choose materials with

grades HB' > 75, K_{Ic}^* > 20, S_y^* > 50 and E* > 30, trying to maximize K_{Ic}^*.

The best materials are then AerMet 100 (K_{Ic}^* = 43), 18Ni(300) (K_{Ic}^* = 35) and AerMet 310 (K_{Ic}^* = 34). Other good choices are S7 steel at 54HRc, 18Ni(350) and AerMet 340, but only if the disk, shell or drum is not heavily notched, because of their lower $K_{Ic}^* \leq 30$. Note that these materials are exactly the same as the ones from the one-piece bar analysis, which is a coincidence, since different grades were used.

Blunt one-piece spinning disks and shells

Blunt one-piece spinning drums are not a good idea, they will have a hard time grabbing and launching the opponent. Sharpness is important for drums. But blunt one-piece spinning disks may be useful, in special if the one-piece disk has large teeth, and if it spins horizontally. Blunt one-piece spinning shells might also be useful.

We can use the same material selection criteria from blunt one-piece spinning bars, as long as we change beam grades to plate grades, so we need HB' > 45 and K_{Ic}^* > 30, ordered by the average $X^* = (K_{Ic}^* + HB') / 2$. Interestingly, these are exactly the same criteria that optimize traditional armor plates, resulting in almost the same materials.

So, the best materials are then K12 Dual Hardness, AerMet 100, HP-9-4-30, AerMet 310, 18Ni(250), 18Ni(300), Ti-6Al-4V ELI, 18Ni(200) and Ti-6Al-4V, all of them with X^* > 50. Tempered 4340 or 5160 steels are not bad options, but not one of the best. Avoid low hardness steels such as 1020 (which will yield very easily, as seen on the spinning disk tooth on the right), and medium-low toughness steels such as S7.

Note that the K12 Dual Hardness steel is probably a bad idea for disks. Although one of the disk surfaces would have HB' = 100, the other one, which is also exposed, would only have HB' = 73. K12 would be fine for shell spinners, as long as the harder surface is facing outwards.

Spinning disks, shells and drums with inserts

If your disks, shells or drums have inserts (as pictured below), then they can have a lower hardness to improve fracture toughness.

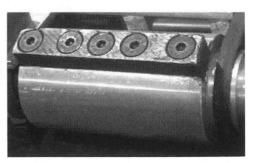

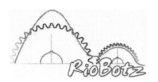 

We can use material selection criteria similar to the ones for spinning bars with inserts, if the beam grades are changed to plate grades. So, we'll choose materials with $K_{Ic}^* > 50$, $E^* > 70$ and $S_y^* > 70$, and order them by the average grade $X^* = (K_{Ic}^* + E^* + S_y^*) / 3$. These are the same criteria used for integrated structure-armor plates, resulting in the same material choices: ZK60A-T5, AZ31B-H24, Elektron 675-T5, Elektron WE43-T5, 2324-T39 Type II, 7475-T7351, 7175-T736, 7055-T7751 or T74 and 7050-T7451, all of them with $X^* > 73$. The aluminum alloys 7075-T73 or T6 and 2024-T3 would be the next options. As we did with integrated structure-armor plates, avoid Ti alloys, steels and polymers.

Note that Ti-6Al-4V ELI is as good as most aerospace aluminum alloys for spinning bars with inserts, but it is significantly worse for spinning disks and drums with inserts. This is not an obvious conclusion, only after the above analyses we were able to show that.

So, the best spinning disks, shells and drums with inserts are made out of high toughness Mg and Al alloys. But these materials result in a high maintenance disk, shell or drum, due to ablation. The picture to the right shows that the aluminum drum of our featherweight Touro Feather, although still functional, has suffered a lot of ablation. This loss of material ends up unbalancing the drum, requiring it to be changed after a few tournaments. On the other hand, its tempered S7 steel teeth are high hardness inserts, with very little loss of material (despite some brittle chipping that can be noticed around their countersunk holes).

Low maintenance spinning disks, shells and drums with inserts

If you want a low maintenance disk, shell or drum, such as the ones pictured below, then you need to select a material with a higher hardness grade, for instance HB' > 45, in addition to the high average grade for plates $X^* = (K_{Ic}^* + E^* + S_y^*) / 3$.

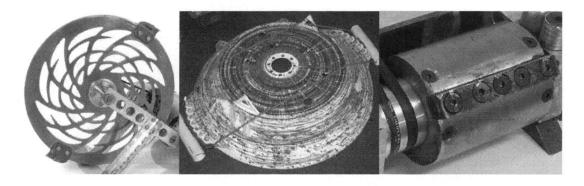

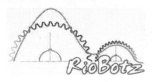

If we also only choose materials with $K_{Ic}* > 25$, $E* > 35$ and $S_y* > 40$, to avoid distortions in the $X*$ average, then we end up with exactly the same criteria for wedges. So, after having ruled out Mg and Al alloys due to their low hardness, the best materials would be Ti-6Al-4V ELI and Ti-6Al-4V, with $X*$ equal to 64 and 63, respectively. The next choices would be AerMet 100, followed by 18Ni(250), HP-9-4-30, 18Ni(200), 18Ni(300), tempered 4340, AR400 and tempered 5160, all of them with $X* > 42$.

3.9.12. Minimum Weight Weapon Inserts

The most important properties of weapon inserts are high impact and fracture toughnesses. If they must remain sharp, then high hardness is also important. Three types of inserts are studied below.

Sharp plate-like weapon inserts

Clampers, lifters and launchers usually use sharp inserts at the tip of their arms to help scoop the opponent. These scoops are basically plates under bending, working as wedges. Therefore, the wedge design analysis can be used here, resulting in Ti-6Al-4V ELI and Ti-6Al-4V scoops. But the lower HB' between 49 and 50 of these alloys might require high maintenance to keep them sharp at every combat.

So, for a low maintenance scoop, a higher hardness is desired, for instance HB' > 75. With this new restriction, the same analysis used for sharp one-piece spinning disks can be applied for these scoops, resulting in AerMet 100 as the best material, followed by 18Ni(300), AerMet 310, S7 steel at 54HRc, 18Ni(350) and AerMet 340. Alternatively, 5160 tempered at 46HRc can be used, but it has a lower HB' = 65 than the other high strength steel options.

Sharp beam-like weapon inserts

Most weapon inserts, such as drum, disk, shell or bar teeth (pictured to the right), spear tips, or sharp thwack, overhead thwack or crusher tips, can be modeled as beams. This is because they usually have both their width and thickness (or their diameter) as free parameters.

To retain their sharpness, we need to select high hardness materials. Forget about Ti-6Al-4V inserts, its grade HB' = 50 won't stand a chance to keep sharp against hard traditional armor or wedge materials such as AR400, with HB' = 60. You'll need something harder than that, preferably with HB' > 75. These hardness and toughness requirements are then similar to the ones used for sharp one-piece spinning bars, resulting in the same optimal materials: AerMet 100 is the best, followed by 18Ni(300), AerMet 310, S7 steel at 54HRc, 18Ni(350) and AerMet 340, coincidently the same materials selected for sharp plate-like inserts.

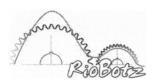

Inserts usually have complex geometries, such as the puzzle-like fitting between the aluminum spinning bar from The Mortician and its hardened steel inserts, pictured to the right. These fittings are essential to guarantee a strong connection during high energy impacts, helping to avoid sheared bolts. Note, however, that sharp notches should be avoided in the insert, because the high hardness steel in general does not have a very high fracture toughness, typically K_{Ic}** < 33. So, if intricate geometries are

necessary, use large notch radii (typically of a few millimeters) to avoid high stress concentration factors (denoted by K_t), which might lead to the fracture of the insert.

For instance, the K_t of an 8mm deep notch with a sharp 0.5mm radius can be roughly estimated by $K_t = 1 + 2 \cdot (8mm / 0.5mm)^{0.5} = 9$, meaning that any stresses near this notch will be locally multiplied by 9. For very ductile and low hardness metals (such a 304 stainless steel) this may not be a problem, because these 9 times higher stresses will probably cause the notch to plastically deform and get blunt. This would increase the notch radius and thus decrease the K_t from 9 to much lower values, even lower than 2. But high hardness metals usually don't have enough ductility to blunt the sharp notch, keeping in this example the K_t in its original high value equal to 9. This is why we have to worry much more with sharp notches in high hardness metal components than in soft metals.

Even though the bar is usually made out of a softer material than its insert, it is also a good idea to avoid sharp notches as well in the bar (or disk, drum, handle). In special if the notch is very close to a threaded hole, which can have K_t of up to 7, because both K_t would get multiplied. For instance, if the sharp notch from the previous example (with $K_t = 9$) was very close to a threaded hole, the resulting K_t could be as high as $9 \times 7 = 63$. The amazing 63 times higher notch root stress would certainly break a very hard low ductility material. Tougher materials such as the ones used in

the bars would be able to lower this K_t through blunting, but maybe it would still be high enough to cause fracture. So, avoid sharp notches at all costs. And never thread any hole from high hardness inserts, always leave the threads, if necessary, to the lower hardness bars, disks, drums and handles.

Using plain through holes can be a good option to avoid the stress concentration from the threads, as seen on the bar to the right, where nuts are used to hold the two bolts from the insert. Note, however, that through holes such as the ones shown in the

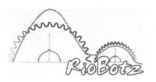 

picture should only be drilled in a very thick bar, otherwise they'll significantly lower its cross section area, compromising strength.

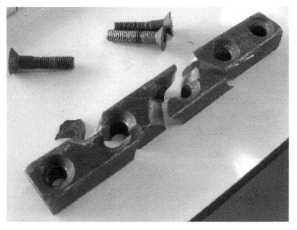

If your inserts are still breaking even after removing all sharp notches from their geometry, then you'll probably need to sacrifice hardness a little bit to improve the impact and fracture toughnesses, by changing the material or the heat treatment. Earlier versions of the S7 steel drum teeth of our middleweight Touro had been tempered to hardnesses between 57 and 59 Rockwell C (HRc), to guarantee their sharpness. However, as seen to the right, this led to their premature fracture in combat. The newer teeth now have a slightly lower 54 HRc hardness, but with a much better impact toughness.

Blunt beam-like weapon inserts

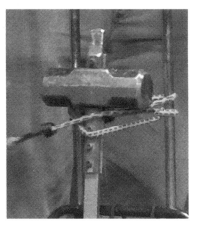

Hammer heads (as pictured to the right), which may be used in hammer, thwack or overhead thwack robots, are usually made out of blunt inserts, which do not need to be sharp. Medium-high hardness is still important, but now we can lower our minimum HB' grade requirement from 75 (for sharp inserts) to, for instance, 45, allowing us to have more material options and also improve other properties such as fracture toughness.

Since the diameter (or other cross-section dimensions) of hammer heads is a design parameter that can be varied, we conclude that these blunt inserts can also be modeled as beams. It is easy to see that these blunt inserts with HB' > 45 have basically the same requirements of blunt one-piece spinning bars, resulting in the same material choices: AerMet 100 as the best, followed by HP-9-4-30, 18Ni(250), Ti-6Al-4V ELI, 18Ni(200), 18Ni(300) and AerMet 310. Other good choices are Ti-6Al-4V, 4340 (at 34HRc) and 4340 (at 39HRc). So, if sharpness is not important, then high toughness titanium alloys are also good choices for inserts, together with high toughness hardened steels.

3.9.13. Minimum Weight Clamper and Crusher Claws

Since clamper and crusher robots have relatively slow active mechanisms, acting without impacts, it is more important to have claws with high S_u and S_y than high K_{Ic}. Claws are basically beams working under bending, therefore to minimize their weight it is important to choose materials with high S_u** and S_y** grades.

If high hardness inserts are used at the tips of the claws, as pictured to the right, then the choices for the claw material can include low hardness alloys. We then select the materials with a high average grade $X^{**} = (Su^{**} + Sy^{**}) / 2$, resulting in Elektron 675-T5 as the best, followed by 7055-T7751, 7175-T736 and 7075-T6, all of them with $X^{**} > 85$. Other reasonable choices are, in that order, 18Ni(350), 7050-T7451, Ti-6Al-4V, 7055-T74, ZK60A-T5, AerMet 340 and 7075-T73, all of them with $X^{**} > 80$.

Note that these material choices didn't consider the possibility of the claws receiving impacts, in special from spinners. They assume that your robot will have some shield or wedge to be able to slow down the spinner before clamping or crushing it. Otherwise, you'll need to choose claw materials with high K_{Ic}^{**} as well. So, if the claw must withstand high impacts as well, the best material choices would be the same as for spinning bars with inserts, such as the Mg alloys ZK60A-T5 and AZ31B-H24, which have high K_{Ic}^{**} grades for beams.

Note also that the Mg and Al alloy options may require high maintenance, due to their low hardness. If low maintenance is also desired, then the best material choices would be the same as for low maintenance spinning bars with inserts, such as Ti-6Al-4V ELI.

Finally, if you want to use one-piece claws (as pictured to the right), without tip inserts, then very high hardness is required. The best choices would then be the same as for sharp one-piece spinning bars, such as AerMet 100 or 18Ni(300).

3.9.14. Minimum Weight Trusses

The weight optimization analysis of trussed elements, which can be used in the structure of trussed robots (as pictured to the right, using welded steel tube trusses), is relatively simple. Besides composites and Be alloys, we've shown that all steels, Al, Ti and Mg alloys result in similar stiffness-to-weight ratios for trusses, which depend on E/ρ. To find the best materials, it is then a matter of looking for the ones that maximize S_y/ρ, S_u/ρ and K_{Ic}/ρ at the same time, since a trussed robot structure needs high S_y not to get bent, high S_u to bear static loads and avoid

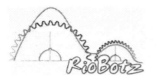 

fatigue, and high K_{Ic} to withstand impacts and cracks. The previous table shows the minimum weight truss grades E***, S_y***, S_u*** and K_{Ic}*** for several materials, obtained by normalizing the E/ρ, S_y/ρ, S_u/ρ and K_{Ic}/ρ ratios using the best materials in the table.

Most trussed robot designs have their trusses exposed or only partially covered by armor plates, allowing them to take direct hits. So, we'll give more importance to K_{Ic}*** than to S_u***. We choose then the materials that have higher average grade X*** = (K_{Ic}*** + S_y***) / 2, requiring as well that K_{Ic}*** > 40 and S_y*** > 50.

The best truss material from the table is Ti-6Al-4V ELI, followed by Ti-6Al-4V, AerMet 100, 18Ni(250) and 18Ni(200), all with X*** > 60. Other good choices are HP-9-4-30 and 4340 tempered to 39HRc, both with X*** > 50. Note that 304 stainless steel wasn't chosen, despite its high toughness, due to its low S_y that will allow the frame to easily get warped. And S7 and 18Ni(350) weren't chosen either, because of their insufficient fracture toughness.

The above options (with X*** > 50) are fine if the robot trusses will be bolted together. But if you want to use welds to join them, then their material will also need to have a high weldability. The above Ti alloys are still the best choice, but you might have trouble welding the presented high strength steel options, even the 4340 steel. There's no point in having high strength trusses if they're joined by weak welds. So, a good alternative to 4340 would be 4130 steel which, despite its lower S_y and S_u, results in stronger welds due to its lower 0.3% carbon content, instead of 0.4% from 4340. But note that, after welding the 4130 steel trussed frame, it needs to be heat treated to relieve residual stresses from the welds and to achieve higher S_y and S_u through temper. This means that repaired 4130 welds at the pits during a competition will probably be weak spots in the following fights, unless they're somehow heat treated.

Ti-6Al-4V and Ti-6Al-4V ELI, on the other hand, in addition to their highest X*** grades, don't need to be hardened after welding, resulting in a much better choice for repairs if proper welding equipment is available at the pits. Their only disadvantage is the higher cost.

Note that bamboo is not represented in the table, however it has a great stiffness-to-weight ratio for trusses, comparable to the performance of metals, grading E*** = 94. Its fracture toughness grade is not too bad, K_{Ic}*** = 31, comparable to low toughness aluminum alloys. The strength grades for trusses S_u*** = 19 and S_y*** = 19 are relatively low compared to high strength metals, however they are actually higher than the grades from 1020 steel, used in civil engineering. So, it's not a surprise to see that China is using bamboo-reinforced concrete to build high-rises. Note however that bamboo has only about half the K_{Ic}*** of 1020 steel, which might mean bad news during an earthquake. For combat, bamboo is not a good option, because most high strength metals will result in much better strength and toughness grades. Also, bamboo trusses, despite being light, would end up with a very high volume, leaving a limited room inside the robot to mount its components. Not to mention the challenge in putting together the bamboo trusses with strong and light joining elements.

3.10. Minimum Volume Design

All minimum weight problems presented in the previous section can be used to choose materials that make your robot lose weight without losing stiffness, strength or toughness. But what if a component needs to have its mechanical properties improved? Well, if there's space to increase the volume of the component (such as in most structural or armor elements), then this shouldn't be a problem, it is just a matter of weight optimization, using the grades from minimum weight design. If the material has not yet been optimized, then you should first change it depending on the functionality of the component, as explained before. After that, you only need to increase the component thickness (for plate elements) or cross section area (for beam and truss elements) until the desired improved properties are obtained.

But there are other components that cannot (or should not) have their volume increased. For instance, a gear that keeps breaking cannot be replaced by a thicker one without modifying the gearbox. A shaft that keeps yielding cannot have its diameter increased without changing its bearings and collars. These examples are volume optimization problems, requiring a minimum volume design, instead of minimum weight, as described next.

Minimum volume design has the goal to find the best materials to optimize the performance of a component while minimizing its volume. It assumes that the weight of the component can be increased, if necessary, but in most cases its dimensions cannot be changed. The idea is to design a component for a desired functionality with the lowest possible volume. Alternatively, if a component is failing in combat and its dimensions should not be changed, then the idea is to improve its mechanical properties only by switching its material, without changing its dimensions, while adding as little weight as possible.

Minimum volume design is quite straightforward in the case where the dimensions must be kept constant. Since the volume is not changed, it is just a matter of directly comparing the material properties. This is more easily performed if we normalize the mechanical properties using the best materials from the presented table, resulting in a system of grade points between 0 and 100 for minimum volume. These grades are represented by the property followed by the ' symbol, namely E', S_y', S_u' and K_{Ic}', shown in the previous table.

Note that minimum volume grades are completely different than minimum weight grades. Polymers are always the worst choice, by far, for minimum volume parts, with grades lower than 4 (out of 100). Al and Mg alloys are also very bad options, none of their grades is higher than 35. Even Ti alloys are not good, their highest grades only go up to 53.

Therefore, steels are always the best choice if you need to minimize volume. Their E' grades are always between 88 and 100. The best material from the table for a minimum volume component with high yield and ultimate strengths is the 18Ni(350) maraging steel, with $S_y' = 100$ and $S_u' = 100$. The best choice to maximize fracture toughness is the 304 stainless steel, with $K_{Ic}' = 100$.

Note that, except for titanium alloys, almost all S_u' grades are only within 4 points from HB' grades. This is no surprise, because there are good correlations between S_u and hardness for different metal alloy families, as mentioned before. For instance, you can estimate $S_u \cong 3.4 \cdot HB$ for steels within a few percent. So, in most metals, high hardness and high S_u usually come together for

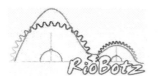

minimum volume design. This trend is also true for titanium alloys, but the correlations are not as good.

Note also that the S_u grades are very different than HB grades for minimum weight design, because the strength of the component depends not only on its material but also on its shape and functionality, while hardness is a local property that only depends on the material.

The main minimum volume design problems in combat robots are presented next.

3.10.1. Compact-Sized Internal Mounts

Very compact robots sometimes need to minimize the volume of internal mounts so they can fit inside. For instance, an internal mount attached to the face plate of a drive system motor may have thickness limitations, in special if the motor cannot be shifted too much inside the robot or if its output shaft is not too long, as pictured to the right. If the 6061-T6 aluminum motor mount in the picture didn't have enough stiffness or strength, you'd probably have to switch its material. Minimum weight design would tell you to use magnesium alloys, but their higher thickness (despite their lower weight) would make it hard to mount any wheel or pulley in such short output shaft. This is a problem of minimum volume design.

So, to improve the properties of the internal mount, we'd need to choose materials with better minimum volume grades. Since internal mounts should have high stiffness as well as high ultimate and yield strengths, we'll select materials with highest average grade $X' = (E' + S_u' + S_y') / 3$.

The best material to minimize volume would then be 18Ni(350) maraging steel, followed by AerMet 340, AerMet 310, 18Ni(300), AerMet 100 and S7 tempered at 54HRc, all of them with average grade $X' > 80$. Other options would be 18Ni(250), HP-9-4-30, and 4340 tempered to 43HRc, all with $X' > 70$. But note that all these steels will result in heavier mounts. This is the price you pay to minimize volume.

If you're also concerned with weight, then you'll have to compromise a little the minimum volume requirement. The idea is to find the lowest density material that will satisfy your X' requirement, without significantly increasing the volume of the component.

For instance, in the above example, the 6061-T6 aluminum has $X' = 19$ and density $\rho = 2.7$. We won't even bother looking for polymers or Mg alloys, because mounts with same volume would have much worse mechanical properties since their X' is always below 3 and 17, respectively. For polymer or Mg alloy mounts to have similar or better mechanical properties than our 6061-T6 mount, they would need to have a much higher thickness, due to their lower density.

We'll then start looking among all aluminum alloys, which have approximately the same density ρ, between 2.66 and 2.86. The highest X' among them is 28.5, for the 7055-T7751 alloy, which would result in a better mount with about the same weight and volume as the 6061-T6 version.

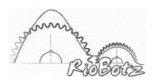

If this X' is still low for your application, or if you still need to decrease the thickness (and therefore the volume) of the mount, then look into Ti alloys. Their density ρ between 4.4 and 4.6 is not too much higher, and you'll be able to achieve X' = 43.7 for the alloy Ti-6Al-4V.

Finally, if you still need a higher X' or a lower thickness, look into steels. You'll end up with a heavier mount, due to their higher density ρ between 7.7 and 8.0, but you'll be able to achieve up to X' = 98.9 for high strength alloys such as 18Ni(350). And you'll be able to get a much thinner mount, such as the steel motor mount pictured to the right.

3.10.2. Compact-Sized Drums

Spinner bars and disks naturally have a high moment of inertia. Therefore, usually the inertia and strength requirements can be met just by switching the material and changing the bar or disk thickness, which was already studied in minimum weight design. On the other hand, designing a compact drum is a little trickier, as seen next.

Compact-sized drums with inserts

If we assume that the drum thickness can be changed (changing the drum internal diameter), then we're facing a minimum weight design problem. It was already seen that Mg and Al alloys with high strength and toughness are the best options for drums with inserts, optionally using Ti-6Al-4V for low maintenance versions.

This could be fine for very fast spinning drums, such as the aluminum drum of our featherweight Touro Feather, which compensates its low moment of inertia with a high spin. Or it could be fine for drums with large outer diameter, such as the aluminum drum from the middleweight Stewie, which can reach a large moment of inertia despite its low density material.

But our low profile drumbot Touro has limited values for the drum outer diameter D (about 5", without its S7 teeth inserts), as well as for its mass m (about 11.6kg or 25.6lb) and length L (180mm, a little over 7"). These values cannot be arbitrarily increased without changing the design of the rest of the robot. A minimum weight design would select the low density Mg or Al alloys, which would result in a drum with high thickness and therefore low internal diameter d. For a drum material with density ρ, it is easy to show that $d^2 = D^2 - 4m/\rho\pi L$, which lowers as ρ decreases.

But, for a given drum mass m and outer diameter D, a very low internal diameter d would lower the moment of inertia $I_{zz} \cong m\cdot(D^2 + d^2)/8$. For Touro, we decided that the resulting lower I_{zz}, combined with a moderate 6,000RPM drum speed, would make an Al or Mg drum have low energy. To maximize I_{zz}, we had to use a material with the highest possible density. A natural choice was steel, due to its ρ between 7.7 and 8.0. Denser metals would be either too expensive (such as tungsten, tantalum, silver, gold, platinum), too brittle if unalloyed (such as molybdenum, cobalt), or too soft (such as lead).

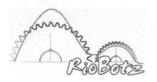

By selecting steels, we were able to get a 1" thick drum wall, which was thick enough to mill channels to hold the teeth without compromising its strength. If the resulting thickness was too thin, we would probably need to use the lower density Ti-6Al-4V for the drum, reducing somewhat the I_{zz} but adding enough thickness to be able to mill the channels without compromising strength.

But now, which steel should we use? We once tried a hardened 410 stainless steel for the drum body, trying to achieve a low-maintenance drum. This was not a good choice, because this high strength steel has only a moderate fracture toughness, while the drum body has several stress risers such as milled channels and threaded holes. Not surprisingly, the drum not only fractured along the threaded holes, but also sheared at the notch root of its tooth channel (as seen on the right), all at the same time during its first impact test against a dead weight.

We had to change the material to improve the impact toughness. Since it is a minimum volume problem, we had to look at the E', S_y', S_u' and K_{Ic}' grades of the material candidates. We know that the drum does not lose functionality if it yields locally, so S_y' is not that important for drums that have teeth inserts (unless there's some major yielding that might compromise the tooth support or unbalance the drum). The grade E' is almost the same for all steels, so it doesn't need to be considered.

Most of the loads on the drum are due to impacts, related to the grade K_{Ic}'. The static loads, related to the grade S_u', are relatively small, they're mostly due to the centrifugal forces of the drum teeth. For instance, each of the 0.63kg (1.39lb) teeth from the 2007 version of Touro's drum, which spin at 6,000RPM (628 rad/s) from a radius 0.065m (2.56"), generates a centrifugal force equal to $0.63 \cdot 628^2 \cdot 0.065 = 16{,}150$N, equivalent to 1,646 kgf or 3,629 lbf. This might seem a large static force, but it is small compared to the dynamic loads generated when hitting a stiff opponent. So, the grade S_u' is not as important as K_{Ic}'.

So, this is a volume optimization problem to maximize K_{Ic}'. The best option among the studied materials is the 304 stainless steel, with K_{Ic}' = 100. Other good options are, in that order, 4340 tempered at 34HRc, 18Ni(200), 1020 steel, HP-9-4-30, 18Ni(250), 4340 at 39HRc and AerMet 100, all of them have K_{Ic}' > 50.

Avoid using S7 steel, its K_{Ic}' at 54HRc is only 25, it could break in a similar way as shown above, near the notches. If the drum has inserts, there is no need to make the drum body out of a very hard material, lowering its impact toughness. Use hard materials only where they are needed, such as on the drum teeth, which must remain sharp. Avoid as well using other steels that might have K_{Ic}' < 50.

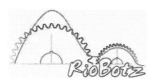

Avoid using as well polymers, magnesium, aluminum and titanium alloys, not only due to their low density, but also due to their $K_{Ic}' < 50$.

This is why Touro's drum body is made out of 304 stainless steel ($K_{Ic}' = 100$), to hold the sharp tempered S7 steel teeth. The only downside is the low yield strength of 304, but this is not much of a problem for the body of a drum. It easily yields, but it also withstands huge impacts, such as the one that broke the spinning bar of Terminal Velocity at Robogames 2007 (the resulting indentation is pictured to the right).

The drum body of our lightweight Touro Light had already been machined using 410 stainless steel, the same material from Touro's fractured drum, before that fracture happened. Instead of machining another drum out of 304 stainless steel, we've decided to save money by keeping the 410 version, but without hardening it through heat treatment. Without tempering the 410 steel, it ended up with a much higher impact toughness than the tempered version, and the much lower yield strength wouldn't be a problem for the drum body. The 304 steel would be a better choice, but the 410 steel without temper was almost as good, surviving Robogames 2007 while leading Touro Light to a gold medal.

Compact-sized sharp one-piece drums

Sharp one-piece drums have their body and teeth milled out of a single bar or block, as a single piece. They are not very popular with the heavier robots, because they're not easy to machine, and if a tooth breaks they need to be entirely replaced.

But they're a good option for lighter robots, such as insects. Teeth inserts for insect classes are very delicate to machine and temper, and they're not simple to attach to the tiny drum body. Teeth made out of hardened flat-headed allen bolts are a popular choice, but their hardness never exceeds 44 Rockwell C, even if using class 12.9 bolts. A one-piece drum can be hardened between 51 and 55 Rockwell C and still have a high toughness, if its material is wisely selected.

To maximize the one-piece drum moment of inertia, it is important to choose a dense material such as steel. So, if the drum outer diameter and width cannot be changed, then we end up facing a minimum volume problem.

We'll choose then steels that have minimum volume grades $HB' > 70$ to guarantee tooth sharpness, $K_{Ic}' > 20$ to avoid drum body or tooth fracture, and $S_y' > 60$ to avoid bent teeth. By choosing the average $X' = (K_{Ic}' + HB') / 2$ as a grading criterion, the best materials are AerMet 100, 18Ni(250), 18Ni(300), and AerMet 310, all of them with $X' > 55$. Another good choice is S7 steel tempered at 54HRc. Avoid using polymers, Mg, Al or Ti alloys, or steels with low hardness or with $X' < 45$.

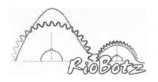

3.10.3. Compact-Sized Shafts, Gears and Weapon Parts

Shafts, gears and most weapon parts are subject to impacts. If their volume cannot be increased but their mechanical properties must be improved, then we face again a minimum volume problem.

For instance, the weapon system of our middleweight spinner Titan originally used a TWM 3R2 gearbox. This nicely crafted sturdy gearbox is made out of a solid aerospace aluminum block, with tempered steel gears and a special titanium shaft adapted for spinning weapons. After very severe

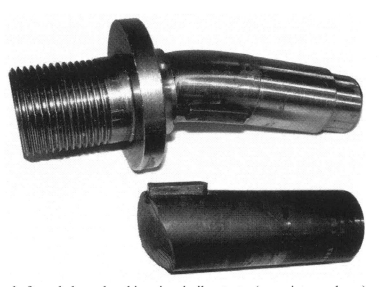

tests and a lot of abuse, we ended up bending the titanium shaft (pictured to the right). The shaft dimensions cannot be increased, otherwise it won't fit in the gearbox, so it's a minimum volume problem. We thought that an S7 steel shaft tempered to 54 Rockwell C (HRc) would be a good replacement, despite its higher weight. It has more than twice the ultimate and yield strengths of Ti-6Al-4V, which would certainly prevent it from getting bent. But our S7 steel shaft ended up breaking in similar tests (see picture above). Experiments don't lie. So why did it happen?

S7 steel already has lower toughness than Ti-6Al-4V in specimens without notches, as seen from their minimum volume grades K_{Ic}' equal to 25 and 33, respectively. It is not much of a difference, but this difference is exacerbated if sharp notches are present. By notch we mean any change in the geometry of the part, such as holes, grooves, fillets. Sharp notches should always be avoided because they are stress risers. But sometimes they are inevitable, such as in keyways.

Our titanium shaft was so ductile (with $\varepsilon_f = 45\%$) that it was able to blunt its sharp notches during the impact and avoid the effects of stress concentration. It's true that it got bent, but at least it withstood the impact without breaking (a broken shaft in your bot will award your opponent many more damage points from a judge than a bent one). But the lower ductility of the S7 steel wasn't able to blunt the sharp notches from its keyway, concentrating the impact energy on that point and making it break. In summary, S7 is notch-sensitive, it exhibits relatively high impact strength in the unnotched condition or if it has notches with generous radii, but it has a very severe degradation in its impact absorbing ability if it has sharp notches. So, S7 steel was a poor choice.

One alternative would be to change the keyway geometry to increase the notch radius, switching the square key to a round one. Another option would be to change the material to 4340 steel tempered to 43 Rockwell C, which would result in about the same ductility as Ti-6Al-4V ($\varepsilon_f = 45\%$) and a much higher $S_u = 1450$MPa. This combination would result in a much better impact strength

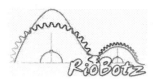

than both S7 and Ti-6Al-4V, even in the presence of notches, due to 4340's relatively high ductility, preventing it from breaking. And its higher ultimate and yield strengths would prevent it from getting permanently bent as it happened with the Ti-6Al-4V shaft.

But there are even better options than 4340 steel. Let's look for materials with minimum volume grades $K_{Ic}' > 30$ to avoid fracturing, and $S_y' > 50$ to avoid getting bent, ordering them by their average grade $X' = (K_{Ic}' + S_y') / 2$. Hardness and stiffness are also important in shafts, as discussed before, so let's look as well for $HB' > 45$ and $E' > 85$.

So, the best materials for compact-sized shafts are, in that order, 18Ni(250), AerMet 100, and 18Ni(200), all of them having $X' > 60$. Other good options are 18Ni(300), AerMet 310, HP-9-4-30, 4340 (tempered to 40-43HRc), and 5160 (tempered to 44-46HRc), all of them with $X' > 45$. Avoid using polymers, Mg, Al or Ti alloys, medium-low strength steels (such as 1020, 1045 and 304), and medium-low toughness steels (such as S7, 18Ni(350) and AerMet 340)

The above material choices are also applicable to compact-sized gears, which must also have high K_{Ic}' and S_y', together with high HB' to prevent tooth wear, and high E' to prevent excessive deflections.

Compact-sized weapon parts that do not get in touch with the opponent can also be included in this category.

But if the weapon part touches the opponent, then its material choice should be the same one used for compact-sized one-piece drums, to retain sharpness due to $HB' > 70$ instead of only requiring $HB' > 45$, resulting in AerMet 100, 18Ni(250), 18Ni(300), AerMet 310 and S7 steel tempered at 54HRc, and excluding lower hardness options such as 4340 and 5160.

3.11. Conclusions on Materials Selection

The main conclusions from the material optimization analyses presented in this chapter are:

- aluminum and magnesium alloys in general, especially the high strength ones, are a very good choice for protected structural walls, integrated structure-armor, structural top covers, bottom covers, ablative armor and internal mounts. They're also a good choice for the body of weapons that have inserts, such as spinning disks, shells, drums, bars and eggbeaters with inserts, hammer, thwack or overhead thwack handles, spears with inserts, lifter and launcher arms with inserts, and clamper and crusher claws with inserts. In all the above applications, avoid using steels, even high performance steel alloys.

- Ti-6Al-4V titanium, in special the ELI version, is the best material for wedges and for minimum weight shafts, gears and trusses. It is a very good option for the body of low maintenance weapons that have inserts, such as low maintenance spinning disks, shells, drums, bars and eggbeaters with inserts, and low maintenance hammer, thwack or overhead thwack handles. It is very good as well for blunt weapons such as blunt one-piece spinner disks, shells, bars and eggbeaters, blunt hammer heads, and blunt thwack or overhead thwack tips. It is also a good option for traditional armor, rammer shields and non-structural top covers.

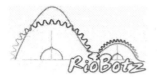 

- steels that combine high toughness, high strength and high hardness are the best materials for traditional armor, rammer shields, non-structural top covers, as well as compact-sized shafts, gears and weapon parts. They are also the best option for sharp weapon parts such as one-piece spinning disks, shells, drums, bars and eggbeaters, one-piece spears, and sharp weapon inserts (such as teeth, spear tips, clamper, lifter or launcher scoops). Together with Ti-6Al-4V, they're also a good option for blunt one-piece spinner disks, shells, bars and eggbeaters, blunt hammer heads, blunt thwack or overhead thwack tips, and minimum weight trusses.

- because shafts are basically beams under bending (and torsion), theoretically aluminum and magnesium alloys would be better candidates than steels for minimum weight shafts. However, an aluminum shaft would need to have a much larger diameter than an equivalent one made out of steel, which would imply in larger bearings, larger gears and pulleys, and larger gearboxes, increasing the robot weight. Therefore, hardened steel shafts are better than aerospace aluminum ones. Ti-6Al-4V titanium is the best option for minimum weight shafts, as long as the weight saved by the shaft is not gained by the slightly larger bearings, gearboxes, etc, when compared to the ones from a steel shaft. As a reference, steel shafts that drive the wheels of robust middleweights usually have diameters between 15 and 20mm (0.59" and 0.79", but it depends on the robot type and number of wheels it uses), and the main steel shafts for spinning weapons may vary from 25 to 40mm in diameter (0.98" to 1.57", but it depends, of course, on the weapon type).

- avoid using plastics such as Lexan in the robot structure, even as armor plate. Even relatively tough polymers such as Lexan and UHMW are not recommended to be used as structural parts. Structural plates made out of the best polymers can achieve higher stiffness-to-weight, strength-to-weight and toughness-to-weight ratios than most steels, however high strength aluminum and magnesium alloys are much better in all those cases. In addition, polymer plates need to be very thick to outperform steels, which will significantly increase the volume of the robot. Lexan used to be an attractive armor material due to its transparency to radio signals, because an all-metal robot used to suffer from the Faraday cage effect. However, the high frequency radios used nowadays, such as the 2.4GHz ones, do not have much problem with all-metal robots. So, plastics should be avoided in structural parts. Plastics are a good option, however, for internal mounts that do not have volume restrictions (such as UHMW motor mounts), or for other specific applications described in section 3.7.

The table in the next page summarizes all the weight and volume optimization analyses performed in this chapter.

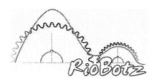 

	Applications	Grading criteria	Best material choices (in order of preference)	Good material choices (in order of preference)	Avoid using
minimum weight plates	Internal mounts; **or** Protected structural walls; **or** Bottom covers	$E^* > 70$ order by $X^* = (E^* + S_u^* + S_y^*)/3$	Mg alloys ($E^*>90$, $X^*>85$): Elektron 675-T5, ZK60A-T5, AZ31B-H24, Elektron WE43-T5	7000 series Al alloys ($E^*>70$, $X^*>77$): 7055-T7751, 7175-T736, 7075-T6, 7050-T7451, 7055-T74	Ti-6Al-4V ($E^*=54$, $X^*<67$), steels ($E^*<40$, $28<X^*<55$), polymers ($48<E^*<56$)
	Integrated structure-armor; **or** Structural top covers; **or** Spinning disks, shells or drums with inserts	$K_{Ic}^* > 50$, $E^* > 70$, $S_y^* > 70$, order by $X^* = (K_{Ic}^* + E^* + S_y^*)/3$	ZK60A-T5, AZ31B-H24, Elektron 675-T5, Elektron WE43-T5, 2324-T39 Type II, 7475-T7751, 7175-T736, 7055-T7751 or T74, 7050-T7451 (all of them with $X^*>73$)	7075-T73 or T6, 2024-T3, Mg and Al alloys with high strength and toughness ($X^*>63$)	Ti-6Al-4V ($E^*=54$, $S_y^*<70$), steels ($E^*<40$, $S_y^*<70$), polymers ($K_{Ic}^*<45$)
	Wedges; **or** Low maintenance spinning disks, shells or drums with inserts	$HB' > 45$, $K_{Ic}^* > 25$, $S_y^* > 40$, $E^* > 35$, order by $X^* = (K_{Ic}^* + E^* + S_y^*)/3$	Ti-6Al-4V ELI ($X^*=64$), Ti-6Al-4V ($X^*=63$)	AerMet 100, 18Ni(250), HP-9-4-30, 18Ni(200), 18Ni(300), tempered 4340, AR400, tempered 5160 (all with $X^*>42$)	Mg, Al, polymers, low strength steels, low strength Ti alloys
	Traditional armor; **or** Rammer shields; **or** Non-structural top covers; **or** Blunt one-piece spinner disks or shells	$HB' > 45$, $K_{Ic}^* > 30$, order by $X^* = (K_{Ic}^* + HB')/2$	K12 Dual Hardness (except for disks), AerMet 100, HP-9-4-30, AerMet 310, 18Ni(250), 18Ni(300), Ti-6Al-4V ELI, 18Ni(200), Ti-6Al-4V (all of them with $X^*>50$)	Ti and steels with both high toughness and high hardness (such as $X^*>37$); possibly use tempered 4340 or 5160 steel	Mg, Al, polymers, low hardness steel and Ti alloys (1020 steel, Ti grade 2), medium-low toughness steels (S7, 18Ni(350), AerMet 340)
	Ablative armor	$HB' < 30$, $K_{Ic}^* > 55$, order by K_{Ic}^*	ZK60A-T5, AZ31B-H24, 5086-H32, 2324-T39 Type II, 7475-T7351, 7055-T74 (all of them with $K_{Ic}^*>69$)	most Mg alloys, high toughness Al alloys ($K_{Ic}^*>55$)	polymers ($K_{Ic}^*<45$), Ti alloys and steels (not ablative because of high melting point and $HB' >> 30$)
	Sharp one-piece spinning disks, shells or drums; **or** Sharp plate-like weapon inserts (sharp clamper, lifter or launcher scoops)	$HB' > 75$, $K_{Ic}^* > 20$, $S_y^* > 50$, $E^* > 30$, order by K_{Ic}^*	AerMet 100, 18Ni(300), AerMet 310 (all of them with $K_{Ic}^*>33$)	S7 steel at 54HRc, 18Ni(350), AerMet 340; possibly use 5160 at 46HRc (but this leads to only $HB' = 65$)	Mg, Al, Ti, polymers, steels with hardness lower than 45HRc
minimum weight beams	Sharp one-piece spinning bars or eggbeaters; **or** Sharp one-piece spears; **or** Sharp beam-like weapon inserts (teeth; spear or other sharp tips)	$HB' > 75$, $K_{Ic}^{**} > 20$, $S_y^{**} > 50$, $E^{**} > 30$, order by K_{Ic}^{**}	AerMet 100, 18Ni(300), AerMet 310 (all of them with $K_{Ic}^{**}>38$)	S7 steel at 54HRc, 18Ni(350), AerMet 340; possibly use 5160 at 46HRc (but this leads to only $HB' = 65$)	Mg, Al, Ti, polymers, steels with hardness lower than 45HRc
	Blunt one-piece spinning bars and eggbeaters; **or** Blunt beam-like weapon inserts (hammer heads; blunt thwack or overhead thwack tips)	$HB' > 45$, $K_{Ic}^{**} > 30$, order by $X^{**} = (K_{Ic}^{**} + HB')/2$	AerMet 100, HP-9-4-30, 18Ni(250), Ti-6Al-4V ELI, 18Ni(200), 18Ni(300), AerMet 310 (all $X^{**}>60$)	Ti-6Al-4V, 4340 (at 34HRc), 4340 (at 39HRc), Ti and steels with both high toughness and high hardness ($X^{**}>37$)	Mg, Al, polymers, low hardness steel and Ti alloys (1020 steel, Ti grade 2), medium-low toughness steels (S7, 18Ni(350), AerMet 340)
	Spinning bars and eggbeaters with inserts; **or** Hammer, thwack or overhead thwack handles; **or** Spears with inserts; **or** Lifter and launcher arms with inserts	$K_{Ic}^{**} > 50$, $E^{**} > 60$, $S_y^{**} > 60$, order by $X^{**} = (K_{Ic}^{**} + E^{**} + S_y^{**})/3$	ZK60A-T5, AZ31B-H24, Elektron 675-T5, 2324-T39 Type II, 7475-T7351, 7055-T7751 or T74, 7175-T736, 7050-T7451 ($X^{**}>75$)	Elektron WE43-T5, 7075-T73 or T6, Ti-6Al-4V ELI, Ti-6Al-4V, 2024-T3 ($X^{**}>70$), high strength Mg, Al and Ti alloys with $X^{**}>60$	polymers ($X^{**}<40$), steels ($X^{**}<58$)
	Minimum weight shafts and gears; **or** Low maintenance spears, lifter and launcher arms, or spinning bars or eggbeaters with inserts; **or** Low maintenance hammer, thwack or overhead thwack handles	$HB' > 45$, $K_{Ic}^{**} > 25$, $S_y^{**} > 40$, $E^{**} > 40$, order by $X^{**} = (K_{Ic}^{**} + E^{**} + S_y^{**})/3$	Ti-6Al-4V ELI ($X^{**}=73$), Ti-6Al-4V ($X^{**}=72$)	AerMet 100, 18Ni(250), 18Ni(200), HP-9-4-30, 18Ni(300), tempered 4340, AR400, tempered 5160 (all of them with $X^{**}>50$)	Mg, Al, polymers, low strength steel and Ti alloys
	Clamper and crusher claws with inserts	high S_u^{**} and S_y^{**}, order by $X^{**} = (S_u^{**} + S_y^{**})/2$	Elektron 675-T5, 7055-T7751, 7175-T736, 7075-T6 (all of them with $X^{**}>85$)	18Ni(350), 7050-T7451, Ti-6Al-4V, 7055-T74, ZK60A-T5, AerMet 340, 7075-T73 (all of them with $X^{**}>80$)	Al and Mg alloys with low Sy, most steels, polymers
trusses	Minimum weight trusses	$K_{Ic}^{***}>40$, $S_y^{***}>50$, order by $X^{***} = (K_{Ic}^{***} + S_y^{***})/2$	Ti-6Al-4V ELI, Ti-6Al-4V, AerMet 100, 18Ni(250), 18Ni(200) ($X^{***}>60$)	HP-9-4-30, 4340 tempered at 39HRc (all with $X^{***}>50$); possibly use tempered 4130	Mg, Al, polymers, low strength steel and Ti alloys
minimum volume design	Compact-sized internal mounts	$E' > 85$ order by $X' = (E' + S_u' + S_y')/3$	18Ni(350), AerMet 340, AerMet 310, 18Ni(300), AerMet 100, S7 tempered at 54HRc (all of them with $X'>80$)	18Ni(250), HP-9-4-30, 4340 tempered at 43HRc (all of them with $X'>70$)	polymers, Mg, Al, Ti alloys, steels with $X' < 50$
	Compact-sized drums with inserts	$\rho > 7.5$, $K_{Ic}' > 50$, order by K_{Ic}'	304 stainless ($K_{Ic}' = 100$)	4340 tempered at 34HRc, 18Ni(200), 1020 steel, HP-9-4-30, 18Ni(250), 4340 at 39HRc, AerMet 100	polymers, Mg, Al, Ti alloys; steels with $K_{Ic}' < 50$
	Compact-sized sharp one-piece drums	$\rho > 7.5$, $HB' > 70$, $K_{Ic}' > 20$, $S_y' > 60$, order by $X' = (K_{Ic}' + HB')/2$	AerMet 100, 18Ni(250), 18Ni(300), AerMet 310 (all of them with $X' > 55$)	S7 tempered at 54HRc	polymers, Mg, Al, Ti alloys; steels with low hardness or with $X' < 45$
	Compact-sized shafts, gears and weapon parts	$HB' > 45$, $K_{Ic}' > 30$, $S_y' > 50$, $E' > 85$, order by $X' = (K_{Ic}' + S_y')/2$	18Ni(250), AerMet 100, 18Ni(200), all of them with $X' > 60$	18Ni(300), AerMet 310, HP-9-4-30, 4340 (tempered to 40-43HRc), 5160 (tempered to 44-46HRc), all of them with $X' > 45$	Mg, Al, Ti, polymers, steels with medium-low strength (1020, 1045, 304) or toughness (S7, 18Ni(350), AerMet 340)

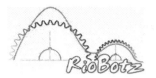 

It is interesting to note that the know-how of experienced builders, coupled with the "survival of the fittest" principle from the theory of evolution, has made several combots converge to very good if not the best material choices studied in this chapter, for instance:

- AR400 (or 4340 steel) for very hard wedges, or Ti-6Al-4V for not-so-hard wedges, both used by Devil's Plunger;

- spinner bars made out of aerospace aluminum and lightly notched S7 steel inserts, used by Last Rites and The Mortician;

- shock mounted Ti-6Al-4V top covers against vertical spinners, used by Pipe Wench;

- Ti-6Al-4V for very light shafts, such as the ones used in the TWM 3M gearboxes;

- aluminum alloys as integrated structure-armor elements, such as Team Plumb Crazy's 6061-T6 extrusions for the unprotected walls (which in theory are unprotected, as long as we do not define red wheels as armor elements!); and

- trussed robots made out of welded and tempered 4130 steel tubes, as in Last Rites and The Mortician (noting that 4130 steel is not the best truss option, but it is the cheapest and most easily weldable among the good ones).

On the other hand, this chapter has shown that there's still a lot to evolve, such as:

- making more use of high strength magnesium alloys for structural parts;

- using high toughness magnesium and aluminum alloys as ablative armor plates;

- using AerMet and maraging alloys in weapon inserts, replacing S7 steel, as well as in compact-sized shafts, replacing 4340 steel, and even in traditional armor plates, replacing Ti-6Al-4V; and

- replacing Ti-6Al-4V with Ti-6Al-4V ELI to improve impact toughness.

Finally, one might think that several high performance materials discussed in this chapter aren't used in combat because of high cost, but this is not entirely true. Most of them are not used because they're not very well known or understood. Magnesium alloys are the third most used structural metal in the world, it is not difficult to find high strength Mg alloys in surplus dealers at a low cost. The ELI version of Ti-6Al-4V, with improved impact properties, is not too difficult to find either. Maraging steels are expensive, but they can be bought in small quantities, they're worth using in critical compact shafts. AerMet alloys such as AerMet 100, on the other hand, are not only expensive, they're difficult to buy in small quantities, they're difficult to machine, and their heat treatment is very complicated, but it is worth the trouble if you want a steel as hard as S7 with 2.15 times higher impact toughness.

I hope that this chapter will, among other things, help making these high end materials more popular in combat robot design.

In the following chapter, the joining elements are studied, necessary to join the presented materials with enough stiffness and strength.

Chapter

4

Joining Elements

Joining elements are used to keep the robot parts held together in a rigid and strong bond. The main types are described below.

4.1. Screws

Screws are joining elements, almost always cylindrical, which have helical threads around their perimeter with one or more entries. Screws are used in countless applications to apply forces, to fasten joints, to transmit power (in worm gears) or to generate linear motion. The helical threads, in general wrapped around according to the right hand rule, are inclined planes that convert the applied torques in the screws into axial forces. The main types of screws are presented to the right.

The screws used in the robot structure should have hex (hexagonal) or Allen head, because they are the ones that allow the highest tightening torques. Screws used in the electronics can be of the flathead or Phillips types.

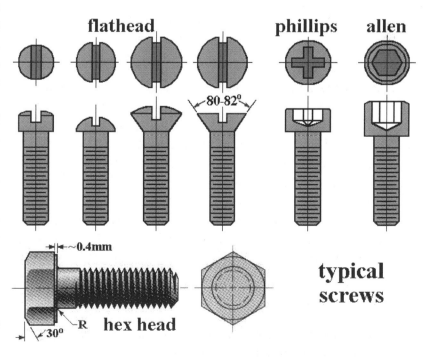

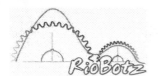

	Hex head – easily tightened with open-ended wrenches. Always use the 8.8 or 10.9 class types (made out of hardened steel), they have twice the strength of regular (mild steel) screws. Except for the few extra cents they'll cost, high class screws mean a free gain in strength, since their weight is the same as low class ones. Stainless steel screws have higher strength than mild steel ones, but much lower than hardened steel screws, so they should not be used in structural parts (besides, they are much more expensive).
 0.02in minimum	Allen – the highest strength screws, use the 12.9 or 10.9 class types (made out of hardened alloy steel), they have 3 times the strength of regular screws. Despite their higher impact toughness, don't use stainless steel screws: their low yield strength will let them easily bend during combat, making it difficult to disassemble the robot. Stainless steel Allen bolt heads are also easier to strip than hardened steel ones. The figure shows, from left to right, the button, standard and flat head (flush head) types. The flat head types are good for thick plates used in the robot's exterior, because they are embedded flush to the plate surface, with less chance of being knocked off by spinners. Avoid using flathead screws to fasten thin sheets, in this case the button head ones should be used, they also work well against spinners. Flat head screws require that the plates are countersunk, which reduces joint strength. To avoid this, do not countersink too deep to create a knife-edge condition in the countersunk member. A knife-edge creates a significant stress riser, as well as it allows the fastener to tilt and rise up on the countersunk surface. As a general rule, at least 0.5mm (0.02in) of the plate thickness should not be countersunk, as pictured to the left.
	Self-drilling – these screws don't require tapped holes since they cut their own thread as they're fastened, being very practical. They're good for wood and sheet metal, but they're a bad option to fasten thicker sheets and plates in the robot structure: they're made out of low strength steel, and they're easily knocked off due to the lack of nuts or properly threaded holes.
	Sandwich mounts – they are basically 2 screws held together by a piece of rubber or neoprene. Besides the male-male version from the figure, there are also threaded ones such as the female-female and male-female. They are excellent dampers to mount the electronics into the robot, leaving it mechanically and electrically isolated from the structure. Note that velcro is also a good choice for light parts, such as the receiver.

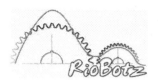

A few robots have the outer armor separated from the inner structure, held together using several sandwich mounts to absorb impacts (usually from spinners). The launcher Sub-Zero uses this damping technique: in the picture to the right, 4 of its sandwich mounts can be seen mounted to the robot structure. However, half of its armor was pulled out by the spinner The Mortician during Robogames 2006 – rubber and neoprene are not very resistant, in special to traction, so use several of these mounts.

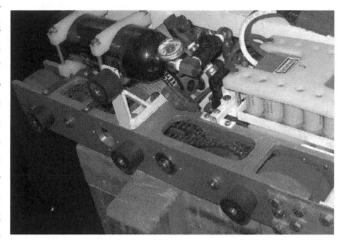

To hold the screws, nuts and washers are used in general. Washers are important to evenly distribute the force of the screws onto the part. Nuts have the inconvenience of needing 2 wrenches to be tightened, one open-ended to hold the nut, and another open-ended or Allen for the screw head. To avoid that, several robots make use of threaded holes. A hole is drilled in the piece with diameter a little smaller than the one of the screw (there are specific tables for that), and a tap (figure to the right) is used to generate threads, guided by a tap wrench  (figure below). Such threaded holes make the robot assembly much easier, without having to deal with nuts, which can be hard to reach and secure  during a quick pitstop, or might fall inside the robot. The mechanical structure of our middleweight Touro has more than 400 screws but no nuts.

The thickness to be tapped in the piece should be at least equal to the thickness of the nut that would be used with the screw, to avoid stripped threads. In addition, avoid tapping low strength aluminum (such as 6063-T5) and Lexan, their threads will have relatively low resistance. Also, avoid tapping deep holes in titanium by hand: besides being tough to tap, there's a good chance that the tap will break inside the piece.

A rule of thumb for a good screw diameter is to make it a little smaller than the sum of the thicknesses of the parts being joined. For instance, to fasten a 5mm thick plate to a 4mm one (totaling 5 + 4 = 9mm), an M8 screw (with 8mm diameter) is a reasonable choice.

And what about the number of screws? In robot combat, the word "overkill" doesn't exist, it is just a matter of your opponent super sizing his/her weapon for your armor to suddenly become undersized. Therefore, the most critical parts should have the largest possible number of screws, using common sense. If you drill too many holes to use more screws, your plates will look like Swiss cheese and they will be weakened. A rule of thumb is to leave the distance of at least one washer diameter between the washers of 2 consecutive screws, in other words, the distance between the centers of the holes should be at least twice the diameter of the washer.

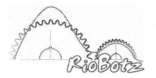

Screws shear much more easily than they break due to traction forces. Therefore, pay attention to the forces that would most likely act on each part of your robot. For instance, in the figure to the right, two parts are joined using a screw to transmit a vertical force. The configuration with the horizontally mounted

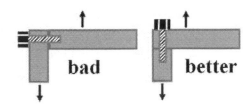

screw is a bad idea, since it will be loaded in shear. Change the design so that the screw will be under traction, as in the right figure. In this way, the screw will be able to take up to twice the load.

Another important thing is the tightening torque of the screw. Impact forces are transmitted entirely to a screw that is loosely tightened, which will end up breaking. A well tightened screw, on the other hand, distributes the received impact loads evenly through the surrounding material, receiving just a smaller portion of the impact force, resulting in a structure with greater stiffness and strength. Always check for loose screws during a pitstop. Usually, open-ended and Allen wrenches have an appropriate length (lever arm) for a single person to be able to manually generate appropriate tightening torques without leaving it loose or breaking the screw. A torquemeter can be used to deliver a higher precision when tightening bolts.

A great investment is to buy a power drill/screw driver. It makes all the difference during a pitstop, removing or tightening screws quickly. An 18V version is a good choice, avoid the cheaper versions with 12V or under (at least for use in a lightweight or heavier robot). We have been using an 18V DeWalt for 5 years, and it still works very well even after all the abuse. They are so reliable that we use 18V DeWalt motors to power the drum of our hobbyweight *Tourinho*, as well as to drive our retired middleweight *Ciclone*, with great results.

Now, how do we guarantee that a screw won't get loose during a match? The tightening torque by itself is not enough to hold the screws in robot combat. Vibration and impacts are very high and they end up making them become loose. A well tightened screw from the top cover of our spinner *Titan* ended up getting unscrewed after 4 full turns, until it was knocked off by our own weapon bar. To avoid this, there are 5 methods:

- spring lock or Belleville washers: they guarantee that the screw remains tightened, working as a spring to load them in the axial direction. Most of the times you can tighten the screw until these washers become flat.

- locknuts: they have a nylon insert that holds well onto the threads of the screws, resisting vibration and holding in place anywhere along the threads of the mating part. The locking element also limits fluid leakage and it won't damage or distort threads.

- counter nuts: if in the middle of a frantic pitstop you don't find any spring lock washers or locknuts, simply add a second nut to the screw (the counter nut), and tighten it well on top of the other one. The pressure between the 2 nuts will help preventing the screw from becoming unfastened.

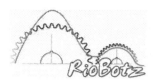

- threadlockers: they literally glue the screw onto the nuts or tapped holes. We use Loctite 242 (blue), it has medium strength and it holds very well. It is enough to use a single drop on the screw thread before tightening it. There are also the Loctite 222 (purple), which is relatively weak for combat, and Loctite 262 (red), with high strength for a permanent bond. But high strength threadlockers could be a problem if you need to disassemble the robot: you might need to heat up the part and deliver a great blow with a hammer to break the Loctite 262 bond. It is a good idea to clean up the screw and the nut or threaded hole with alcohol or acetone before using the threadlocker, to improve bonding. But in practice we always end up forgetting to do that, especially in the rush of a pitstop. Don't use threadlockers in Lexan, because they react with it and weaken the material.

- threaded shaft collars: they work as nuts with a small screw to lock them in place, as pictured to the right. They are the safest way to tighten a screw or threaded shaft. The spinning bars of our middleweights *Ciclone* and *Titan* are attached to their weapon shaft using threaded collars. Plain shaft collars, used in plain shafts, are also very useful, as discussed below.

4.2. Shaft Mounting

To attach pulleys, gears, sprockets and wheels to the robot shafts, keys and keyways are the best option. Keys are usually square steel bars that are inserted between the shaft and another component inside channels called keyways. They are an efficient way to deliver high torques. They also work

as a mechanical fuse, breaking as a result of overloads and saving the shaft and the other component. The keyway channels can be tricky to machine, especially the ones from the shaft (left figure). The internal keyways (right figure) are easier to make using a keyway broach and a collared bushing.

Avoid attaching components to shafts using pins or set screws. Set screws are tightened in the radial direction through the component (such as in the sprocket pictured to the right). Avoid pins and set screws, they are not a good option in the presence of impacts, they usually get loose or break. If a set screw must be used, then at least make sure you apply some threadlocker in it.

Another solution for shaft mounting is to use of a keyless bushing (pictured to the right), such as Trantorque or Shaftloc. You only need to tighten the collar nut to torque up these bushings in a few seconds, without keys or cap screws. As you tighten the collar nut, the inner sleeve contracts onto the shaft while the outer

sleeve expands to hold your component. Just match the bushing internal diameter to your shaft diameter, and the bushing outer diameter to the bore of your sprocket, pulley, or gear.

To guarantee that the attached components won't slide along the axial direction of the shaft, you can use retaining rings or plain shaft collars. Retaining rings (left figure below) are mounted in such a way to wrap a groove lathed in the shaft

with an external diameter A equal to their internal diameter. You must be careful with the shaft groove, it is a stress riser that could make the shaft break under severe bending stresses. Shaft collars are more resistant than rings, however they are much heavier. They are easy to install, it is enough to insert the shaft in the collar and tighten the locking screw(s), generating a holding force of the order of several tons in some cases. The two main types are the one-piece collars (middle figure above), more difficult to install, and the two-piece collar (right figure), which can be separated into two parts for easier installation.

As discussed before, threaded shaft collars are also a great option. The threads guarantee that the collar won't move axially even during huge impacts. This is important to maintain, for instance, a constant pressure against other components without letting them get loose. The spinner Hazard uses threaded collars to hold the weapon bar onto the shaft, as pictured to the right. The motor torque is transmitted to the bar through the friction forces caused by a large washer. This washer is pressed against the bar using a spring element (in green in the figure), held by the threaded collars. Note that two collars are used in this case, to improve strength and to act together as counter nuts. Note also that flat surfaces were machined on the collars to make it easier to get tightened, with an open-end wrench.

Another joining element is the worm-drive clamp (pictured to the right). It is practical, easy to assemble, and it works well to clamp cylindrical objects, such as motors and air tanks, or even components with different shapes such as batteries. In the case of batteries, it is advisable to wrap electrical tape all over the clamp to avoid shorts. Use several clamps to distribute well the load and to avoid breakage. Always perform several tests to guarantee that the clamps will resist to impacts during combats.

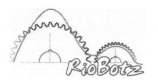

Notice however that collars are much more resistant and reliable than clamps to hold cylindrical objects, especially motors. The weapon motor of our drumbot *Touro*, for instance, is held by two aluminum collars, with two locking screws each (as pictured to the right). These specific motor mounts for use with the S28-series Magmotors

can be found at www.robotmarketplace.com. This method, besides much more resistant, is as practical as using clamps, because to switch the motor it is enough to deal with only 4 screws.

4.3. Rivets

Never use rivets! They are easy to install and they are used in aircrafts, but they are not suitable for the direct hits found in combat.

Spinners love riveted robots. Rivets are only applicable to join metal sheets, in special aluminum ones, not plates. They're a bad choice even for steel sheets, which would be better off if they were welded together.

Remember that the small flange that holds the rivets is made out of a material that you were able to deform by hand, using a manual tool, as pointed out in Grant Imahara's book [10]. Besides, to disassemble the robot, you would need to drill the rivet to remove it, which consumes as much or more time than unfastening a screw.

4.4. Hinges

Hinges are an excellent solution to attach wedges. Articulated wedges have the advantage of always being in contact with the arena floor, with zero clearance. This avoids other wedges from getting underneath them. They're also a great solution to make an invertible wedge: if the robot is flipped over, the wedge's own weight will make it rotate and keep touching the arena floor from its other surface.

Two important types are the hinges with lay flat leaves and the ones with tight-clearance leaves. The tight-clearance type can be seen in the figure to the right, in its open and closed configurations. Their disadvantage is that the hinged part cannot be laid flat on a surface, which might be a problem to mount them to the robot's walls.

The hinge with lay flat leaves, pictured to the right in its open and closed configurations, is usually a better choice.

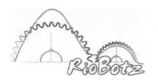

Piano hinges are a very popular and inexpensive choice to hold articulated wedges. Because they are continuous, they have the advantage of distributing the loads evenly along the entire wedge. Avoid using versions with brass leaves or pins, which have low strength. Steel and stainless steel versions are better, even though most of them still have low yield strength S_y. A yielded hinge will get stuck and it won't work efficiently. It is not easy to find a piano hinge that is both light weight and suitable for combat.

Another option is the use of door hinges (pictured to the right). In this case, it is a good idea to use only two of them for each wedge. Any misalignment when using 3 or more of those hinges might make the wedge articulation become stuck in certain positions. In this case, the wedge's own weight won't be enough to guarantee that it will touch the floor. The highest strength door hinges found in the market are made out of stainless steel (SS) type 304, which still has low S_y. Make sure that the pins are also made out of SS, not brass. Use oversized door hinges, remembering that the loads will be taken by only two of them.

Another option is to machine your own hinges out of titanium or hardened steel, integrated with the robot's structure and wedge plates. This is the solution adopted by the lifter Biohazard, pictured to the right. These integrated hinges are not easy to get knocked off in battle.

4.5. Welds

Many robots are made out of welded structures. Their main advantage is the short building time, without worrying about the high precision required to align holes in bolted components. The pitstop repairs are also faster, the weld filler works as a glue-all to hold parts together, even in the presence of misalignment, and to fill out holes and voids resulting from battle. Welds can be very resistant if well made, and they are a good option for mild and stainless steels. The figure to the right shows an oxy-acetylene system.

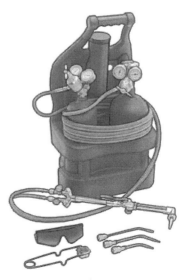

However, welded structures present a few problems. The welds or the surrounding heat affected zone tend to be the weakest point of the structure. To compensate for that, a lot of filler material is needed, increasing the robot weight. Note also

that in several competitions the access to welding equipment during the pitstops may be limited.

Also, the welds are deposited at a much higher temperature than the one of the base material. As they cool down, thermal effects make the welds contract and compress the base material. As the base material resists this compression, the weld ends up with residual stresses in tension. These tension stresses are so high that the welds usually end up yielding, beyond S_y. These residual tensile stresses decrease a lot the fatigue strength of the material. A few ways to reduce these stresses are to pre-heat the base material before depositing the filler weld material, so that the temperature difference between them is not too high, or to perform heat treatment after welding to decrease the residual stresses. Grinding and polishing the weld surface is a good idea, it generates a good surface finish that increases a lot the fatigue life of the component, because cracks usually initiate at the badly finished asperities of the welds, which locally concentrate stresses.

Another problem is that several high strength materials are either non-weldable, or they present problems if welded during a pitstop. For instance, most high strength aerospace aluminum alloys cannot be welded, and welded 4130 steel structures only acquire high strength after heat treatment (HT) – therefore, if some weld breaks and it needs to be repaired during a pitstop, the strength of the surrounding material will be compromised because there won't be time to perform another HT.

The welds in aluminum alloys need to be made using MIG equipment (Metal Inert Gas, seen in the left figure), and in titanium preferably using TIG (Tungsten Inert Gas, right figure). The use of such equipment requires some skill and experience in order not to 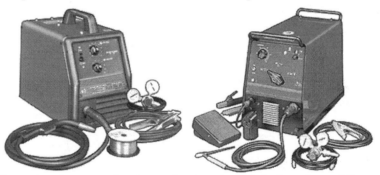 compromise strength. These equipments rely on an inert gas that is released during the welding process, shielding the heated part from the surrounding atmospheric gases, which would react to and weaken the weld.

Finally, it is important to clean the parts before welding, and to chamfer thicker plates to guarantee a through-the-thickness weld, increasing strength. The choice of filler material is also important. As mentioned before, grade 2 (commercially pure) titanium makes a great filler for grade 5 titanium (Ti-6Al-4V) – although grade 2 has lower strength, its higher ductility prevents cracking from the thermal effects during the welding process, resulting in better impact toughness during battle.

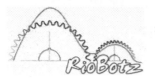 

Chapter

5

Motors and Transmissions

Motors are probably the combat robot's most important components. They can be powered either electrically, pneumatically, hydraulically, or using fuels such as gasoline. One of the most used types is the brushed direct current (DC) electrical motor, because it can reach high torques, it is easily powered by batteries, its speed control is relatively simple, and its spinning direction is easily reversed. Brushless DC motors are also a good choice, in special because they're not as expensive as they used to be. However, most brushless motor speed controls don't allow them to be reversed during combat, limiting their use to weapon systems, not as drivetrain motors. There are also other types of electrical motors, but not all of them are used in combat. For instance, step motors have in general a relatively low torque compared to their own weight. And the speed of AC motors is more difficult to control when powered by batteries, which can only provide direct current. In the next sections, we'll focus on both brushed and brushless speed motors, as well as on their transmission systems.

5.1. Brushed DC Motors

The three main types of brushed DC motors are the permanent magnet (PM), shunt (parallel), and series. The series type motors are the ones used as starter motors, they have high initial torque and high maximum speed. If there is no load on their shaft, starter motors would accelerate more and more until they would self-destruct, this is why they're dangerous. In a few competitions they can be forbidden for that reason. They are rarely used in the robot's drivetrain because it is not easy to reverse their movement, however they are a good choice for powerful weapons that spin in only one direction.

The PM DC motors and the shunt type have similar behavior, quite different from the starter motors. The PM ones are the most used, not only in the drive system but also to power weapons. They have fixed permanent magnets attached to their body (as pictured in the next page, to the left), which forms the stator (the part that does not rotate), and a rotor that has several windings (center figure in the next page). These windings generate a magnetic field that, together with the field of the PM, generates torque in the rotor. To obtain an approximately constant torque output, the winding contacts should be continually commutated, which is done through the commuter on the rotor and the stator brushes (pictured in the next page, to the right). Electrically, a DC motor can be modeled

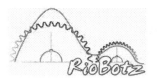

as a resistance, an inductance, and a power source, connected in series. The behavior as a power source is due to the counter electromotive force, which is directly proportional to the motor speed. The choice of the best brushed DC motor depends on several parameters, modeled next.

To discover the behavior of a brushed DC motor (permanent magnet or shunt types), it is necessary to know 4 parameters:

- V_{input} – the applied voltage to the terminals (measured in volts, V);

- K_t – the torque constant of the motor, which is the ratio between the torque generated by the motor and the applied electric current (usually measured in N·m/A, ozf·in/A or lbf·ft/A);

- R_{motor} – electric resistance between the motor contacts (measured in Ω); a low resistance allows the motor to draw a higher current, increasing their maximum torque;

- I_{no_load} – electric current (measured in ampères, A) drawn by the motor to spin without any load on its shaft; small values mean small losses due to bearing friction.

The equations for a brushed PM DC motor are:

$$\tau = K_t \times (I_{input} - I_{no\_load})$$
$$\omega = K_v \times (V_{input} - R_{motor} \times I_{input})$$

where:

- τ – applied torque at a given moment (typically in N·m, ozf·in or lbf·ft);

- ω – angular speed of the rotor (in rad/s, multiply by 9.55 to get it in RPM);

- I_{input} – electric current that the motor is drawing (in A);

- K_v – the speed constant of the motor, which is the ratio between the motor speed and the applied voltage, measured in (rad/s)/V; it can also be calculated by $K_v = 1 / K_t$;

Although neglecting the motor inductance, the above equations are good approximations if the current doesn't vary abruptly. The consumed electric power is $P_{input} = V_{input} \times I_{input}$, and the generated mechanical power is $P_{output} = \tau \times \omega$. We want the largest possible mechanical power output while spending the minimum amount of electrical power, which can be quantified by the efficiency $\eta = P_{output}/P_{input}$, which results in a number between 0 and 1. Since $K_t \times K_v = 1$, the previous equations result in:

$$\eta = \frac{P_{output}}{P_{input}} = \frac{(I_{input} - I_{no\_load}) \cdot (V_{input} - R_{motor} \cdot I_{input})}{V_{input} \cdot I_{input}}$$

In an ideal motor (which doesn't exist in practice), there would be no mechanical friction losses, resulting in $I_{no\_load} = 0$, and the electrical resistance would be zero, resulting in $R_{motor} = 0$, and in that case $\eta = 1$ (100% efficiency). Real motors have $0 \leq \eta < 1$ (efficiency between 0 and 100%).

The curves showing drawn current I (I_{input}), angular speed ω, output power P_{output}, and efficiency η as a function of the torque τ applied to the motor at a certain moment are illustrated below.

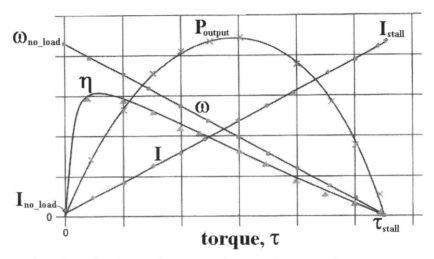

The above plots show that the maximum speed ω_{no_load} happens when the motor shaft is free of external loads, with $\tau = 0$, resulting in $I_{input} = I_{no\_load}$, and therefore

$$\omega_{no\_load} = K_v \times (V_{input} - R_{motor} \times I_{no\_load})$$

The maximum current I_{stall} happens when the motor is stalled, with speed $\omega = 0$, therefore $I_{stall} = V_{input} / R_{motor}$, generating the maximum possible torque for that motor $\tau_{stall} = K_t \times (I_{stall} - I_{no\_load})$. In practice, your motor won't see that much current, because in addition to the winding resistance from the motor, there will be the resistances from the battery and electronic system. The actual maximum current must be calculated from the system resistance R_{system}

$$I_{stall} = V_{input} / R_{system} = V_{input} / (R_{motor} + R_{battery} + R_{electronics})$$

The previous equations should also have their R_{motor} value switched to the actual R_{system}. Several manufacturers publish their motor datasheets based on values calculated using R_{motor}. This can be deceiving, because the actual (lower) performance the motor will have is obtained from R_{system}.

As seen in the plot, the maximum value of the mechanical power P_{output} happens when ω is approximately equal to half of ω_{no_load}. More precisely, differentiating the previous equations, it can be shown that the maximum P_{output} happens when

$$I_{input} = V_{input}/(2 \times R_{system}) + I_{no\_load}$$

On the other hand, the highest efficiency happens in general between 80% and 90% of ω_{no_load}, more precisely when $I_{input} = \sqrt{V_{input} \cdot I_{no\_load} / R_{system}}$.

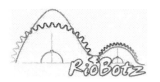

5.1.1. Example: Magmotor S28-150

We will now work out an example using the presented equations. Consider the motor Magmotor S28-150 (pictured to the right) connected to one NiCd 24V battery pack. Therefore $V_{input} = 24V$, while the motor has $K_t = 0.03757 N \cdot m/A$, $R_{motor} = 0.064\Omega$, and $I_{no\_load} = 3.4A$. Also, $K_v = 1/K_t = 26.62$ (rad/s)/V = 254 RPM/V. The motor resistance, in fact, needs to be added to the battery resistance (0.080Ω in this example) and the electronics resistance (about 0.004Ω, but it depends on the speed controller), resulting in $R_{system} = 0.064 + 0.080 + 0.004 = 0.148\Omega$.

The top speed of the motor (without loads on the shaft) is $\omega_{no\_load} = 254 \times (24 - 0.148 \times 3.4) = 5,968 RPM$. The maximum current (with the motor stalled) is $I_{stall} = 24 / 0.148 = 162A$, generating the maximum torque $\tau_{stall} = 0.03757 \times (162 - 3.4) \cong 6.0 N \cdot m$. In this case, with the motor stalled, the mechanical power is zero and therefore the efficiency is zero, however the electric power is maximum, $P_{input\_max} = V_{input} \times I_{stall} = 24 \times 162 = 3,888W = 5.2HP$, remembering that 1HP (horsepower) is equal to 745.7 W (Watts). Note that this does not mean that you have a 5.2HP motor. All this power is wasted when the motor is stalled, converted into heat by the system resistance. Therefore, avoid leaving the motor stalled for a long time during a match, it can end up overheating.

The maximum mechanical power happens when $I_{input} = 24 / (2 \times 0.148) + 3.4 = 84.5A$, and it is worth $P_{output\_max} = (84.5 - 3.4) \times (24 - 0.148 \times 84.5) = 932W = 1.25HP$. Notice that the manufacturer says that the maximum power is 3HP for that motor, you would only get that if the battery and electronic system resistances were zero, leaving only the motor resistance 0.064Ω. Recalculating using only the motor resistance 0.064Ω instead of 0.148Ω, P_{output_max} would result in 3HP, but this value is just theoretical.

The maximum mechanical power of 1.25HP happens when $\omega = 254 \times (24 - 0.148 \times 84.5) = 2,919 RPM$, very close to half the ω_{no_load} of 5,968RPM, as expected. Note however that this P_{output_max} happens for $P_{input} = 24 \times 84.5 = 2,028W = 2.72HP$, with an efficiency of only $\eta = 1.25HP/2.72HP = 0.46 = 46\%$. As it can be seen in the graph for this motor, pictured to the right, the maximum mechanical power happens at speeds that are not necessarily efficient.

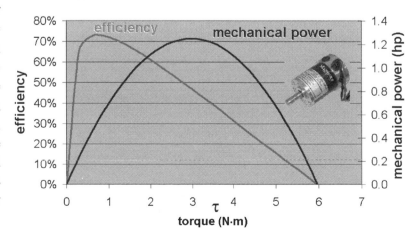

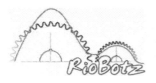

The maximum efficiency happens if $I_{input} = \sqrt{24 \cdot 3.4 / 0.148} = 23.5A$, associated with the speed $\omega = 254 \times (24 - 0.148 \times 23.5) = 5,213RPM$ (about 87% of ω_{no_load}). From the previous equations, we get a maximum efficiency of 73%. If theoretically the battery and electronics didn't have electrical resistance, the maximum efficiency would go up to 82%, the value that the manufacturer displays, which is just an upper limit of what you'd be able to get in practice.

5.1.2. Typical Brushed DC Motors

The above example can be repeated for several other motors. The next page shows a table with a few of the most used motors in combat robots, and their main parameters. Several parameters are based only on R_{motor}, their actual values would not be as good after recalculating them using R_{system}.

The Bosch GPA and GPB, shown in the table, have been extensively used in Brazil to drive middleweights, however they have a low ratio between maximum power and their own weight. In addition, the GPA generates a lot of noise, which can reduce the range of 75MHz radio control systems. This problem can be minimized using capacitors between the motor brushes, or switching to, for instance, 2.4GHz radio systems.

The DeWalt 18V motor with gearbox is a good choice for the drive system, we've used it in our middleweight *Ciclone*. It has an excellent power-to-weight ratio. Its main disadvantages are that it is not easy to mount to the robot structure, the gearbox casing is made out of plastic, and its resulting length including gearbox ends up very high to fit inside compact robots. Note that some older discontinued DeWalt cordless drills had other disadvantages, using Mabuchi motors instead of the higher quality DeWalt ones, and using a few plastic gears among the metal ones in their gearbox.

The NPC T64 already includes a gearbox with typically a 20:1 reduction, which is already embedded in the values of K_t (already multiplied by 20 with respect to the motor values without gearbox), in K_v (already divided by 20), and in its weight. The data in the table already include the power loss and weight increase due to the gearbox, which explains the relatively low power-to-weight ratio. But, even disregarding that, the performance of this motor is still not too high. The reason many builders use it is due to its convenience, it is easily mounted to the robot and it is one of the few high power DC motors that come with a built-in gearbox. There is also a version of that motor with almost the same weight but twice the power, the NPC T74, however this version is not so easy to find. Care should be taken with the NPC T64 and NPC T74 gears (pictured to the right, with red grease), they might break under severe impacts if used to power weapons. As recommended by the manufacturer, use them only as drive motors. Our middleweight overhead thwackbot *Anubis* is driven by two NPC T74, but their gears

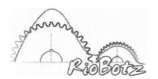 

ended up breaking after extreme impacts. This was no surprise, since we were indirectly using these drive motors to power the overhead thwackbot weapon. After replacing the gears, we've shock-mounted the motors to the robot structure, which solved the problem. Therefore, in overhead thwackbots, which have weapons that are powered by the drive system, it is a good idea to shock-mount the gearmotors.

Name	Bosch GPA	Bosch GPB	D-Pack	DeWalt 18V
Voltage (V)	24	12	12 (nominal)	24
P_{output_max} (W)	1,175	282	3,561	946
Weight (kg)	3.8	1.5	3.5	0.5
Power/Weight	309	188	1,017	1,892
I_{stall} / I_{no_load}	23	25	63	128
K_t (N·m/A)	0.061	0.042	0.020	0.0085
K_v (RPM/V)	167	229	485	1,100
R_{motor} (Ω)	0.13	0.121	0.00969	0.072
I_{no_load} (A)	8.0	3.9	19.6	2.6

Name	Etek	Magmotor S28-150	Magmotor S28-400	NPC T64 (w/gearbox)
Voltage (V)	48	24	24	24
P_{output_max} (W)	11,185	2,183	3,367	834
Weight (kg)	9.4	1.7	3.1	5.9
Power/Weight	1,190	1,284	1,086	141
I_{stall} / I_{no_load}	526	110	127	27
K_t (N·m/A)	0.13	0.03757	0.0464	0.86
K_v (RPM/V)	72	254	206	10
R_{motor} (Ω)	0.016	0.064	0.042	0.16
I_{no_load} (A)	5.7	3.4	4.5	5.5

An excellent motor for driving middleweights is the Magmotor S28-150, it is used in our robots Titan and Touro. A good weapon motor for a middleweight would be the Magmotor S28-400, with

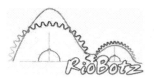 

higher torque and power, which we use to power Touro's drum. Using a single S28-150 to power the weapon of a middleweight is not a good idea, there's a good chance that it will overheat.

Because of that, to spin the bar of our middleweight Titan, we use 2 Magmotors S28-150 mechanically connected in parallel by acting on the same gear of the weapon shaft. Note that the two S28-150 motors result together in a higher top speed (6,096RPM instead of 4,944RPM at 24V), stall torque (about 28 instead of 26.5Nm, in theory) and power (6HP instead of 4.5HP) than a single S28-400, weighing only a little more (7.6lb instead of 6.9lb). But we're considering switching to a single S28-400 for three reasons: two S28-150 motors electrically connected in parallel will draw much more current than a single S28-400 if the weapon stalls (which might damage the batteries), the S28-400 can be overvolted more than the S28-100 because it better dissipates heat (compensating for the lower resulting speed, torque and power), and the 6.7" length of the S28-400 will save space inside the robot if compared to the 8" combined length of both S28-100.

The D-Pack motor is a good candidate to replace the Magmotors, besides being much cheaper. However, its electrical resistance is so low that it almost shorts the batteries and electronics. Because of that, its current must be limited if used with speed controllers, otherwise there's a good chance of damaging the electronics, in special since this motor is usually overvolted to 24V instead of powered by its nominal 12V. If used with solenoids to power weapons, make sure that they can take the high currents involved. This motor is difficult to find even in the US.

The Etek motor is really impressive. It may deliver up to 15HP (1HP = 745.7W), and it can deliver high torque and high speed at the same time. It is a little too heavy for a middleweight: we ended up using it in our spinner *Ciclone* but we had to power it at only 24V, because the additional battery packs that would be needed to get to 48V would make the robot go over its 120lb weight limit. The super-heavyweight shell spinner Super Megabyte only needs one of these motors, powered at 48V, to spin up its heavy shell. A few daring builders have overvolted it to 96V, but current limiting is highly recommended.

Besides the maximum power-to-weight ratio, a parameter that indicates the quality of a motor is the ratio I_{stall} / I_{no_load} between the maximum and no-load currents. The higher the ratio, the higher the current and therefore the torque the motor can deliver, with lower friction losses associated with I_{no_load}. Excellent motors have a ratio above 50. The NPC T64 only has 27, but you must take into account that its I_{no_load} was measured including the gearbox, which contributes with significant friction losses. Without the gearbox, this I_{stall} / I_{no_load} ratio for the NPC T64 would probably reach 50. The Bosch GPA and GPB are not very efficient, their ratio is around 24. The best motors are the D-Pack (with ratio 63), Magmotors and DeWalt (around 110 to 130), and Etek, with the astonishing ratio of 526 (which is just a theoretical value, since it assumes that the batteries have zero resistance and that they can dish out an I_{stall} of 3,000A at 48V).

A few DC motors allow the permanent magnets fixed in their body to be mounted with an angular offset with respect to their brush housings (typically about 10 to 20 degrees, it depends on the motor), which allows you to adjust their phase timing. If the motor is used in the robot drive system, it should have neutral timing, in other words, it should spin with the same speed in both directions, helping a tank steering robot to move straight. But if it is used to power a weapon that only spins in one direction, you can advance the timing to typically get a few hundred extra RPM

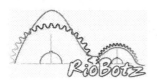

(on the other hand, in the other direction the motor speed would decrease). To advance the timing, loosen the motor screws that hold its body, power it without loading its shaft, and slightly rotate its body (where the permanent magnets are attached to) until the measured I_{no_load} current is maximum, and then fasten the body back in place. For neutral timing, rotate the body until I_{no_load} is identical when spinning in both directions.

Regarding hobbyweights (12lbs, about 5.4kg), a few inexpensive gearmotor options for the drive system are the ones from the manufacturers Pittman and Buehler, which can be found in several junk yards. Our hobbyweight drumbot *Tourinho* originally used, in 2006, 2 Buehler gear motors (with 300 grams each, about 0.66lb), and our hobbyweight wedge Puminha used 4 Pittmans (with 500 grams each, about 1.10lb). We've bought used ones in Brazil for about US$10 to US$15 each (after bargaining). Most of them have nominal voltage 12V, however we've used them at 24V for 3 minute matches without overheating problems. Remember that by doubling the voltage the power is multiplied by four. The only problem is that the small gears can break due to the higher torques at 24V – we've broken quite a few 12V Pittmans after abusing them in battle at 24V. The only way to know whether they'll take the overvolting is by testing them. It's also a good idea to always have spare motors.

There are much better gearmotor options for hobbyweights, and even heavier robots, than the ones from Pittman and Buehler, however they usually need some modifications to get combat-ready. We've been using, for the drive system in our hobbyweights, Integy Matrix Pro Lathe motors, as pictured to the right, adapted to Banebots gearboxes that were  modified following Nick Martin's recommendations, described in the March 2008 edition of Servo Magazine.

A good combination for the drive system of a featherweight is a larger Banebots gearbox connected to Mabuchi's RS-775 motor. For a lightweight, it might be a good idea to go for 18V DeWalt motors, either connected to DeWalt gearboxes or to custom-made ones.

For middleweights, S28-150 Magmotors are usually a good choice for the drive system, connected for instance to Team Whyachi's famous TWM 3M gearbox (pictured below to the left), or to the newer TWM 3M12 version (pictured below in the middle). The S28-400 Magmotors are more appropriate for the drive system of heavyweights and super heavyweights, connected for instance to the TWM3 gearbox (pictured below to the right).

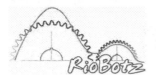 

A good option for the drive system of beetleweights is the Beetle B16 gearmotor (shown in the left picture), sold at The Robot Marketplace (www.robotmarketplace.com). For antweights and fairyweights, the Sanyo 50 micro geared motor (shown in the right picture) is a very popular choice.

There are several other good brushed DC motors besides the ones presented above, not only for the drive system, but also to power the weapon. Brushless motors, studied in section 5.2, have been successfully used as weapon motors in several weight classes. It is useful to do a research on which motors have been successfully used in combat. Several motors can be found at The Robot Marketplace (www.robotmarketplace.com), and much more information can be obtained, for instance, in the RFL Forum (http://forums.delphiforums.com/therfl).

5.1.3. Identifying Unknown Brushed DC Motors

If you bought your motor from a junkyard, or if you found it forgotten somewhere in your laboratory, and you don't have any clue about its characteristics, you can follow the steps below:

- Seek any identification on the motor, and look for its datasheet over the internet.
- Make sure it is a DC motor. If there are only 2 wires connecting it, there is a good chance it is DC, otherwise it could be an AC, brushless or step motor.
- Measure the electrical resistance between the terminals, obtaining R_{motor}.
- Apply increasingly higher voltages, such as 6V, 9V, 12V, 18V, 24V, waiting for a few minutes at each level, while checking if the motor warms up significantly. If it gets very hot even without loads, you're probably over the nominal voltage, so reduce its value.
- Most *high quality* motors can work without problems during a 3 minute match with twice their nominal voltage, this is a technique used in combat (such as the 48V Etek powered at 96V). The 24V Magmotors are exceptions, they are already optimized for this voltage, tolerating at most 36V, and even so the current should be limited in this case.
- Once you've chosen the working voltage V_{input}, connect the motor (without loads on its shaft) to the appropriate battery, the same that will be used in combat, and measure I_{no_load}. Note that the value of I_{no_load} does not depend much on V_{input}, however it is always a good idea to measure it at the working voltage. If you have an optical tachometer (which uses strobe lights, such as the one to the right), you can also measure the maximum no-load motor speed ω_{no_load}. A cheaper option is to attach a small spool to the motor shaft, and to count how long it takes for it to roll up, for instance, 10 meters or 30 feet of nylon thread – the angular speed in rad/s will be the length of the thread divided by the radius of the spool, all this divided by the measured time (the

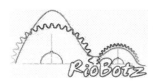 

thread needs to be thin, so that when it's rolled up around the spool the effective radius doesn't vary significantly).

- Attach the motor shaft to a vise grip, holding well both the motor and the vise grip, and connect the battery. Be careful, because the torques can be large. The measured current will be I_{stall}, associated with the circuit resistance $R_{system} = R_{battery} + R_{motor}$, so $I_{stall} = V_{input} / R_{system}$ and then calculate $R_{battery} = (V_{input} / I_{stall}) - R_{motor}$. Do not leave the motor stalled for a long time, it will overheat and possibly get damaged. Also, take care not to dent the motor body while holding it, for instance, with a C-clamp, as pictured below.

- Repeat the procedure above, but supporting one end of the vise grip by a scale or spring dynamometer (with the vise grip in the horizontal position, see the picture to the right). Then, measure the difference between the weights with the motor stalled and with it turned off, and multiply this value by the lever arm of the vise grip to obtain the maximum torque of the motor, τ_{stall}. For instance, if the scale

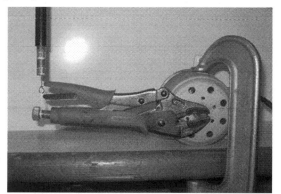

reads 0.1kg with the motor turned off (because of the vise grip weight) and 0.8kg when it is stalled, and the lever arm (distance between the axis of the motor shaft and the point in the vise grip attached to the scale) is 150mm, then $\tau_{stall} = (0.8\text{kg} - 0.1\text{kg}) \cdot 9.81\text{m/s}^2 \cdot 0.150\text{m} = 1.03\text{N·m}$.

- Because $\tau_{stall} = K_t \times (I_{stall} - I_{no\_load})$, you can obtain the motor torque constant by calculating $K_t = \tau_{stall} / (I_{stall} - I_{no\_load})$.

- Alternatively, if you were able to measure ω_{no_load} with a tachometer or spool, then you can calculate the motor speed constant using $K_v = \omega_{no\_load} / (V_{input} - R_{system} \times I_{no\_load})$. Check if the product $K_t \times K_v$ is indeed equal to 1, representing K_t in N·m/A and K_v in (rad/s)/V. This is a redundancy check that reduces the measurement errors. If you weren't able to measure ω_{no_load}, there is no problem, simply calculate $K_v = 1 / K_t$, taking care with the physical units.

- Finally, once you have the values of V_{input}, K_t (and/or K_v), R_{system} and I_{no_load}, you can obtain all other parameters associated with your motor + battery system using the previously presented equations (don't forget to later add the resistance of the electronics as well).

5.2. Brushless DC Motors

A brushless DC motor is a synchronous electric motor powered by DC current, with an electronically controlled commutation system instead of a mechanical one based on brushes. Similarly to brushed DC motors, current and torque are linearly related, as well as voltage and speed.

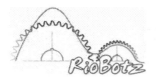 

In a brushless DC motor, the permanent magnets rotate, while the armature windings remain static. With a static armature, there is no need for brushes. The commutation is similar to the one in brushed DC motors, but it is performed by an electronic controller using a solid-state circuit rather than a commutator/brush system.

Compared with brushed DC motors, brushless motors have higher efficiency and reliability, reduced noise, longer lifetime due to the absence of brushes, elimination of ionizing sparks from the commutator, and reduction of electromagnetic interference. The stationary windings do not suffer with centrifugal forces. The maximum power that can be applied to a brushless DC motor is very high, limited almost exclusively by heat, which can damage the permanent magnets. Their main disadvantage is higher cost, which has been decreasing due to their mass production, as the number of applications involving them increases.

The better efficiency of brushless motors over brushed ones is mainly due to the absence of electrical and friction losses due to brushes. This enhanced efficiency of brushless motors is greatest under low mechanical loads and high speeds. But high-quality brushed motors are comparable in efficiency with brushless motors under high mechanical loads, where such losses are relatively small compared to the output torques.

Their kV rating is the constant relating the motor RPM at no-load to the supply voltage. For example, a 1,000 kV brushless motor, supplied with 11.1 volts, will run at a nominal 11,100 RPM.

Most brushless motors are of the inrunner or outrunner types. In the inrunner configuration, the permanent magnets are mounted on the spinning rotor, in the motor core. Three stator windings are attached to the motor casing, surrounding the rotor and its permanent magnets. The picture to the right shows a brushless inrunner of the KB45 series, used to power the spinning drum of our featherweight Touro Feather.

In the outrunner configuration, the windings are also stationary, but they form the core of the motor (as it can be seen in the Turnigy motor in the left picture), while the permanent magnets spin on an overhanging rotor (the "spinning can") which surrounds the core. Outrunners typically have more poles, set up in triplets to maintain the three groups of windings, resulting in a higher torque and lower kV than inrunners. Outrunners usually allow direct drive without a gearbox, because of their lower speed and higher torque. Due to their relatively large diameter, they're not a good option to be horizontally mounted inside very low profile robots. Remember to leave a generous clearance all around an outrunner, to prevent its outer spinning can from touching any structural part of the robot that could be bent during combat. Popular brushless outrunners are the ones from Turnigy and the more expensive ones from the famous Czech Republic company AXi, pictured above. We've also

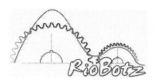

tested very good outrunners from E-Flite (such as E-Flite's Park 250) and Little Screamers (such as the "De Novo" model).

One important thing about outrunners is that they should be mounted "behind the firewall" for combat applications. Firewall is the flat panel, cross-shaped mount or standoff at the front of a model airplane where the motor is attached to. Supporting the motor in front of the firewall, as

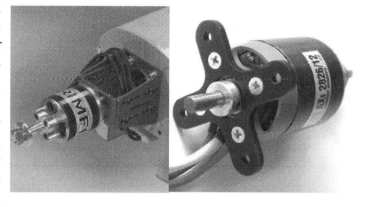

shown in the left picture, is a good idea in model airplanes to help the motor cool down with the aid of the propeller air flow. The motor shaft mostly sees axial loads in this case.

But pulleys used to power robot weapons put large bending forces on the motor shaft. So, for combat applications it is important to support the motor by mounting it as close to the output shaft as possible, behind the firewall, as shown in the right picture. To mount outrunner motors behind the firewall, you might need to reverse the position of the output shaft, for it to stick out from the face where the firewall is attached to, which can be done through the repositioning of the shaft retaining clips or screws.

Since most brushless speed controllers do not allow the motor to reverse its spin direction during combat, the use of brushless motors in combots is usually restricted to weapons that only spin in one direction. But reversible brushless speed controllers will soon become cheap and small enough to allow their widespread use in the robot drive system as well.

More information on brushless motors can be found, for instance, in the wikipedia link http://en.wikipedia.org/wiki/Brushless_DC_motor.

5.3. Power Transmission

To transmit power from the motors to the wheels or weapon, it is necessary to use gears, belts or chains. Each one of those elements is described next.

5.3.1. Gears

There are 3 main types of gears, as pictured in the next page:

(a) cylindrical gears, with straight or helical teeth;

(b) conical gears, which have perpendicular and convergent axes; and

(c) worm gears, consisting of a worm (which is a gear in the form of a screw) that meshes with a worm gear (which is similar in appearance to a spur gear, and is also called a worm wheel), where the worm and worm gear have perpendicular axes that do not converge.

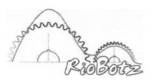

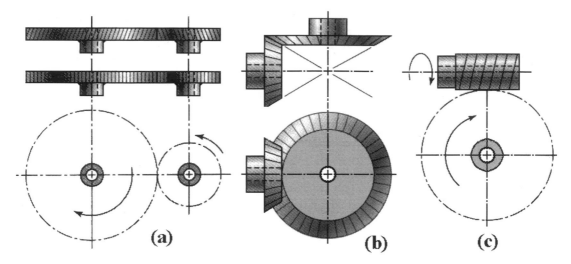

(a) (b) (c)

Among the cylindrical gears, the straight-toothed ones don't generate axial forces, but they are noisier than the helical ones. The helical-toothed gears are more resistant, however they generate axial forces, except for the double helical ones, which cancel these loads. Grease them well before use, to increase their service life. The TWM 3M gearboxes that drive our middleweights Titan and Touro only use straight-toothed cylindrical gears, in two stages. The gears are made out of hardened steel to resist impacts. Avoid using cast iron or mild steel gears, they might not resist the rigors of combat, as seen in the figure to the right.

Conical gears are an efficient option to transmit power at 90 degrees. The gearbox of the weapon system of our spinner Titan uses a large conical gear attached to the weapon shaft, powered in parallel by two S28-150 Magmotors, each one with a smaller conical gear. In the same way as with cylindrical gears, the reduction ratio between two conical gears only depends on the ratio between the number of teeth of each of them. For instance, if the motor gear has 20 teeth and the weapon gear 30 teeth, then the reduction ratio is $30/20 = 1.5$, meaning that the torque of each motor will be multiplied by 1.5, and the weapon speed will be 1.5 times slower than the motor speed.

Worm gears are used in several gearmotors, because they can have a large reduction ratio with a single stage. This ratio is equal to the number of teeth of the driven worm gear, which can be a large number. Most of them are self-locking, meaning that the driven worm gear can be designed so that it can't turn the worm. This can be dangerous in combat, because a large impact can cause the worm gear to break its teeth due to self-locking. Another disadvantage is due to the low efficiency (high power loss) caused by the functional sliding between the worm and worm gear. Because of that, avoid using electric windshield wiper motors, they have low power-to-weight ratios, and the power losses due to the worm gears are high.

Our first combat robots, the middleweight overhead thwackbots *Lacrainha* and *Lacraia*, use worm gearboxes driven by Bosch GPB motors. Besides the low power of the GPB, their cast iron

gearboxes are very heavy. A good option for the drive system is to mill a gearbox out of a solid block of aerospace aluminum, and to use straight-toothed cylindrical gears made out of hardened steel. Milling such solid block is not easy without a CNC system, because any small error (of the order of 0.1mm) may cause misalignment between the shafts and then compromise the service life of the gears, not to mention the reduction in efficiency due to the added friction losses. Besides, any error during the milling process may mean the waste of an expensive block of aerospace aluminum.

After a few tries with our manual mill, we've realized that the TWM 3M gearbox (pictured to the right), sold by Team Whyachi (www.teamwhyachi.com), is worth every penny. It is milled out of a solid block of aerospace aluminum, with hardened steel gears, and the output wheel shaft is made out of grade 5 titanium (Ti-6Al-4V). We've used the TWM 3M gearboxes together with the Magmotor S28-150 to drive the wheels of our middleweights Titan and Touro.

5.3.2. Belts

Belts are flexible machine elements used to transmit force and power to relatively long distances, driven by pulleys. These elements can replace gears in many cases, with several advantages: besides being relatively quiet, belts help to absorb impacts and vibrations through their flexibility, and they tolerate some misalignment between the pulleys.

The main types of belts are the timing belts (a.k.a. synchronous or toothed belts, see the left figure) and the V-belts (right figure, showing quadruple-sheave pulleys), manufactured in standard sizes in rubber or polymeric base, in general reinforced with high resistance fibers.

Timing belts (pictured to the right) keep the relative position between the pulleys, synchronizing the movements and preventing sliding. They can be used to transmit power to the drive system. They can also be used in the robot weapon system, but in this case it is recommended to use some type of torque limiter (discussed ahead) to bear impact loads.

V-belts (pictured to the right), on the other hand, allow the pulleys to have some relative sliding, working as a clutch. This is very useful in combat robot weapons, allowing some sliding at the moment of impact against the opponent, which is good not to stress too much the motor or to rupture the belt. Touro uses a pair V-belts to power its drum.

For small diameter pulleys, use cogged V-belts (pictured to the right), they are more flexible and dissipate heat better because of the cogged design. Note that they're not timing belts, the cogs are not used as teeth.

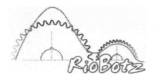

If your V-belt broke off in combat and you don't have time during a pitstop to open up your robot to install a new one, then a good alternative is to use an adjustable-length V-belt (pictured to the right). Sold by the foot, it is perfect for making replacement V-belts, easily installed by simply twisting its sections for coupling or uncoupling. Its only problem is that it tends to stretch with use, so standard or cogged V-belts are better if you have time to install them.

There are still round belts (with circular cross section), but in general they are only used in low power applications, such as in sewing machines, or in lighter combat robots such as insects.

The calculation of the nominal length L of the circumference of the belt is made starting from the distance C between the centers of the pulleys and from the primitive diameters of the smaller pulley (d) and of the larger one (D), using the equations below.

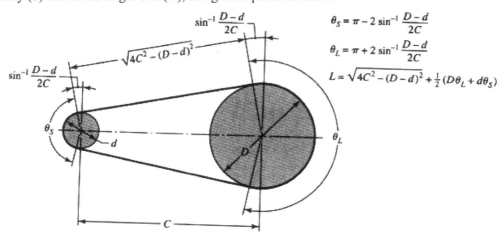

$$\theta_S = \pi - 2\,\sin^{-1}\frac{D-d}{2C}$$

$$\theta_L = \pi + 2\,\sin^{-1}\frac{D-d}{2C}$$

$$L = \sqrt{4C^2 - (D-d)^2} + \tfrac{1}{2}(D\theta_L + d\theta_S)$$

Note that θ_S and θ_L above need to be calculated in radians. Also, note that primitive diameters cannot be measured with a caliper, they are "imaginary" nominal values that need to be obtained from specific tables or from the manufacturer's catalog. Since the belts are only sold in standard sizes, you'll probably have to round up or down the calculated L. To prevent slack, you'll need to be able to slightly adjust the pulley distance C, or to install a belt tensioner, which can be easily made out of a small ball bearing fastened along the path of the belt between both pulleys.

An important parameter in the choice of timing belts and pulleys is the pitch, which is the distance between the tips of two consecutive teeth, as pictured to the right. The larger the pitch, the larger the tooth and the torque it  can handle. A few common belt denominations and pitches in the US are the MXL (2/25" pitch) and XL (1/5") for extra light duty, L (3/8") for light, H (1/2") for heavy, and XH (7/8") and XXH (1-1/4") for extra heavy duties. The metric denominations are 3M, 5M, 8M, 14M and 20M, where each number is the pitch in mm. The metric timing belts have high strength versions that are good for combat, such as the Optibelt Omega A, B and HP, with increasing strength. To have an idea of scale, our middleweight *Ciclone* uses 8M (8mm pitch), our hobbyweight *Tourinho* uses 5M (5mm pitch), and our beetleweight *Mini-Touro* uses 3M (3mm pitch) timing belts to power their spinning bar and drums.

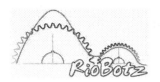 

5.3.3. Chains

Chains are also flexible elements used to transmit force and power. They are a good option because they are cheap and they can have any length, you only need to custom define their size using specific tools. Their disadvantages are: they are less efficient than belts, which results in certain power loss; they are noisy; they need tensioners to keep the chains stretched; and they can come out from the sprocket due to misalignments or other deformations, or due to large impacts. Since combat robots will suffer several impacts, care should be taken with such transmission type.

To avoid these problems, it is a good idea to use short chains, eliminating the need for tensioners, and to protect them very well. This can be seen in the picture to the right, which shows a great modular drive unit sold at www.battlekits.com, designed by the famous BioHazard builder Carlo Bertocchini.

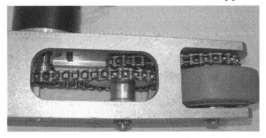

5.3.4. Flexible Couplings

Flexible couplings allow a shaft to efficiently transmit power to another one, even in the presence of misalignments. They consist of 2 rigid coupling hubs, usually made out of cast iron, fixed to each shaft usually using keyways, and of an elastic element (rubber spider) between them, see the picture to the right. They are used in general to connect the motor shaft to the wheel shaft. Besides

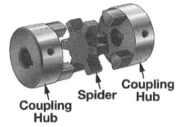 

tolerating misalignments, they absorb impacts and vibrations, which is highly advisable if your drive system gears aren't very resistant.

Our middleweight *Ciclone* uses such couplings between its wheels and the 18V DeWalt gearmotors that drive them, which helps to prevent an infamous plastic gear inside very old versions of the DeWalt gearbox from breaking. An inconvenience of flexible couplings is their overall length (right figure above), which is relatively large, increasing the distance between the motors and the wheels, which can make your robot become too wide.

Avoid using these couplings to power impact weapons, it is likely that the rubber spider won't take the high impact torques.

Another method to couple misaligned shafts is through universal joints (a.k.a. universal or U-joint, pictured to the right). Avoid using them: they are heavy, their strength is relatively low (their pins, which have a much smaller diameter than the joint itself, are the weakest point), and the energy efficiency is low, getting worse if the shafts have large misalignments. In combat robots, always try to replace universal joints with belt or chain transmissions.

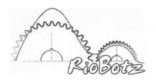

5.3.5. Torque Limiters

Torque limiters are power transmission elements that allow relative sliding between the coupled components, acting as a clutch. They are an important component in impact weapons that do not use V-belts or some other element that acts as a clutch to limit the torques transmitted back to the robot.

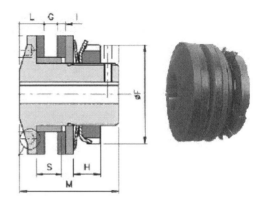

The figure to the right shows the torque limiter used by our middleweight bar spinner *Ciclone*, the DSF/EX 2.90, manufactured by the Italian company *Comintec*. The spinning bar is sandwiched between 2 flanges, one fixed and the other movable. The movable flange is fastened onto the bar, applying a constant pressure with the aid of a Belleville washer (see chapter 4), in such a way to transmit friction torques to accelerate the weapon bar. The flanges allow the bar to slide in the event of an impact, acting as a clutch.

It is not necessary to buy an off-the-shelf torque limiter. It is possible to build yourself much smaller, lighter, more resistant and cheaper versions. You basically need two flanges attached to a shaft, which can be two sturdy hardened steel shaft collars – a plain one and a threaded one to be tightened with the aid of a Belleville washer against the driven element such as the weapon bar. Phenolic laminates such as garolite are a good clutch material to be inserted in between the collars and the bar. The torque limiter from our spinner Titan is much smaller than *Ciclone*'s, because the lower flange is already embedded onto the weapon shaft, saving weight and increasing strength. It is then enough to use a Belleville washer and a threaded collar to attach its weapon bar.

5.4. Weapon and Drive System Calculations

Using the above information on DC motors and power transmission elements, we can already design a typical robot weapon and drive system. We will present next a few examples.

5.4.1. Example: Design of Touro's Drive System

We will calculate the acceleration time and final speed of our middleweight Touro. It uses two Magmotors S28-150, one for each of its two wheels. The used TWM 3M gearbox has a reduction ratio of n = 7.14, in other words, the wheel spins 7.14 times slower than the motor, and with 7.14 times more torque. Touro's mass is about 55kg (120lb), however we estimated that the 2 wheels support about 50kg or less (roughly 110lb), because they are not perfectly aligned with the robot center of mass. The two skids beside the drum support the remaining 10lb. Therefore, each wheel supports static loads of about 25kg

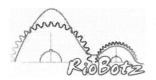 

(55lb), the equivalent to $25 \times 9.81 = 245N$. Note however that when the robot is accelerating, the active wheels might see a larger normal force, in special if the robot tilts back (as it happens when dragster cars accelerate) without having their rear structure touch the ground.

We will assume that the friction coefficient between the wheels and the arena floor is 0.9. This is a good number for rubber wheels with 65 Shore A hardness (see chapter 2) on a steel floor of a clean arena. This value might drop to 0.8 or even 0.6 when the arena is dirty, covered with dust and debris. The largest traction force that each wheel can generate without skidding is then $0.9 \times 245N = 220.5N$.

Touro's wheels have 6" diameter, therefore their radius is $r = 76.2mm$. The torque that makes the wheels skid is then $220.5N \times 0.0762m = 16.8N\cdot m$, and the torque that the Magmotor needs to deliver to the gearbox is $\tau_{max} = 16.8N\cdot m / 7.14 = 2.35N\cdot m$. As we've seen in section 5.1, this motor generates torques of up to $6.0N\cdot m$, therefore Touro should skid in the beginning of its acceleration and only later stop slipping. The maximum electric current in each motor while the robot is skidding is $I_{max} = \tau_{max} / K_t + I_{no\_load} = 2.35/0.03757 + 3.4 = 66A$. If you are an aggressive driver and spend 50% of a match accelerating at full throttle, then in 3 minutes (0.05 hours) you would spend in both motors $2 \times 66A \times 0.05h \times 50\% = 3.3A\cdot h$ (approximately, ignoring the consumption when the robot is not accelerating). Therefore, for the drive system, 1 battery pack with 24V and 3.6A·h would probably be enough (more details on batteries can be seen in chapter 8).

But be careful with the limit of validity of the calculation: we had calculated $I_{stall} = 162A$ for each motor, but a single NiCd pack would not be able to supply $2 \times 162 = 324A$ for both motors. But, actually, we use two 24V battery packs in parallel in Touro, so there is no problem, both are able to generate together 324A (at least with the weapon turned off). Note also that if we used two packs in parallel to power a single motor, we would have an equivalent electrical resistance of half the one of a single pack, $R_{battery} = 0.080\Omega / 2 = 0.040\Omega$, which would change all the previous calculations due to the new value of R_{system}. However, because we use 2 packs to drive 2 motors, the calculations using 0.080Ω are still valid.

The maximum theoretical speed of Touro, if there were no friction losses in the gearbox, would happen with the motor spinning at $\omega_{no\_load} = 5,968RPM = 625rad/s$, generating a top speed $v_{max} = (\omega_{no\_load} / n) \times r = (625 / 7.14) \times 0.0762m = 6.67m/s = 24km/h$ (almost 15 miles per hour), a relatively high speed for a middleweight.

While Touro is skidding, the current on each motor is $I_{max} = 66A$, which only happens for low speeds of the motor, from zero up to $\omega_1 = K_v \times (V_{input} - R_{system} \times I_{max}) = 254 \times (24 - 0.148 \times 66) = 3,615RPM = 379rad/s$. The robot speed when the wheels stop skidding is given by $v_1 = (\omega_1 / n) \times r = (379 / 7.14) \times 0.0762m = 4.04m/s = 14.5km/h$ (9.04mph).

During this period, when the wheels are skidding, the robot's acceleration would be equal to the friction coefficient times the acceleration of gravity, worth $a_1 = 0.9 \times 9.81 = 8.83m/s^2$. Actually, this value would be true for any all-wheel-drive robot, but Touro has two skids beside its drum that take together about 10lb of the robot weight. With all active wheels taking only 50kg (110lb), the acceleration would then be $a_1 = 0.9 \times 9.81 \times 50kg / 55kg = 8.03m/s^2$.

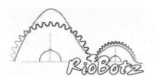 

But because the skids are in front of Touro, when it accelerates they are almost lifted off from the ground, making almost the entire robot weight go to the active wheels, as discussed in chapter 2. Thus, the previously calculated $a_1 = 8.83 \text{m/s}^2$ is a better approximation. Note that this assumption regarding the skids almost lifting off would slightly change the values of τ_{max}, I_{max} and v_1, however we will keep their previously calculated values for the sake of simplicity.

The resulting movement while the robot is skidding is a uniformly accelerated one, which happens during a time interval of $\Delta t_1 = v_1 / a_1 = 4.04 / 8.83 = 0.46 \text{s}$.

After that time, the current in each motor starts to decrease, getting below 66A. The instantaneous current delivered to each motor is then $I_{input} = [V_{input} / R_{system}] - [\omega / (K_v \times R_{system})] = I_{stall} - [\omega / (K_v \times R_{system})]$, and the motor torque results in

$$\tau = K_t \cdot (I_{input} - I_{no\_load}) = K_t \cdot (I_{stall} - I_{no\_load} - \frac{\omega}{K_v \cdot R_{system}})$$

Therefore, the torque at each wheel is $\tau \times n$, which generates a traction force of $\tau \times n / r$. The 2 wheels generate together twice that force, and then from Newton's second law we obtain the equation $2 \times \tau \times n / r = 55 \text{kg} \times a_2$. This robot's acceleration a_2 varies because it depends on the motor speed ω:

$$a_2 = \frac{2 \cdot n}{r \cdot 55} \cdot K_t \cdot (I_{stall} - I_{no\_load} - \frac{\omega}{K_v \cdot R_{system}}) = \frac{2 \cdot 7.14}{0.0762 \cdot 55} \cdot 0.03757 \cdot (162 - 3.4 - \frac{\omega}{26.62 \cdot 0.148})$$

resulting in $a_2 = 20.3 - 0.0325 \cdot \omega$. Be careful with these calculations, because K_v needs to be represented in (rad/s)/V, and not in RPM/V. Because the wheels are not slipping anymore, the robot speed can be obtained directly from the motor speed, $v = (\omega / n) \times r = \omega / 93.7$, resulting in an acceleration $a_2 = 20.3 - 3.04 \cdot v$.

The robot never achieves the theoretical maximum speed, because the behavior is asymptotic. But the time interval between the moment the robot stops skidding (when $v = 4.04 \text{m/s}$) and the moment it reaches, for instance, 95% of its maximum speed ($v = 0.95 \times 6.67 \text{m/s} = 6.34 \text{m/s}$) can be calculated:

$$\Delta t_2 = \int dt = \int_{4.04}^{6.34} \frac{dv}{20.3 - 3.04 \cdot v} = \frac{1}{3.04} \cdot \ln\left(\frac{20.3 - 3.04 \cdot 4.04}{20.3 - 3.04 \cdot 6.34}\right) = 0.68 \text{s}$$

where ln stands for the natural logarithm function.

Thus, the total acceleration time of Touro, from its resting position up to 95% of its maximum speed, is $\Delta t = \Delta t_1 + \Delta t_2 = 0.46 + 0.68 = 1.14 \text{s}$, a very close value to the measured one in our tests. The graph to the right shows the results. If your robot doesn't have enough torque to skid

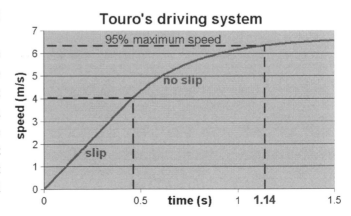

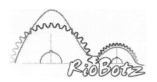

(slip) during its acceleration, then it is enough to make calculations based on the integral above using the initial speed $v = 0$ (in other words, $\Delta t_1 = 0$, therefore $\Delta t = \Delta t_2$).

Notice that, for the robot to be agile, it is important that such acceleration time Δt is short, such as in Touro. It is not a good idea to have a very high maximum speed if the robot can't achieve it quickly enough, without the need to cross the entire arena.

It is important to emphasize that the above calculations would also be valid if the robot had 4 active wheels powered by the same 2 motors. The torque from each motor would be distributed to the 2 wheels it drives, however the combined traction force of these 2 wheels would be added up, resulting in practically the same acceleration and time intervals calculated above.

But if there were 4 motors for the 4 wheels, then the calculation results would definitely change, because we would be multiplying by 2 the system power. Probably Δt_1 would remain the same, since it is mainly determined by the tire coefficient of friction, but Δt_2 would certainly decrease. These calculations would not be difficult to perform using the above methodology.

5.4.2. Example: Design of Touro's Weapon System

We will calculate the acceleration time of Touro's drum, and the kinetic energy it stores. Touro's drum can be approximately modeled as a steel cylinder with external radius $R = 65mm$ and internal radius $r = 40mm$, with length $L = 180mm$. The density of steel is roughly 7.8, therefore the drum mass is $m = \pi \cdot (65^2 - 40^2) \cdot 180 \cdot 7.8 \cdot 10^{-6} kg/mm^2 = 11.6kg$ (about 25.6lb). The rotational moment of inertia with respect to the horizontal spin axis is $I_{zz} = m \cdot (R^2 + r^2)/2 = 11.6 \cdot (65^2 + 40^2)/2 = 33785 kg \cdot mm^2 = 0.0338 kg \cdot m^2$.

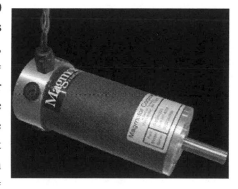

The weapon motor is one Magmotor S28-400 (pictured to the right) connected to 2 NiCd battery packs in parallel, therefore $V_{input} = 24V$, $K_t = 0.0464 N \cdot m/A$, $R_{motor} = 0.042\Omega$, and $I_{no\_load} = 4.5A$. We have then $K_v = 1/K_t = 21.55$ (rad/s)/V = 206 RPM/V. The motor resistance needs to be added to the resistance of the electronics and solenoid (about 0.004Ω) and of the batteries, which for being in parallel have an equivalent resistance of half of a single pack ($0.080\Omega / 2 = 0.040\Omega$ in this case), resulting in $R_{system} = 0.042 + 0.004 + 0.040 = 0.086\Omega$. Note that those 2 packs are the same as the ones used in the drive system of Touro, therefore we will assume in the following calculations that the robot is not being driven around during the weapon acceleration.

The 2006 version of Touro had V-belt pulleys used in the weapon system with same diameter, therefore there was no speed reduction ($n - 1$). The theoretical top speed of the drum is then $\omega_{no\_load} = 206 \times (24 - 0.086 \times 4.5) = 4,864 RPM = 509 rad/s$ (in 2007, this speed was increased to about 6,000RPM by reducing the diameter of the drum pulley). In practice, because of the friction losses,

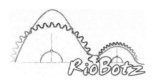 

the drum (from the 2006 version of Touro) spins at a little more than 4,750RPM, which was measured using a strobe tachometer.

The peak current at the beginning of the acceleration is I_{stall} = 24 / 0.086 = 279A. Note that, ideally, the V-belts should not slide during the drum acceleration, they should only slip at the moment of impact against the opponent. This is why they need to be well tensioned. Assuming that they don't slide during the acceleration, the only other thing we need to know is whether the batteries are able to supply the required 279A.

If this weren't true, we would need to split the calculations into 2 parts: an initial acceleration period when the batteries would be supplying their maximum current (which would be a certain value smaller than I_{stall}), and another period when the batteries would be able to supply the motor needs. The solution of this problem would not be difficult, the calculations would be similar to those made for the design of the drive system, adding up the time intervals from both parts.

In the case of Touro, the batteries are able to supply together the required 279A, which simplifies the calculations. As studied before, the motor torque is a function of its angular speed ω:

$$\tau = K_t \cdot (I_{input} - I_{no\_load}) = K_t \cdot (I_{stall} - I_{no\_load} - \frac{\omega}{K_v \cdot R_{system}}) = 12.74 - 0.025 \cdot \omega$$

Because the gear ratio is n = 1 (same diameter pulleys), this torque is applied directly to the drum to accelerate it:

$$\tau = I_{zz} \cdot \frac{d\omega}{dt} \implies 12.74 - 0.025 \cdot \omega = 0.0338 \cdot \frac{d\omega}{dt}$$

It would not be difficult to include the effect of a gear ratio n different than one, the procedure would be similar to the one used in the drive system calculations.

The acceleration (spin up) time of the drum from zero speed up to, for instance, 90% of its maximum speed (0.90 × 509 = 458rad/s), is then

$$\Delta t = \int dt = \int_0^{458} \frac{0.0338 \cdot d\omega}{12.74 - 0.025 \cdot \omega} = \frac{0.0338}{0.025} \cdot \ln\left(\frac{12.74 - 0.025 \cdot 0.0}{12.74 - 0.025 \cdot 458}\right) = 3.1s$$

The graph to the right summarizes the drum spin up results. Considering the friction losses, it would be expected in practice that the actual value would be slightly above 3.1s. On the other hand, fully charged 24V NiCd batteries are able to deliver up to 28V, which would more than compensate for these friction losses.

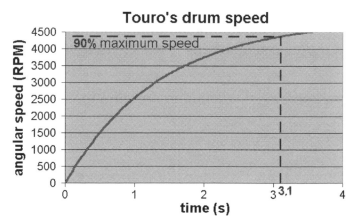

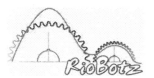 

As a result, the above approximation ends up quite reasonable: the experimentally measured spin up time until 90% of the maximum speed was about 3s. In general, it is a good idea that the spin up time of a weapon is shorter than 4 seconds (see chapter 2), therefore 3s is a good value.

Note that those calculations assumed that the robot was not moving around, and therefore the 2 battery packs were used exclusively to accelerate the weapon. If the robot was driving around during the weapon acceleration, then naturally the actual spin up time would be longer than 3s.

The accumulated kinetic energy by the drum after these 3.1s would be $E = I_{zz} \cdot \omega^2 / 2 = 0.0338 \cdot 458^2 / 2 = 3{,}545J$ (for 90% of its maximum speed), the equivalent to about 10 caliber 38 shots, or 1 rifle shot. The actual maximum kinetic energy, from the measured speed 4,750RPM (497rad/s), is $E = I_{zz} \cdot \omega^2 / 2 = 0.0338 \cdot 497^2 / 2 = 4{,}174J$.

Theoretically, this energy would be able to fling a middleweight opponent to a height of $h = E / (m \cdot g) = 4174 / (55 \cdot 9.81) \cong 7.7$ meters (more than 25 feet into the air). In practice, the height is much lower because the impact is not entirely transmitted to the opponent, and a lot of the energy is dissipated in the form of heat and deformation. The equations to estimate the actual height will be presented in chapter 6.

Finally, note that the above calculations can be applied to horizontal and vertical spinners as well, not only to drumbots, as long as the weapon inertia I_{zz} is known. For instance, a flat bar with mass m, length 2·a and width 2·b, spinning around its center of mass, has $I_{zz} \cong m \cdot (a^2 + b^2)^2 / 3$. And a solid disc, with mass m and radius a, would have $I_{zz} \cong m \cdot a^2 / 2$. More details can be seen in chapter 6.

5.4.3. Energy and Capacity Consumption of Spinning Weapons

It is very important to calculate the energy consumption of an electrical motor from a spinning weapon, in order to evaluate battery requirements. The weapon consumption can be divided into a portion needed to spin up the weapon after each impact, and another one from friction losses.

It is possible to estimate the energy and capacity consumption of the battery during the spin up of a weapon with moment of inertia I_{zz} in the spinning direction, assuming that I_{no_load} is much smaller than I_{stall} (which is true for all good quality DC motors). In this case, if we approximate $I_{no\_load} = 0$, the motor torque is simply $\tau = K_t \times I_{input}$. The torque transmitted to the weapon after a reduction ratio of n:1 is $\tau_{weapon} = \tau \times n$, and the weapon angular speed is reduced to $\omega_{weapon} = \omega / n$.

In the equation $\omega = K_v \times (V_{input} - R_{system} \times I_{input})$, the only variable terms are ω and I_{input}, all others are constant, therefore the angular acceleration is $d\omega/dt = -K_v \times R_{system} \times dI_{input}/dt$. The dynamic equation of the system is then:

$$\tau_{weapon} = I_{zz} \cdot \frac{d\omega_{weapon}}{dt} \quad \Rightarrow \quad \tau \cdot n = \frac{I_{zz}}{n} \cdot \frac{d\omega}{dt} \quad \Rightarrow \quad K_t \cdot I_{input} \cdot n = -\frac{I_{zz} \cdot K_v \cdot R_{system}}{n} \cdot \frac{dI_{input}}{dt}$$

and therefore

$$I_{input} dt = -\frac{I_{zz} \cdot K_v \cdot R_{system}}{K_t \cdot n^2} \cdot dI_{input}$$

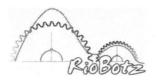 

The *capacity consumption* of a battery, which is its *energy consumption* divided by its voltage, is then obtained by integrating the current with respect to time, from its initial value I_{stall} (in the start of the weapon acceleration) until its final zero value (because it approaches $I_{no\_load} = 0$), therefore

$$\text{Capacity Consumption} = \int I_{input} dt = -\frac{I_{zz} \cdot K_v \cdot R_{system}}{K_t \cdot n^2} \cdot \int_{I_{stall}}^{0} dI_{input} = \frac{I_{zz} \cdot K_v \cdot R_{system}}{K_t \cdot n^2} \cdot I_{stall}$$

The above equation is valid for any spinning weapon powered by PM DC motors. But be careful with the units, K_v should be in (rad/s)/A and the resulting capacity consumption is in A·s. In the 2006 version of Touro, the capacity consumption during each spin up of its drum was

$$\text{Capacity Consumption} = \frac{0.0338 \cdot 21.55 \cdot 0.086}{0.0464 \cdot 1^2} \cdot 279 = 377 \text{A} \cdot \text{s} = 0.104 \text{A} \cdot \text{h}$$

However, we still need to consider the capacity consumption due to the friction losses of the weapon. This is very hard to model theoretically, but it can be easily measured experimentally. To do that, we've powered Touro's drum and, at its maximum speed, we've measured the electrical current going through the motor, which was about 40A. Be careful when testing weapons, safety always comes first! This average 40A value is continuously consumed while the drum is powered, to compensate for friction losses from the motor, drum bearings and V-belts, as well as the aerodynamic losses due to the high tangential speed of the drum teeth.

We will consider that the drum is powered during an entire 3 minute match, and that it delivers about 10 large blows against the opponent (therefore needing to fully accelerate 10 times). The total capacity consumption of the weapon motor in 3 minutes (180 seconds) is then approximately:

Weapon Capacity Consumption $= 40\text{A} \times 180\text{s} + 10 \times 377 \text{A} \cdot \text{s} = 7200 \text{A} \cdot \text{s} + 3770 \text{A} \cdot \text{s} = 10970 \text{A} \cdot \text{s} \cong 3.1 \text{A} \cdot \text{h}$

Note that most of the weapon consumption (almost 66% in this case) is used up to compensate for friction and aerodynamic losses. Therefore, you should always use well lubricated ball, roller or tapered bearings. Shielded and sealed bearings are a good option to avoid debris, but the sealed type usually results in higher friction. Plain bronze bearings can also be used, but the friction losses are usually higher, and they can heat up a lot at high speeds unless they are of the plugged type.

The total energy consumption of Touro in 3 minutes, adding the contributions of the drive system (with an aggressive driver accelerating half of the match, as previously considered) and the weapon system (with the drum turned on during the whole time and delivering 10 great blows) is

Total Capacity Consumption = 3.3A·h + 3.1A·h = 6.4A·h

therefore two 24V battery packs with 3.6A·h each (totaling 7.2 A·h) would be enough.

These calculations can also help to define the driver's strategy in case of contingency. For instance, if you have available two packs with only 2.4A·h each (totaling 4.8A·h), the driver could accelerate the robot (drive system) 25% of the time, turn off the weapon during 30% of the match (attacking as a rammer during this time), and still be able to deliver 10 great blows, because after recalculating the capacity consumption we would get

Total Capacity Consumption = 1.65A·h (drive) + 2.45A·h (weapon) = 4.1A·h

which would be enough if both batteries can actually deliver 4.8A·h.

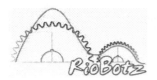

5.5. Pneumatic Systems

Up to now we've basically focused on DC motors, because they're the most used actuators in combat. However, there are other actuation elements that are as good as or even better than electric systems. Their only disadvantage is the higher complexity and, in some cases, reduced reliability.

Pneumatic systems are capable of generating a great amount of energy in a short period, which is fundamental for robots with intermittent weapons such as hammerbots or launchers (such as the lightweight Hexy Jr, from Team WhoopAss, pictured to the right). They are usually powered by high pressure air or nitrogen (N_2), or liquid CO_2.

CO_2 can be stored in reservoirs in the liquid form. This allows tanks to store a great amount of CO_2 in a small space. The storage pressure is about 850 to 1000psi (about 60 to 70 atmospheres). Because it is used in paintball weapons, many components for CO_2 are easily found. The problem with CO_2 is that the phase change from liquid to gas is an endothermic process, which can make the reservoirs freeze during a match.

Air and N_2 can be compressed in gas form to higher pressures, such as 3000psi (about 200 atmospheres). Their advantages over CO_2 are that they do not have the freezing problem and they are lighter (saving about 0.5kg in typical middleweights with a full tank). The disadvantage is in the need for high pressure components, which are more expensive. Besides, a few competitions limit the pressure that can be stored in the robots.

In a simplified way, the pneumatic systems consist of one or more storage tanks, connected to a pressure regulator, accumulator, solenoid valve, and pneumatic cylinder, not to mention the safety valves.

Storage tanks are necessary because it is not practical to use air (or N_2) compressors. Besides being heavy, air compressors would not have enough power to supply the robot's needs in time during an attack, even if some accumulator was present to act as a buffer.

Regulators are components that transform the high pressures of the pneumatic tanks (about 1000psi for CO_2 and 2000 to 3000psi for air or N_2) into lower pressures that can be used in conventional pneumatic systems, typically between 150 to 250psi.

Accumulators are buffers, small reservoirs that store the gas already in the operating pressure of the robot's weapon. They are necessary only if your regulator doesn't generate enough flow for an efficient attack. They usually store enough gas for one attack, guaranteeing the required flow during the entire stroke of the cylinder, without suffering the bottleneck effects of the regulator or safety valves.

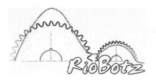 

The cylinders are actuated through solenoid valves, which usually are of two types: two-way two-port, to power single-acting cylinders (which are only powered in one direction, they usually need a spring return), or four-way five-port, for double-acting cylinders (which are pneumatically powered in both directions). Naturally, the larger the piston area, the larger the generated forces by the cylinder.

The picture below shows the super-heavyweight launcher Ziggy, with its high pressure tank, high pressure regulator (a "GO regulator" PR-59 Series), and its cylinder, which powers a 4-bar linkage. The solenoid valves cannot be seen in the picture, because they're on the opposite side. No accumulator is used in this case, therefore the regulator is the system bottleneck, even though it is of a high flow type. The picture shows as well the Magmotor S28-400 motors and TWM 3M gearboxes used in the chained drive system.

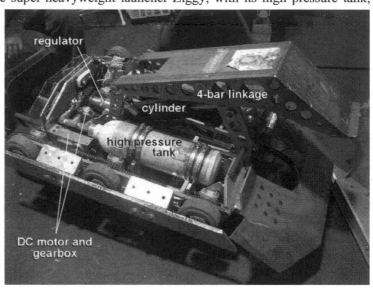

The figure below displays the schematics of a single-acting cylinder. The items A, B, C and D of the diagram represent the high pressure line, and the remaining elements are the operating pressure line.

A. high pressure tank;
B. high pressure purge valve;
C. high pressure gauge;
D. regulator;
E. low pressure gauge;
F. safety valve;
G. accumulator;
H. low pressure purge valve;
I. normally closed two-way valve;
J. normally open two-way valve;
K. single-acting cylinder, with spring return.

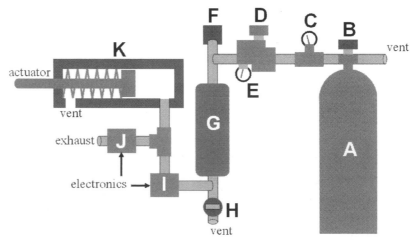

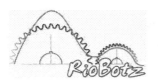

Several daring builders use CO_2 tanks without regulators, sending high pressure directly to the cylinders. For that it is necessary to eliminate any bottlenecks in the pressure line, removing any needle valves and avoiding turns and sharp corners in the pipeline. If the entire flow is free of bottlenecks, it is not necessary to use an accumulator. The schematics would be similar to the one above, except that the items D, E, F, G and H would be eliminated. But be careful: instead of working with 150 to 250psi, the items I, J and K would be submitted directly to about 1000psi. They might not tolerate such pressure level.

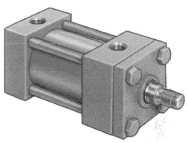

To tolerate such unregulated pressure, you would need hydraulic components, especially hydraulic cylinders, as pictured to the right. A few of them are rated to up to 2500psi. However, they would be pneumatically powered. Be careful with these systems, because they are potentially self-destructive! Hydraulic cylinders are not designed for the high-speeds of the pneumatic systems, thus there is a chance that the piston will break due to the impact at the end of its stroke. Use certified systems for 2500psi hydraulic if you plan to power them at 1000psi pneumatic.

Even so, as pointed out by Mark Demers, builder of Ziggy, an unmodified 2500psi cylinder which is not designed for impact loading at stroke end doesn't guarantee it will hold up at 1000psi pneumatic: "Impact loads are dramatically higher than static loads. I recommend some sort of external constraint to eliminate the impact load which occurs at the end of the stroke. The higher the launching force, the more sense it makes to add a limiting constraint. Back in the days of BattleBots, the Inertia Labs launcher robots (T-Minus, Toro, Matador) used nylon strap restraints to limit the extension of the arm and relieve the cylinder from the shock loading. Nylon straps are also used in Monster trucks to prevent over-extension in the suspension. Ziggy's 4-bar system limits the stroke of the cylinder by design – the cylinder has an 8" stroke but the linkage does not allow extension of more than 7.75."

In addition to end-of-stroke restraints, most cylinders used in launchers need modifications to take the impact loads. Team Hammertime's famous launchers, such as Bounty Hunter (pictured to the right) and Sub-Zero, use cylinders powered by high-pressure CO_2. Their builder, BattleBots veteran Jerry Clarkin, has modified his cylinders for the additional load. As pointed out by Mark Demers, "the cylinders Jerry is using have an

air cushion at the end of their stroke. Additionally, Jerry has added high strength steel tie rods and steel containment plates to dramatically increase the axial strength of the cylinder."

But be careful, do not try using unregulated systems unless you already have a lot of experience with conventional regulated pneumatics. Also, don't forget to check the competition rules to see whether the use of such unregulated pressures is allowed.

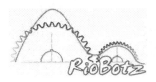

Back to regulated systems, to power a double-acting cylinder, it is necessary to use a slightly different schematics, shown below.

A. high pressure tank;
B. high pressure purge valve;
C. high pressure gauge;
D. regulator;
E. low pressure gauge;
F. safety valve;
G. accumulator;
H. low pressure purge valve;
I. four-way valve;
J. double-acting cylinder.

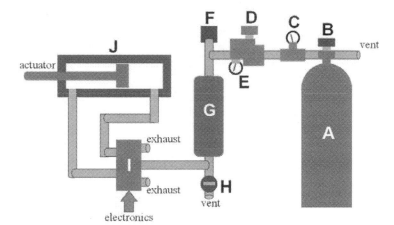

A few tips to increase the speed of your cylinder are: use a larger accumulator; use hoses and connections with the largest possible diameter; avoid sharp corners in the path of the hoses and pipes; leave the gas exhaust path as free as possible, directed towards outside the robot. More information can be found at www.teamdavinci.com/understanding_pneumatics.htm, and in the references [4] and [10].

5.6. Hydraulic Systems

Among weapon system actuators, hydraulic cylinders are the ones capable of generating the largest forces. Their inconvenience is in the low speed of the weapon, which is a big issue in combat. A two-stage hydraulic system would solve this issue, however its implementation is very complex. The hydraulic cylinder is powered by hydraulic servo-valves through solenoids. These systems also require a compressor (hydraulic pump), which needs to be powered either electrically or using an internal combustion engine (ICE). Hydraulic fluid leakage is also a common problem.

Hydraulic weapon systems were only successfully used in crusher bots. The picture below, from www.boilerbots.com, shows the weapon system from the famous super heavyweight Jaws of Death. Note the need for an electric system (for the servo-valves and drivetrain), hydraulic system (weapon), as well as an ICE to power the hydraulic pump. There are so many heavy components required in the weapon system, that usually only a super heavyweight is able to use them without compromising drivetrain speed or armor. Few hydraulic robots are still active, mainly due to their complexity.

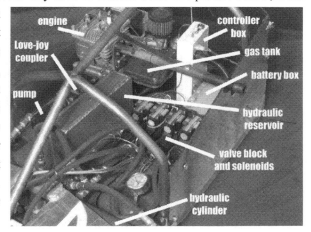

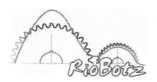

5.7. Internal Combustion Engines

Internal combustion engines (ICEs) are capable of storing a great amount of energy. The energy density of gasoline, for instance, is about 100 times larger than that of NiCd batteries. They deliver more power to weapon systems than an electric motor would. Another advantage is that their torque increases (up to a certain point) with speed, unlike PM DC motors, which tend to zero torque at high speeds. Internal combustion engines also do not lose power when the tank is almost empty, as opposed to DC motors, which start to run slow as the batteries drain. In addition, the loud noises can impress well the judges during a match.

The ICE system design is relatively simple, you just need a good quality servo-motor and a centrifugal clutch (such as the ones used in go-karts). These clutches guarantee that the weapon will not spin until the beginning of the match, as required by the competition rules, even with the ICE turned on.

A great challenge is to guarantee that the ICE works upside down, in case the robot is invertible, guaranteeing that the fuel flow remains constant and without leakage. Chainsaw motors are good candidates, because their carburetor can operate in any orientation. ICEs used in airplanes also work upside down. Jet engines have also been used to power spinners, however they would usually be too heavy for a middleweight, sometimes even for a heavyweight.

The ICEs only spin in one direction, therefore they are only used to power combat weapons. To power a drive system, the ICE would need a complex gear system to reverse the wheel spin.

A serious problem with an ICE is the large radio interference that the spark plugs can cause. Therefore, place the receiver and electronics as far away as possible from the motor. To eliminate this problem, you can also use resistor spark plugs, which cause ignition through electric resistance, not causing any radio interference. Or you can use, for instance, 2.4GHz radio systems, which do not suffer from such ICE noise problems.

The greatest disadvantage of an ICE robot is its low reliability. The technique to turn it on in the beginning of a match is known as "pull and pray": you pull-start it, and pray for it to keep running. If the ICE dies during a match, it will be impossible to start it again, unless it has its own onboard starter, controlled by an additional channel of the radio. This system adds weight to the robot and it also suffers reliability problems. Besides, you will end up needing to use 4 radio channels to power a single ICE.

In summary, ICE robots are extremely powerful and dangerous, but due to their low reliability they depend a lot on luck to win a competition without technical problems.

A curiosity: the robot Blendo (pictured to the right) was the first ICE spinner, using a lawnmower motor. It was built by Jamie Hyneman, and its electronic system was wired by Adam Savage. Jamie and Adam's appearances in BattleBots called the attention of producer Peter Rees, leading to their debut hosting the famous Discovery Channel TV show MythBusters.

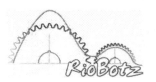

Chapter

6

Weapon Design

In this chapter, we'll address several specific issues concerning weapon design. The idea is to show that mechanics calculations based on basic physics concepts and common sense can help a lot in the design of powerful and robust weapon systems.

6.1. Spinning Bar Design

It is not difficult to specify the dimensions of the bar of a spinner robot using basic stress analysis and a very simplified impact model. Consider the bar on the right, made out of hardened steel with length $2 \cdot a$, width $2 \cdot b$ and thickness t, with a central hole of radius r'. During the impact, the angular momentum of the left and right hand sides of the spinning bar, L_L and L_R, will cause an average reaction force F from hitting the opponent, as seen in the figure. Since the bar is symmetric, it is easy to see that $L_L = L_R$.

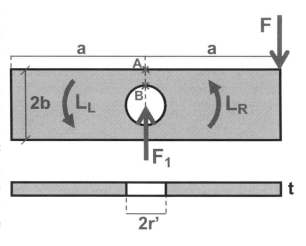

If we assume that the chassis of the spinner robot is much heavier than its bar, we can say that the average reaction force F_1 from the weapon shaft is approximately equal to F, therefore $F_1 \cong F$ (we'll see later a better model that will allow different values for F_1 and F). We will also assume that the opponent is much heavier than the bar, and that the impact is inelastic, making the bar stop spinning after the brief time interval Δt of the impact. Therefore, the average torque $F \cdot a$ with respect to the weapon shaft, caused by the force F, must be able to bring the initial value of the angular momentum $(L_L + L_R)$ of the bar to zero during this Δt, resulting in $F \cdot a = (L_L + L_R) / \Delta t$, which gives $L_L / \Delta t = L_R / \Delta t = F \cdot a/2$. The average bending moment M_{max} in the middle of the bar, the region where bending is maximum, can then be calculated, $M_{max} = L_L / \Delta t = F \cdot a - L_R / \Delta t = F \cdot a/2$.

The stress at point A (see figure) due to bending is $\sigma_A = 3 \cdot M_{max} \cdot b/[2 \cdot (b^3 - r'^3) \cdot t]$. The stress at point B, theoretically, would be smaller than in A, however the hole acts as a stress raiser. It

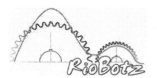 

amplifies the stresses close to its border. In the geometry and loading of this example, it multiplies the stress by a factor of approximately 2, in other words, the stress concentration factor is 2 (this value is obtained from specific stress concentration factor tables [8]). Therefore, the stress acting at B is $\sigma_B = 2 \times 3 \cdot M_{max} \cdot r' / [2 \cdot (b^3 - r'^3) \cdot t]$.

If $\sigma_A > \sigma_B$, then if the bar breaks it will be from the outside in, beginning to fracture in A (where the stress is higher) and propagating a crack abruptly until point B. The bar will then break in two because the residual ligament on the other side of the hole (region below the hole in the figure) will be overloaded and break. All this happens in a split second – in metals, the fracture propagates at a speed of about 2 to 3km/s (1.24 to 1.86 miles per second), therefore a typical middleweight spinner bar would take about 0.01ms to fracture.

On the other hand, if $\sigma_B > \sigma_A$, the bar will break from the inside out, beginning fracturing from point B to point A. This was how the 5160 steel bar from our middleweight spinner *Ciclone* broke during the Winter Challenge 2005 competition, from B to A. That was because the diameter 2·r' of the hole of *Ciclone*'s bar was large with respect to its width 2·b, penalizing point B.

A good design choice would be to try to make point A at least as resistant as point B. For that, it is enough to equate $\sigma_A = \sigma_B$. After a little algebra with the previous expressions, we get b = 2·r'. Therefore, design your spinning bar (and the weapon shaft that will hold it) so that its width 2·b is at least twice the diameter 2·r' of its center hole. The bar from our middleweight spinner Titan was designed having this in mind.

And how much force would the bar support? Consider, for instance, 2·a = 1000mm, 2·b = 80mm, 2·r' = 2·b/2 = 40mm, and the thickness t = 12mm. The steel bar, with average density ρ = 7800kg/m^3, would have a mass of, approximately (without considering the hole), $\rho \cdot (2 \cdot a) \cdot (2 \cdot b) \cdot t$ = 7800kg/m^3 · 1m · 0.080m · 0.012m = 7.5kg (16.5lbs), which is a reasonable value for a middleweight – from the 30-30-25-15 rule, a middleweight would have 16.3kg (36lbs) for the weapon system, leaving in that example 16.3 – 7.5 = 8.8kg (more than 19lbs) for the weapon shaft, bearings, transmission and motor, an also reasonable value. A hardened steel with 45 Rockwell C (unit that measures how hard the material is, see chapter 3) tolerates a maximum stress of about 34 × 45 = 1530N/mm^2 before breaking (this 34 factor is only valid for steels, estimating well the ultimate strength from the Rockwell C hardness).

Making both stresses at A and B equal to 1530N/mm^2, then $\sigma_A = \sigma_B = 3 \cdot M_{max} \cdot b / [2 \cdot (b^3 - r'^3) \cdot t]$ = 1530N/mm^2, where the bending moment M_{max} = F·a/2, resulting in F = 68,544N, equivalent to almost 7 metric tons! Now it is necessary to guarantee that the weapon shaft and the rest of the robot can tolerate such average 7 tons, which can be made using the same philosophy presented above, from basic stress analyses. Approximate calculations can be very efficient if there is common sense and some familiarity with the subject.

Clearly, bars with b > 2·r' will result in even more strength, because a higher width 2·b will decrease both σ_A and σ_B values. But don't exaggerate, otherwise you'll have to decrease too much its thickness t not to go over the weight limit, compromising the strength in the out-of-plane bending direction.

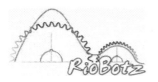

Another solution would be to have a variable width bar, with a wider middle section, as pictured to the right. Note how the bar shape is optimized, with an increasingly wider middle section to withstand high bending moments, and sharp and heavy inserts at its tips to guarantee a high moment of inertia in the spin direction. Note also the ribs milled in the bar to increase its bending strength without adding too much weight.

6.2. Spinning Disk Design

There has always been a great debate whether bars or disks make the best spinning weapons. Consider that the robot design allows a spinning weapon with mass m_b, and its reach from the weapon shaft must have a length a. Then let's compare a bar with length 2·a to a disk with radius a, pictured to the right. Both weapons, with same mass m_b, would be originally spinning until suffering an impact force F from hitting the opponent. If made out of the same material with a mass density ρ, then the thicknesses t and t_d of the bar and disk would be approximately $t \cong m_b/(\rho \cdot 4 \cdot a \cdot b)$ and $t_d \cong m_b/(\rho \cdot \pi \cdot a^2)$.

Concerning moment of inertia (I_b), it is easy to see from the values in the figure that the disk is a better choice, unless the half-width b of the bar is very large, above 0.707·a. The moment of inertia of a narrow bar (with b much smaller than

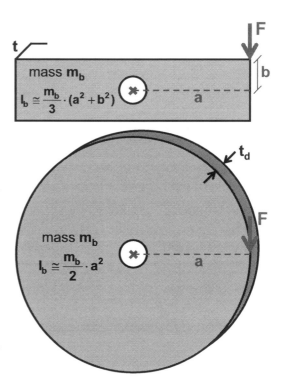

a) would be roughly 66.7% of the value of a disk with same mass m_b and length a.

Let's take a look now at the stresses. If the width 2·b of the bar is much higher than twice the diameter of the center hole, then the maximum stress due to the force F is approximately $\sigma_{bar} \cong 3 \cdot F \cdot a/(4 \cdot t \cdot b^2)$. And assuming the disk diameter is much larger than its hole diameter, then the maximum stress would be $\sigma_{disk} \cong 3 \cdot F/(4 \cdot t_d \cdot a)$. It is easy to show from these equations and the expressions for t and t_d that, assuming b smaller than a, any bar would see higher maximum stresses than the disk.

So, disks are a better choice concerning both stresses (while delivering an impact) and moment inertia. But they have a major drawback. If a vertical disk is hit by a horizontal spinner, or a

horizontal disk is hit by a drum or vertical spinner, it will see a force F* perpendicular to its plane, as pictured to the right, which will cause a maximum out-of-plane bending stress of approximately $\sigma_{disk}* \cong 3 \cdot F*/t_d^2$. The same perpendicular force would cause on the bar a maximum out-of-plane bending stress of $\sigma_{bar}* \cong 3 \cdot F* \cdot a/[(b-r') \cdot t^2]$, where r' is the radius of the hole. It is easy to show that any disk would result in higher maximum out-of-the-plane stresses than a bar.

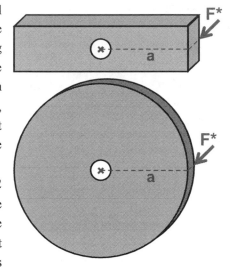

For instance, a bar with b = 2·r' would only see $\pi^2/32 \cong 31\%$ of the stresses found on an equivalent disk. The above equations, together with estimates for F and F*, are very useful to find out values for the bar width 2·b that will meet requirements for maximum allowable stresses σ_{bar} and $\sigma_{bar}*$, as well as to decide whether a disk would be an acceptable option despite its high resulting $\sigma_{disk}*$.

Therefore, we conclude that horizontal bars are a better choice than horizontal disks against drums, vertical spinners, and wedges (which can deflect a hit and also cause high out-of-plane bending stresses). Horizontal disks would be better against all other types of opponents, due to their higher in-plane bending strength and also higher moment of inertia. And vertical disks would be a better choice than vertical bars against most robots, except against horizontal spinners, which will most likely warp or break the disk with a powerful out-of-plane hit. You can get away with a vertical disk against a horizontal spinner, but you should either limit the disk radius (having a lower radius a to decrease $\sigma_{bar}*$), or protect it against out-of-plane hits having it recessed into the chassis or using a wedge, as seen in the lightweight K2 on the right.

Note that the calculations above assumed solid bars and disks, without considering any shape optimization. But the conclusions would still hold if comparing an optimized bar to an optimized disk. Shape optimization can also generate hybrids between disks and bars, trying to get the best of both worlds. The drum teeth from the middleweight Angry Asp (pictured to the right) are a good example of that, with their wide disk-like mid-section and elongated bar-like overall shape.

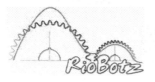 

6.3. Tooth Design

One important issue when designing spinning weapons such as disks, bars, drums and shells is regarding the number of teeth and their height. Too many teeth on a spinning disk, for instance, will make the spinner chew out the opponent instead of grabbing it to deliver a full blow. Everyone who's used a circular saw knows that fewer teeth means a higher chance of the saw binding to the piece being cut, which is exactly what we want in combat.

6.3.1. Tooth Height and Bite

Before we continue this analysis, we need to define the tooth bite d. The tooth bite is a distance that measures how much the tips/teeth of the spinner weapon will get into the opponent before hitting it. For instance, if two robots are moving towards each other with speeds v_{x1} and v_{x2}, one of them having a bar spinning with an angular speed ω_b (in radians per second), as pictured to the right, then the highest bite d = d_{max} would happen if the bar barely missed the opponent before turning 180 degrees to finally hit it. The time interval the bar takes to travel 180° (equal to π radians) is $\Delta t = \pi/\omega_b$, during which both robots would approach each other by $d_{max} = (v_{x1}+v_{x2})\cdot\Delta t = (v_{x1}+v_{x2})\cdot\pi/\omega_b$. So, the tooth bite d could reach values up to d_{max}.

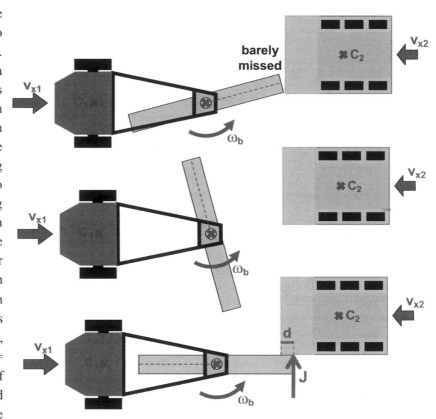

Small values for d mean that the spinner will have a very small contact area with the opponent, most probably chewing its armor instead of binding and grabbing it. So, a spinner needs to maximize d to deliver a more effective blow. This is why an attack with the drive system at full speed is more effective, since a higher speed v_{x1} will result in a higher d. And this is why very fast spinning weapons have a tough time grabbing an opponent, their very high ω_b ends up decreasing the tooth bite d.

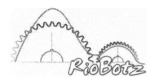

The maximum obtainable tooth bite $d = d_{max}$ can also be generalized for a weapon with n teeth. In this case, the teeth are separated by $2 \cdot \pi/n$ radians, as pictured to the right, resulting in $\Delta t = 2 \cdot \pi/(n \cdot \omega_b)$, and therefore $d_{max} = (v_{x1}+v_{x2}) \cdot \Delta t = (v_{x1}+v_{x2}) \cdot 2 \cdot \pi/(n \cdot \omega_b)$. Since the tooth bite cannot be higher than d_{max}, there is no reason to make the tooth height $y > d_{max}$ (see picture), which would decrease its strength due to higher bending moments. Therefore, the optimal value for the tooth height y is some value $y < (v_{x1}+v_{x2}) \cdot 2 \cdot \pi/(n \cdot \omega_b)$.

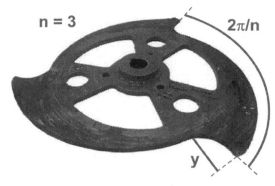

Using the maximum values of both v_{x1} and v_{x2} speeds will probably result in large values for y, so it is a reasonable idea to assume $v_{x2} = 0$. Most attacks will happen at full speed v_{x1} but without the opponent moving towards you. Besides, a spinner doesn't know beforehand the value of v_{x2} of all of its possible opponents. So, a tooth height $y = v_{x1,max} \cdot 2 \cdot \pi/(n \cdot \omega_{b,max})$ is usually more than enough. Note that this height assumes the weapon at full speed, if you want to deal with lower ω_b speeds before it fully accelerates then the value of y should be increased accordingly.

The tooth height calculated above can still be reduced if necessary without compromising much the tooth bite d. This is because the above estimates assumed that one tooth barely misses the opponent, until the next tooth is able to grab it at a distance d. But, if instead of barely missing the opponent, the previous tooth had barely hit it, it would have hit it with a distance much smaller than d. It is a matter of probability, the tooth bite can be any value between 0 and d, with equal chance (constant probability density). So, in 50% of the attacks at full speed the travel distance d will unluckily be between 0 and $d_{max}/2$, and in the other 50% it will luckily be between $d_{max}/2$ and d_{max}. An (unlucky) hit with d very close to zero probably won't grab the opponent, and it will significantly reduce the attacker speed v_{x1} until the next tooth is able to turn $2 \cdot \pi/n$ radians, decreasing the distance d of subsequent hits. If v_{x1} gets down to zero without grabbing the opponent, you'll probably end up grinding it. If this happens, the best option is to back up, and then charge again trying to reach $v_{x1,max}$ and hoping for a high d.

The chance of d being exactly d_{max} is zero, because it is always smaller than that, so if you want you can make the tooth height $y < d_{max}$. If you choose, for instance, $y = d_{max}/2 = v_{x1,max} \cdot \pi/(n \cdot \omega_{b,max})$, your robot won't notice any difference with this lower height in 50% of the hits, when $d < d_{max}/2$, while on the other 50% (where d would be higher than $d_{max}/2$) the opponent will touch the body of the drum/disk before being hit by a tooth, resulting in $d = y = d_{max}/2$. As long as this $d_{max}/2$ value is high enough to grab the opponent instead of grind it, it is a good choice.

For instance, the 2008 version of our featherweight Touro Feather had a drum with $n = 2$ teeth (pictured to the right) spinning up to $\omega_{b,max} = 13{,}500$ RPM (1413.7 rad/s). Since the robot top speed is $v_{x1,max} = 14.5$mph (equal to 23.3km/h or

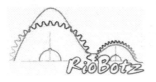 

6.48m/s), then $d_{max} = 6.48 \cdot 2 \cdot \pi/(2 \cdot 1413.7) \cong 0.014m = 14mm$. Since the overall height of the drum needed to be smaller than 4" by design, a tooth height $y = 14mm$ would result in a drum body with low diameter. We then chose $y = 10mm$ for the tooth to stick out of the drum body. This 10mm height is usually enough to grab an opponent. Also, in 10mm/14mm = 71% of the hits at full speed, the tooth height y will be higher than the tooth bite d. The opponent will only touch the drum body in the remaining 29% of the hits, when the next tooth will be able to hit the opponent with its full 10mm height (unless the opponent had bounced off immediately after hitting the drum body).

But beware with a frontal collision between two vertical spinning weapons, because the opponent may be able to grab your drum or disk body with its teeth before you can grab it. In this case, it is a game of chance. The robot with higher teeth will have a better chance of grabbing the opponent, as long as it spins fast enough. Since a vertical spinning bar does not have a round inner body, it basically behaves as if its "tooth height" y was equal to the bar radius. So, usually a powerful vertical bar will have an edge in weapon-to-weapon hits against drums or vertical disks.

6.3.2. Number of Teeth

An important conclusion from the previous analyses is that you must aim for a minimum number of teeth, n. The lower the n, the higher the value of d. Disks with $n = 3$ or more teeth are not a good option. The best choice is to go for $n = 2$, as with bars or two-toothed disks. Even better is to try to develop a one-toothed spinning weapon, such as the disk of the vertical spinner Professor Chaos, but this requires a careful calculation to avoid unbalancing by using, for instance, a counterweight diametrically opposite to that tooth.

Note that a one-toothed weapon does not have to be too much asymmetric, nor will it need heavy counterweights, if you do your math right. For instance, the one-toothed bar pictured to the right can be made out of a symmetrical bar, as long as

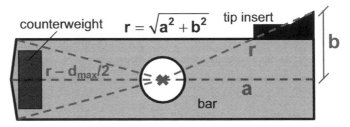

the short end is chamfered to reach a maximum radius $r - d_{max}/2$, where r is the effective radius of the long end including the insert, calculated from a and b as shown in the figure, and the maximum tooth bite d_{max} is calculated for $n = 1$ tooth. In this way, with the bar at full speed, even if the long end barely misses the opponent, the short end won't touch it because during a half turn it would approach at most half of d_{max}. After the full turn it would have approached up to d_{max}, hitting for sure with the long end. With such $n = 1$, it is possible to move twice as much into the opponent before hitting it, transferring more impact energy.

With this proposed one-toothed bar geometry, the counterweight wouldn't have to be much heavy, because its mass would only have to account for the mass of the tip insert plus the removed mass from the chamfers. This bar is also relatively easy to fabricate, with very little material loss. In fact, for wide bars with large inserts, which increase the value of b, it is even possible to design the bar such as $a = r - d_{max}/2$, making it almost symmetrical even after chamfering. In addition, if you perform some shape optimization removing some material from the long end, it is even possible to

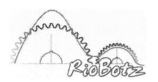 

remove the counterweight, but be careful not to compromise the bar strength at its most stressed region.

In our experience, to bind well to the opponent, the tooth bite should not be below 1/4", no matter if the robot is a hobbyweight or a super heavyweight. We've tested different tooth heights with our drumbot hobbyweight Touro Jr and featherweight Touro Feather, and values below 1/4" made the robot grind instead of grab the used deadweights. With this in mind, it is possible to generate a small table with estimated maximum weapon speeds to avoid the grinding problem. We only need to make sure that in at least 50% of the hits at full speed the tooth will be able to travel at least 1/4" (0.00635m), thus $\omega_{b,max}$ can be found from 1/4" = $d_{max}/2 = v_{x1,max} \cdot \pi/(n \cdot \omega_{b,max})$, see the table to the right. Of course these are just rough estimates, because tooth sharpness and armor hardness also play a role helping or avoiding dents that bind with the opponent.

number of teeth n	drivetrain speed $v_{x1,max}$	maximum $\omega_{b,max}$ to avoid grinding
3	5mph (8km/h)	3520RPM
	10mph (16km/h)	7040RPM
	15mph (24km/h)	10560RPM
2	5mph (8km/h)	5280RPM
	10mph (16km/h)	10560RPM
	15mph (24km/h)	15840RPM
1	5mph (8km/h)	10560RPM
	10mph (16km/h)	21120RPM
	15mph (24km/h)	31680RPM

6.4. Impact Theory

In the previous sections, we've used very simplified models to describe the impact of a spinner weapon on another robot. We'll extend these models here, to get a deeper understanding of the physics behind these impacts, and hopefully design a better spinner.

6.4.1. Impact Equations

We'll consider the problem of a bar spinner hitting a generic opponent (pictured in the next page), during the impact and right after it. The spinning bar has a length $2 \cdot r$, a mass m_b and moment of inertia I_b with respect to its center in the spin direction. It is initially spinning with an angular speed ω_b. The chassis of the spinner robot, without its bar, has a mass m_1 and a moment of inertia I_1 in the spin direction with respect to the chassis center of mass C_1. The opponent has mass m_2 and moment of inertia I_2 in the spin direction with respect to its center of mass C_2. We'll assume that the opponent does not have vertical spinning weapons, which could cause gyroscopic effects.

We'll assume that the impact impulse J that the bar inflicts on the opponent is in a direction perpendicular to their approach speed ($v_{x1} + v_{x2}$), and the distance between C_2 and the vector J is a_2. If the impact force was constant during the time interval Δt of the impact, then the impulse J could be simply calculated multiplying this force times Δt, otherwise we'd have to integrate the force over time to get J. The bar will also generate a reaction impulse J_1 over the weapon shaft of the spinner. This impulse J_1 is at a distance a_1 from the chassis center of mass C_1. The distance a_1 is usually greater than r for an offset spinner when hitting as shown in the picture, or very close to zero for traditional spinners that have their weapon shaft close to C_1.

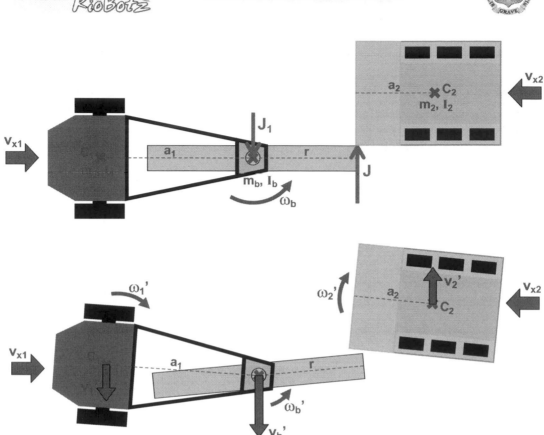

The picture also shows the moment right after the impact, where the opponent will gain a speed v_2' in the direction of J, when it will start spinning with an angular speed ω_2'. Note that both initial speeds v_{x1} and v_{x2} remain unchanged, because we assumed the impulse J perpendicular to them (we're implicitly assuming that there is no friction during the impact). The bar ends up with a slower angular speed ω_b' after the impact, while its center moves with a speed v_b' as a result of the reaction impulse J_1. The spinner robot chassis will gain a speed v_1' in the direction of J_1, and it will start spinning with an angular speed ω_1'. Note that ω_b' is measured with respect to the arena, and not with respect to the (now spinning) chassis.

If we assume that no debris is released from either robot during the impact, that the spinner bar has a perfect clutch system (which does not transmit any torque during the impact), and that the opponent does not have any spinning weapons that might cause some gyroscopic effect (studied later in this chapter), then basic physics equations of conservation of linear and angular momentum can show that

$$v_2' = \frac{J}{m_2}, \quad \omega_2' = \frac{J \cdot a_2}{I_2}, \quad v_1' = \frac{J_1}{m_1}, \quad \omega_1' = \frac{J_1 \cdot a_1}{I_1}, \quad v_b' = \frac{J - J_1}{m_b}, \text{ and } \omega_b' = \omega_b - \frac{J \cdot r}{I_b}$$

To find the values of J and J_1, we need to know the coefficient of restitution (COR) of the impact, defined by e, with $0 \le e \le 1$. The COR is the relative speed between the bar and the opponent after the impact divided by the relative speed before the impact. A purely elastic impact,

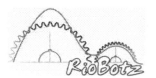 

where no energy is dissipated, would have e = 1. A purely inelastic impact, where a good part of the impact energy (but not all) is dissipated, would have e = 0. All other cases would have 0 < e < 1. Note that this dissipated energy is not only due to damage to the opponent in the form of plastic deformation or fracture, it also accounts for absorbed energy by the opponent's shock mounts, vibrations, sounds, and even damage to the weapon system of the spinner.

For very high speed impacts, as we see in combat, the COR e is usually close to zero, simply because no material can absorb in an elastic way all the huge impact energy. For instance, a bullet (taken from its cartridge) might have up to e = 0.9 when dropped from a 1" height against a metal surface. But the very same bullet, when fired to hit the same surface at very high speeds, will plastically deform and probably become embedded into the surface, resulting in e = 0. So, the value of e depends not only on the materials involved, but also on the impact speed.

In this problem, the approach speed before the impact is only due to the speed of the tip of the spinning bar, namely $v_{tip} = \omega_b \cdot r$. The relative speed between the bar tip and the opponent right after the impact (the departure speed) is due to several terms, such as the linear and angular speeds of the spinner and opponent robots, as well as the remaining angular speed of the bar, resulting in

$$e = \frac{v_{departure}}{v_{approach}} = \frac{v_2' + \omega_2' \cdot a_2 + v_1' + \omega_1' \cdot a_1 - \omega_b' \cdot r}{\omega_b r}$$

The speed v_b' of the center of the bar can be obtained from the linear and angular speeds of the spinner chassis, resulting in $v_b' = v_1' + \omega_1' \cdot a_1$. Solving all the previous equations, we finally obtain the values of the impulses J and J_1

$$J = M \cdot (1 + e) \cdot v_{tip} \qquad \text{and} \qquad J_1 = \frac{J \cdot I_1 m_1}{I_1 m_1 + m_b \cdot (I_1 + m_1 a_1^2)}$$

where M is the effective mass of both robots, obtained from the effective masses M_1 from the spinner and M_2 from the opponent, namely

$$\frac{1}{M} = \frac{1}{M_1} + \frac{1}{M_2}, \quad \text{where} \quad \frac{1}{M_1} = \frac{1}{m_b + m_1 \cdot \dfrac{I_1}{I_1 + m_1 a_1^2}} + \frac{r^2}{I_b} \quad \text{and} \quad \frac{1}{M_2} = \frac{1}{m_2} + \frac{a_2^2}{I_2}$$

With these values of J and J_1, we can now calculate all the speeds after the impact. For instance, let's check a few limit cases to better understand the equations.

6.4.2. Limit Cases

If the spinner, instead of hitting an opponent, hits a very light debris very close to its center of mass (therefore $a_2 \cong 0$), then the debris has an effective mass $M_2 = m_2$. Since m_2 is much smaller than m_1 and m_b, we find that M_2 is much smaller than M_1, which leads to $M \cong M_2$. So, the effective mass of the entire system is only $M = m_2$, resulting in a small impulse $J = m_2 \cdot (1+e) \cdot v_{tip}$ that will accelerate the debris to a speed $v_2' - J/m_2 = (1+e) \cdot v_{tip}$.

This means that if the debris is a little lump of clay (inelastic impact with $e \cong 0$), it will be thrown with basically the same speed v_{tip} of the bar tip. On the other hand, if it is a very tough rubber ball that won't burst due to the impact, its $e \cong 0.8$ will allow it to be thrown at 1.8 times the

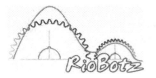 

speed of the bar tip. Also, since m_2 is very small, the equations predict that the speeds of the spinner robot will almost remain unchanged, which makes sense since very little energy was transferred to the debris.

The other limit case is the spinner hitting a very heavy arena wall. The wall is so much heavier than the robot that we can assume that $m_2 \to \infty$ and $I_2 \to \infty$, resulting in $M_2 \to \infty$ and therefore the system effective mass is $M \cong M_1$, the effective mass of the spinner robot. This will result in the maximum impact that the spinner can deliver, $J = M_1 \cdot (1+e) \cdot v_{tip}$. This value is twice the impact that would be delivered to an opponent with $M_2 = M_1$. This is why it is a much tougher test to hit an arena wall than an opponent with similar mass. And, of course, the equations will tell that the speeds of the arena after the impact will be approximately zero.

6.4.3. Impact Energy

Before the impact, we'll assume that the attacking robot (such as a spinner) will have an energy E_b stored in its weapon. For the spinner impact problem presented above, $E_b = I_b \cdot \omega_b^2 / 2$. The impact usually lasts only a few milliseconds, but it can be divided into two phases: the deformation and the restitution phases.

In the deformation phase, a portion E_d of the stored energy E_b is used to deform both robots (such as bending the spinner bar or compressing the opponent's armor), while the remaining portion E_v is used to change the speeds of both robots and weapon. It is not difficult to prove using the presented equations that $E_d = M \cdot v_{tip}^2 / 2$ for the spinner impact problem. Interestingly, this would be the deformation energy E_d that a mass M with a speed v_{tip} would generate if hitting a very heavy wall, as pictured to the right. So, the higher the effective mass M, the higher the E_d. We'll see later in this chapter how an attacking robot can manage to maximize M to increase the inflicted damage to the opponent.

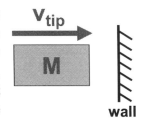

After the deformation reaches its peak, the restitution phase starts. A portion E_k of the deformation energy E_d was stored as elastic deformation, which is then retrieved during the restitution phase to change even more the speeds of both robots. The remaining portion E_c of E_d (where $E_d = E_k + E_c$) is the dissipated energy, transformed into permanent deformations, fractures, vibration, noise, as well as damped by the robot structure and shock mounts. We can show that, for an impact with COR equal to e, $E_k = E_d \cdot e^2$ and $E_c = E_d \cdot (1-e^2)$.

So, a perfectly elastic impact (e = 1) would have no dissipated energy ($E_c = 0$), and a perfectly inelastic impact (e = 0) would dissipate all its deformation energy ($E_c = E_d$). Note that inelastic impact does not mean that the entire energy of the system (which originally is E_b) is dissipated, it only means that the portion E_d is completely dissipated.

In summary, $E_b = E_v + E_d = E_v + E_k + E_c$, where the energy ($E_v + E_k$) will account for the changes in linear and angular speeds of the robots and weapon, and E_c will be dissipated.

6.4.4. Example: Last Rites vs. Sir Loin

Let's solve the impact equations for an example inspired on the heavyweight match between the offset spinner Last Rites (pictured to the right) and the eggbeater-drumbot Sir Loin, at RoboGames 2008. The

mass of each robot is assumed as $m = m_1 + m_b = m_2 = 220$lb. We'll estimate the weight of the bar with the steel inserts as $m_b = 44$lb $= m/5$, therefore $m_1 = 220 - 44 = 176$lb $= 4 \cdot m/5$. The bar is assumed to have a length $2 \cdot r = 40$", spinning at $\omega_b = 2000$RPM (209.4 rad/s) before the impact, with an offset length $a_1 = 30$". If the opponent robot is assumed to have a square shape with side length $2 \cdot a_2 = 30$", then the value of a_2 for the studied impact situation is about $a_2 = 15$".

The speed of the bar tip is $v_{tip} = \omega_b \cdot r = 209.4$rad/s $\cdot 20$" $= 106.4$m/s (equal to 383km/h or 238mph). The moment of inertia of the bar is approximated as $I_b = m_b \cdot r^2/3 = 5867$lb·in^2 (1.72kg·m^2). The moment of inertia I_2 of the second robot is roughly estimated assuming it is a 30" square with uniform density, as seen from above, resulting in $I_2 = 220$lb$\cdot 2 \cdot 15$"$^2/3 = 33{,}000$lb·in^2. The value of I_1 for the bar spinner chassis is roughly estimated as $I_1 = I_2 \cdot m_1/m = I_2 \cdot 4/5 = 26{,}400$lb·in^2. The effective mass of both robots is then

$$M_1 = \{\frac{1}{44 + 175 \cdot \dfrac{26400}{26400 + 175 \cdot 30^2}} + \frac{20^2}{5867}\}^{-1} = 12.1 \text{ lb} \quad \text{and} \quad M_2 = \{\frac{1}{220} + \frac{15^2}{33000}\}^{-1} = 88 \text{ lb}$$

and the effective mass of the system is $M = \{M_1^{-1} + M_2^{-1}\}^{-1} = 10.64$lb (4.825kg). So, even though the spinning bar had a kinetic energy of $E_b = I_b \cdot \omega_b^2 / 2 = 1.72$kg·m$^2 \cdot (209.4rad/s)^2 / 2 = 37{,}654$J, the deformation energy involved in the impact (distributed to both robots) is only $E_d = M \cdot v_{tip}^2 / 2 = 4.825kg\cdot(106.4m/s)^2 / 2 = 27{,}309$J (72.5% of E_b).

If the impulse vector was aligned with the center of mass C_2 of the other robot, then the distance a_2 would be equal to zero. In this case, the effective mass M_2 would be much higher, equal to the robot mass $m_2 = 220$lb. Despite this much higher M_2, the effective M would not increase too much, resulting in $M = \{M_1^{-1} + M_2^{-1}\}^{-1} = 11.47$lb (5.202kg) and $E_d = 29{,}445$J.

For offset spinners such as Last Rites or The Mortician, there's a way to increase even more the deformation energy caused by the impact. For frontal impacts, part of the energy of its bar is wasted making the offset spinner gain an angular speed ω_1', as shown in the next picture for The Mortician. To avoid that, the impulse J should be parallel to the line joining the chassis center of mass C_1 and the center of mass of the bar. In this case, the resulting impulse vector J_1 on the weapon shaft would be aligned with C_1, therefore its distance to C_1 would be $a_1 = 0$.

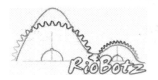

This configuration can be achieved with a special maneuver adopted by Ray Billings, the builder and driver of Last Rites. Last Rites starts facing away from the opponent, and then it turns 90 degrees to hit in the way shown below. In the best case scenario (for Last Rites), the impulse J_1 would be aligned with C_1, making $a_1 = 0$, and J would be aligned with C_2, resulting in

$$M_1 = \{\frac{1}{44+175} + \frac{20^2}{5867}\}^{-1} = 13.7 \text{ lb} \quad \text{and} \quad M_2 = \{\frac{1}{220} + \frac{0^2}{33000}\}^{-1} = 220 \text{ lb}$$

Then, $M = \{M_1^{-1} + M_2^{-1}\}^{-1} = 12.94\text{lb}$ (5.868kg). Note that the angular speed ω_1 of the spinner chassis before the impact would help to slightly increase v_{tip}, but the effect is usually negligible, because ω_b is much higher than ω_1.

So, it is not ω_1 that makes a difference here, but the alignment of the impact, which significantly increases M. If we neglect the effect of ω_1 on v_{tip}, then $E_d = 33,215$J, so the maneuver could increase the impact deformation energy in almost 22%.

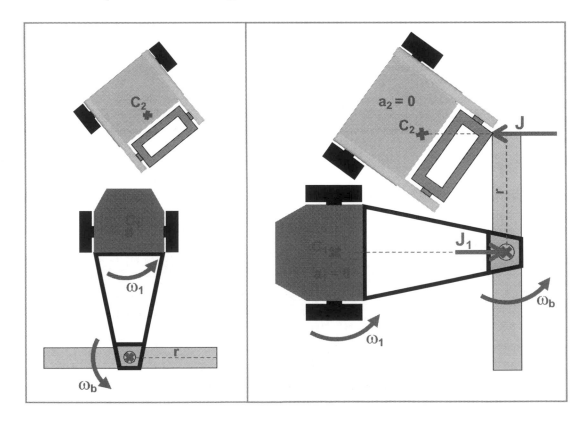

But remember that both robots have to absorb parts of the energy E_d, so you must make sure that the attacker can also withstand the higher impact from the maneuver. In the sequence shown to the right, Last Rites won the fight after performing the described maneuver against Sir Loin, dishing out enough energy to fracture the opponent's chassis, which was already damaged from previous hits, and breaking off the eggbeater.

To calculate the speeds after the impact, we would need to know the COR e. A high speed video of the fight, for instance, could provide an estimate of the angular speed ω_b' of the bar right after the impact. If we assume that ω_b' = 0, and that the impact happened in the ideal way drawn above (with $a_1 = a_2 = 0$), then

$$\omega_b' = 0 = \omega_b - \frac{J \cdot r}{I_b} = \omega_b - \frac{M \cdot (1+e) \cdot (\omega_b \cdot r) \cdot r}{m_b \cdot r^2 / 3} = \omega_b - \frac{12.94 \cdot (1+e) \cdot \omega_b}{44/3} \quad \Rightarrow \quad e \cong 0.13$$

This low value for e is reasonable, considering the high energy of the impact. The impulse values are then J = 705.5Ns and J_1 = 564.4Ns, accelerating both robots to v_1' = v_2' = 7.07m/s (25.5km/h or 15.8mph), while keeping ω_1 unchanged (ω_1' = ω_1). Note that this v_2' value assumed that Sir Loin didn't lose its eggbeater weapon. Since the eggbeater broke off, it is expected that it was flung with a speed above v_2', while the heavier remaining chassis acquired a speed lower than v_2'.

During the first phase of the impact, while the spinning bar was compressing the opponent's chassis, the original energy E_b = 37,654J was used in part to deform both robots, with E_d = 33,215J, and the remaining $E_v = E_b - E_d$ = 4,439J was used to change their speeds. Since the COR is small, a very small part of the deformation energy is elastically stored, namely $E_k = E_d \cdot e^2 = 33,215J \cdot 0.13^2$ = 561J. The remaining energy $E_c = E_d \cdot (1-e^2)$ = 32,654J is dissipated by both robots, either by their structural parts and shock mounts in the form of vibration and sound, or transformed into plastic (permanent) deformations or fractures.

During the second part of the impact, the small stored elastic energy E_k = 561J is restituted to the system, further accelerating both robots. Therefore, from the original E_b = 37,654J, 86.7% (E_c = 32,654J) is dissipated while 13.3% ($E_v + E_k$ = 5,000J) is used to change the speeds of both robots and spinning bar.

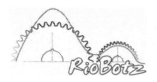

6.5. Effective Mass

Let's study a little bit more the concept of effective masses for an impact problem. As we've seen above, these masses are very important to find out how much energy is dissipated during the impact, potentially causing damages to the opponent. Therefore, an attacking robot should aim for high M_1 values. We'll assume below that the masses of the robot chassis and weapon are m_1 and m_b, respectively. The weapon mass ratio is then defined as $x \equiv m_b/(m_1+m_b)$, it measures how much of the robot mass is spent on its weapon. We'll also define a normalized effective mass, $M_1' \equiv M_1/(m_1+m_b)$, to make it easier to present the results.

6.5.1. Effective Mass of Horizontal Spinners

A spinning bar with length $2 \cdot r$ has a moment of inertia I_b of at least $m_b \cdot r^2/3$. If it has a large width, or if its shape is optimized, as discussed before, then the value of I_b can reach significantly higher values. It is easy to show from this result that a horizontal bar spinner without an offset shaft (therefore $a_1 = 0$) has normalized effective mass $M_1' \geq x/(3+x)$. An offset bar spinner would have a lower M_1' due to its a_1, which is usually greater than zero, unless it performs the maneuver described before to make $a_1 = 0$.

A spinning disk with radius r has I_b of at least $m_b \cdot r^2/2$. Shape optimization can increase this value, concentrating most of the mass m_b on the disk perimeter (as pictured to the right), trying to reach the (unreachable) value of $I_b = m_b \cdot r^2$. A horizontal disk spinner with $a_1 = 0$ has then $M_1' \geq x/(2+x)$, while an offset disk spinner would have a lower M_1'. These values are higher than the ones for bars.

A few robots have successfully implemented a horizontal spinning ring, supported by rollers, such as the hobbyweight Ingor (pictured to the right). The advantage of a ring-shaped weapon is its high I_b, which can reach up to $m_b \cdot r^2$ for a ring with external radius r. This results in M_1' up to $x/(1+x)$, better than bars and disks.

A horizontal shell spinner, on the other hand, can have different shell shapes. If the shell is shaped like a disk, with uniformly distributed mass, then we can estimate $I_b \cong m_b \cdot r^2/2$ and then $M_1' \cong x/(2+x)$. But if its shape is optimized to concentrate most of its mass at its perimeter, then it behaves as a ring, therefore I_b can reach values up to $m_b \cdot r^2$, with M_1' up to $x/(1+x)$.

A disk-shaped thwackbot, with radius r, is basically a full-body spinner. It spins is entire mass (m_1+m_b), therefore its I_b is at least $(m_1+m_b) \cdot r^2/2$, achieving higher values if its weight is more

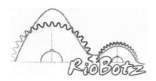

concentrated on its perimeter. Assuming that the spin axis coincides with the robot center of gravity (otherwise it would become unbalanced), then the offset $a_1 = 0$, resulting in M_1' equal to at least 1/3 (33.3%).

6.5.2. Effective Mass of Vertical Spinners and Drumbots

All previous analyses were based on horizontal spinners, which can suffer changes in their angular speed and in the speed in the direction of the impact. But the chassis of vertical spinning robots such as drumbots or vertical spinners does not accelerate during an impact, if the impulse J is vertical in the upwards direction, as pictured below. As long as the spinning drum, disk or bar has solid ground supports that will transmit the entire impulse J without allowing the robot to tilt forward after the attack, the chassis vertical speed v_1' and angular speed ω_1' should remain equal to zero. Obviously, the arena floor won't let the attacking robot move down. Therefore, drumbots and vertical spinners that spin their weapon upwards behave as if they had a chassis with infinite inertia, with $m_1 = \infty$ and $I_1 = \infty$. Note that I_1, I_b and I_2 here are the values in the weapon spin direction, which is horizontal, not vertical. The weapon can change its angular speed, ending up with a slower ω_b' after the impact, so its moment of inertia I_b in the horizontal spin direction is still considered. But the weapon speed v_b' in the vertical direction must remain equal to zero, behaving as if $m_b = \infty$. Note that, since there is no restriction for the opponent to be launched upwards, it will gain a speed v_2' in the direction of J, and it will start spinning with an angular speed ω_2', calculated from the previously presented equations.

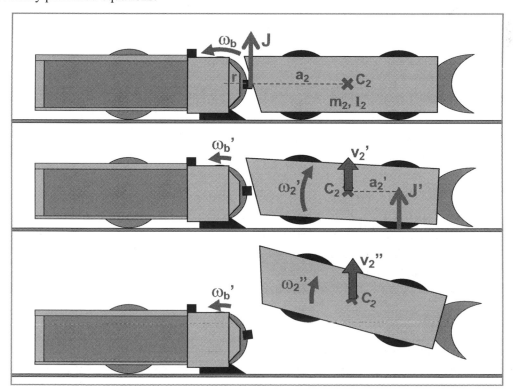

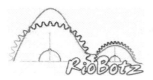 

For m_1, m_b and I_1 tending to infinity, we have $M_1 = I_b/r^2$. So, a vertical bar spinner will have I_b of at least $m_b \cdot r^2/3$, resulting in $M_1 > m_b/3$, therefore $M_1' = M_1/(m_1+m_b) \geq x/3$. Similarly, a vertical disk spinner will have I_b of at least $m_b \cdot r^2/2$, resulting in $M_1' \geq x/2$. And a drumbot, which has I_b between $m_b \cdot r^2/2$ (for a solid homogeneous drum) and $m_b \cdot r^2$ (for a hollow drum with thin walls), ends up with a normalized effective mass M_1' between $x/2$ and x.

But the opponent almost always suffers a second impulse J' immediately after the impact, at the wheel (or skid, or some other ground support) that is farther away from the location of the first impact. This happens because this wheel develops, right after the first impact, a downward speed $v_{wheel}' = \omega_2' \cdot a_2' - v_2'$, where a_2' is the horizontal distance between the wheel and C_2, see the picture above. This v_{wheel}' is almost always positive, pushing the wheel down against the ground, which reacts with the vertical impulse J'. This second impulse makes v_2' increase to a value v_2'', and ω_2' decrease to ω_2'', following the linear and angular momentum equations $J' = m_2 \cdot (v_2'' - v_2')$ and $J' \cdot a_2' = I_2 \cdot (\omega_2' - \omega_2'')$), where I_2 is the moment of inertia of the opponent at C_2 in the spin direction of the weapon. These new values can be calculated if the coefficient of restitution (COR) e' between the wheel and the ground is known. The resulting equations are quite lengthy, but not difficult to obtain, as seen next.

6.5.3. Example: Drumbot Impact

Let's solve an example for a special case where $I_2 = m_2 \cdot a_2^2/3$ and $a_2 = a_2' = 0.3m$, typical of a very low profile opponent with four wheels located near its perimeter, resulting in $M_2 = m_2/4$. Remember that I_2 here is the value in the horizontal spin direction, not in the vertical one as in the horizontal spinner impact calculations. If, for instance, the opponent robot is hit by a solid drum with mass $m_b = m_2/6$ and tip speed $v_{tip} = 32m/s$ (115km/h or 71.6mph), then the drumbot's effective mass for the first impact is $M_1 = I_b/r^2 \cong m_b/2 = m_2/12$, resulting in $M = m_2/16$. Powerful weapon impacts are nearly inelastic so, if we can assume that $e = 0.2$, then $J = M \cdot (1+e) \cdot v_{tip} = m_2 \cdot 1.2 \cdot v_{tip}/16$, leading to $v_2' = J/m_2 = 2.4m/s$ and $\omega_2' = J \cdot a_2/I_2 = 3 \cdot 2.4/a_2 = 24rad/s$ (229RPM). Note that these values are only valid if no debris is released from either robot, and if the opponent does not have any horizontal spinning weapons that might cause some gyroscopic effect (studied later in this chapter).

Right after the first impact, there would be a downward wheel speed $v_{wheel}' = \omega_2' \cdot a_2' - v_2' = 4.8m/s$. The wheel will depart from the ground after the second impact with a speed $v_{wheel}'' = v_{wheel}' \cdot e' = 4.8 \cdot e' = v_2'' - \omega_2'' \cdot a_2'$. The other equation relating the unknowns ω_2'' and v_2'' comes from the second impulse $J' = m_2 \cdot (v_2'' - v_2') = I_2 \cdot (\omega_2' - \omega_2'')/a_2'$, resulting in an increased $v_2'' = (3.6 + 1.2 \cdot e')m/s$ and a decreased $\omega_2'' = 12 \cdot (1-e')rad/s$.

So, a purely inelastic wheel impact ($e' = 0$) would result in $v_2'' = 3.6m/s$ and $\omega_2'' = 12rad/s$, an angular speed more than enough to flip the opponent. And a purely elastic wheel impact ($e' = 1$) would lead to $v_2'' = 4.8m/s$ and $\omega_2'' = 0rad/s$, launching the opponent without flipping it at all. So, interestingly, by using rubber wheels, the opponent makes it more difficult to get flipped over because of their high COR, up to $e' = 0.85$ for relatively slow impacts, with low v_{wheel}'. High speed impacts, however, tend to decrease the value of e'.

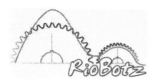

If e' = 0.75 could be used for the considered speed v_{wheel}' = 4.8m/s, then the resulting speeds after the second impact would be v_2" = 4.5m/s and ω_2" = 3rad/s (28.6RPM). These launching speeds would make the opponent reach a height of v_2"^{2}/(2g) = 1.03m, where g = 9.81m/s^2 is the acceleration of gravity. The flight time would be approximately Δt = 2·v_2"/g = 0.92s, but it can be a little less than that if the opponent lands vertically on its nose, instead of flat on the ground. During this flight time, the opponent flips 3rad/s·0.92s = 2.76rad = 158°, more than enough to get flipped over.

Note that it is not unusual to see an opponent being spun, for instance, by 540° before touching the ground after a powerful hit from a horizontal spinner. But it is very difficult for a vertical spinner or drumbot to cause a 540° flip while launching an opponent. The reason for that is the second impact, which only happens in a vertical launch. In the example above, it was able to decrease the opponent's angular speed from ω_2' = 229RPM to only ω_2" = 28.6RPM. If the second impact hadn't happened, the original ω_2' and v_2' after the first impact would have made the opponent flip 673°, instead of only 158°. But, because of the second impact, the drumbot from this example would have to spin its drum with v_{tip} = 59.2m/s to launch the opponent 3.53m into the air to make it flip 540°. Even if the drumbot had the required energy and the arena was tall enough, some part of the opponent would probably break off and prevent it from reaching such height. This is why we don't usually see drumbots or vertical spinners flipping opponents beyond 180°.

6.5.4. Effective Mass of Hammerbots

Technically, hammerbots are not spinners, however their impact behavior can be directly obtained from the previous equations if we consider them as bar spinners that only rotate 180 degrees before hitting the opponent.

If the hammer is a homogeneous bar with length r, without a hammer head, then its I_b is m_b·r^2/3. If it has a hammer head, then part of its mass m_b will be concentrated at its tip, increasing I_b.

Differently from vertical spinners that spin upwards, hammerbots hit downwards, so their chassis is subject to being launched, and m_1, m_b and I_1 cannot be assumed as infinite. So, their model is similar to the one for horizontal (and not vertical) bar spinners, including the 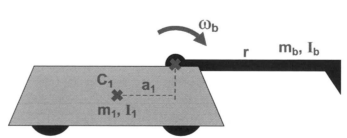 effects of m_1, m_b and I_1. If the hammer pivot coincides with the chassis center of mass, then the offset a_1 is zero, and the resulting normalized effective mass is the same as the one from the bar spinner, M_1' ≥ x/(3+x). Otherwise, if a_1 is different than zero, as in the picture to the right, then it must be modeled as an offset horizontal bar spinner.

Note that, similarly to a drumbot's opponent, the hammerbot will probably receive a second impact immediately after the first impact. This second impact, which happens on its back wheels, will be discussed later in this chapter.

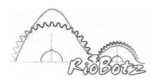

6.5.5. Full Body, Shell and Ring Drumbots

Curiously, shell drums are not very popular, even though they have one of the highest possible effective masses. Shell drums are the vertical equivalent of shell spinners, they spin their entire armor to try to launch the opponents. The heavyweight Barber-ous II (pictured to the right) is an example of a shell drum, it uses two drive motors for the wheels and a separate motor for the drum, which doubles as its armor. The shell drum is supported on a shaft that is aligned with the wheel axis. Alternatively, if the drum was a cylinder mounted on rollers, it should be called a ring drum, the vertical equivalent of a ring spinner.

A robot type that might have never been tried is a full-body drum, which is basically an overhead thwackbot without a long rod. This robot would use the power of its two wheel motors to spin its entire chassis (and not only its armor) as if it were a big drum, maximizing its moment of inertia. It would not need a separate motor for the weapon. The challenge would be to implement at each wheel an independent braking system that would allow the chassis to spin up without moving the robot around. After reaching full speed, the braking system would be released, and the robot would be driven by slightly accelerating or braking each wheel motor. With a clever gearing system to make each wheel turn in the opposite sense of each motor, it would be possible to implement a "kamikaze attack": with the chassis/drum spinning at full speed while facing the opponent, both motors could be shorted out or reversed, directing part of the drum energy straight to the wheels to move the robot towards the opponent with very high acceleration. The combination of high robot speed, high drum angular speed and moment of inertia, and high effective mass, would result in a devastating blow to the opponent. In theory, if its wheels were very light, the weapon mass ratio $x \equiv m_b/(m_1+m_b)$ would be very close to 1 (100%), allowing the normalized effective mass M_1' of full-body drums to get very close to the absolute maximum $M_1' = 1$ (100%).

Unfortunately, all these shell, ring or full-body drumbots have a major drawback: they are easily launched by their own drum energy if attacked from behind, where the spin direction would be downwards. In addition, similarly to shell, ring and full-body horizontal spinners, their internal components need to be very well shock-mounted to avoid self-destruction.

6.5.6. Effective Mass Summary

The table in the next page summarizes the values of the weapon moment of inertia I_b and normalized effective mass M_1', as a function of weapon mass ratio x, for the robot types discussed above. The results are also presented in a graph, which shows a mapping with the ranges of the values of x and M_1' for different robot types.

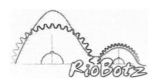 

As seen on the graph, for a given weapon mass ratio x, drumbots have the highest effective mass, while horizontal and offset bar spinners have the lowest. This does not necessarily mean that drumbots are better than horizontal bar spinners, because the impact impulse and energy also depends on the weapon tip speed v_{tip}.

Drums, for instance, cannot have a very large radius r without reducing their thickness and possibly compromising their strength. They also cannot compensate their lower radius r with an arbitrarily high angular speed ω_b to achieve high v_{tip}, because they would lower their tooth bite (as explained before), ending up grinding instead of grabbing the opponent. So, despite their excellent M_1, drumbots have limitations in their achievable v_{tip}.

Spinning bars, on the other hand, can achieve a very large radius r without compromising strength. Their angular speed ω_b does not need to be too high to generate a very fast v_{tip}. So,

robot type	weapon moment of inertia	normalized effective mass
offset bar spinner	$I_b \geq m_b \cdot r^2/3$	$x/4 \leq M_1' \leq x/(3+x)$
bar spinner	$I_b \geq m_b \cdot r^2/3$	$M_1' \geq x/(3+x)$
offset disk spinner	$I_b \geq m_b \cdot r^2/2$	$x/3 \leq M_1' \leq x/(2+x)$
disk spinner	$I_b \geq m_b \cdot r^2/2$	$M_1' \geq x/(2+x)$
shell spinner	$m_b \cdot r^2/2 \leq I_b < m_b \cdot r^2$	$x/(2+x) \leq M_1' < x/(1+x)$
ring spinner	$I_b < m_b \cdot r^2$	$M_1' < x/(1+x)$
vertical bar spinner	$I_b \geq m_b \cdot r^2/3$	$M_1' \geq x/3$
vertical disk spinner	$I_b \geq m_b \cdot r^2/2$	$M_1' \geq x/2$
drumbot	$m_b \cdot r^2/2 \leq I_b < m_b \cdot r^2$	$x/2 \leq M_1' < x$
hammerbot	$I_b \geq m_b \cdot r^2/3$	$M_1' \geq x/(3+x)$
thwackbot	$I_b \geq (m_1+m_b) \cdot r^2/2$	$M_1' \geq 1/3$

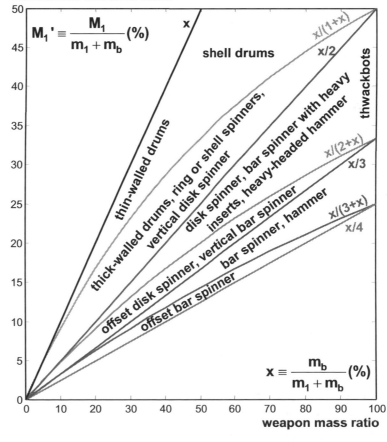

they make up for their poor M_1 with their amazing v_{tip} speeds. In summary, all robot types have their advantages and disadvantages, fortunately there is no single superior design, guaranteeing diversity.

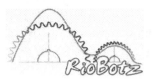

6.6. Effective Spring and Damper

We've learned that the effective mass of the impact determines how much of the weapon energy will be used to compress and deform both robots. But how is this energy distributed between the two robots? To find that out, we need to evaluate the stiffness and damping properties of the system. The stiffness is responsible for storing the elastic energy E_k, while the system damping is related to the dissipated impact energy E_c.

6.6.1. A Simple Spring-Damper Model

A very simple model would consider stiffness and damping coefficients for the structure of the attacking robot (k_1 and c_1), for its weapon (k_b and c_b), for the contact region between the weapon tip and the opponent's armor ($k_{contact}$ and $c_{contact}$), and for the opponent's structure (k_2 and c_2). The picture below schematically shows virtual effective spring-dampers with these stiffness and damping coefficients, for a horizontal spinner impact and for a vertical spinner impact.

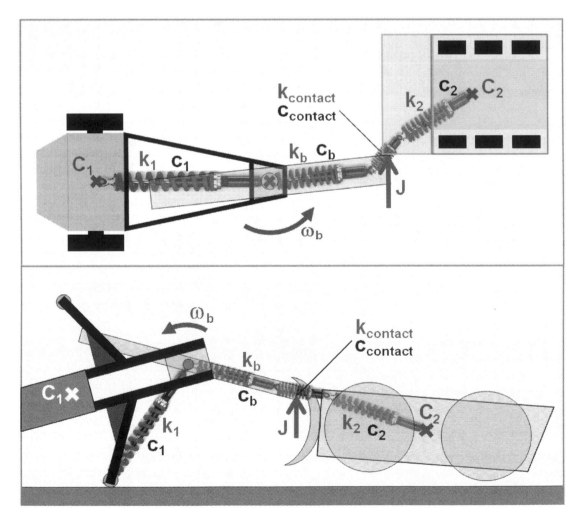

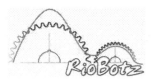 

The k_2 and c_2 coefficients come from the stiffness and damping between the impact point and the center of mass C_2 of the opponent robot. The $k_{contact}$ and $c_{contact}$ coefficients are due to the localized contact (compression) between the weapon tip and the opponent's armor, achieving high values for a blunt tip and low ones for a sharp tip. The k_b and c_b coefficients represent the weapon properties, which in the figure above would be the bending stiffness and damping of the bars.

For a horizontal spinner (or hammer), the coefficients k_1 and c_1 would represent the stiffness and damping properties of the attacking robot between its center of mass C_1 and the weapon shaft (or pivot). On the other hand, for a drumbot or vertical spinner (that does not tilt forward during the impact), these k_1 and c_1 properties would reflect the stiffness and damping of the path between the weapon shaft and the ground supports, not necessarily passing through C_1, because the reaction forces from the opponent are transmitted directly to the ground.

This relatively complex impact problem, which deals with 3 different bodies (the attacker chassis, its weapon, and the opponent), involving translations and rotations, can be analyzed as a very simple impact problem, pictured below. It is equivalent to an effective mass M_1, moving at a speed v_{tip}, hitting an effective mass M_2 through a compliant interface, made out of 4 spring-damper systems in series.

The system can be simplified even more, to a single effective mass M hitting a rigid and heavy wall through an effective spring-damper system with stiffness K and damping C, see the figure to the right. The values of K and C are obtained from the equations of springs in series and dampers in series,

$$\frac{1}{K} = \frac{1}{k_1} + \frac{1}{k_b} + \frac{1}{k_{contact}} + \frac{1}{k_2} \quad \text{and} \quad \frac{1}{C} = \frac{1}{c_1} + \frac{1}{c_b} + \frac{1}{c_{contact}} + \frac{1}{c_2}$$

6.6.2. Spring and Damper Energy

When the springs are fully compressed, we define their elastically stored energies as E_{k1}, E_{kb}, $E_{kcontact}$ and E_{k2}, where $E_{k1} + E_{kb} + E_{kcontact} + E_{k2} = E_k$. Similarly, the energies dissipated by each of the dampers are called E_{c1}, E_{cb}, $E_{ccontact}$ and E_{c2}, where $E_{c1} + E_{cb} + E_{ccontact} + E_{c2} = E_c$. The individual energies are obtained from

$$\begin{cases} E_{k1} = E_k \cdot \dfrac{K}{k_1}, & E_{kb} = E_k \cdot \dfrac{K}{k_b}, & E_{kcontact} = E_k \cdot \dfrac{K}{k_{contact}}, & E_{k2} = E_k \cdot \dfrac{K}{k_2} \\[3mm] E_{c1} = E_c \cdot \dfrac{C}{c_1}, & E_{cb} = E_c \cdot \dfrac{C}{c_b}, & E_{ccontact} = E_c \cdot \dfrac{C}{c_{contact}}, & E_{c2} = E_c \cdot \dfrac{C}{c_2} \end{cases}$$

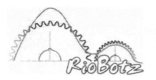

The above equations are such that K is always smaller than the smallest stiffness coefficient, and C is smaller than the smallest damping coefficient. Also, for instance, if k_2 and c_2 are much smaller than the other stiffness and damping coefficients, then $K \cong k_2$ and $C \cong c_2$, and the energies E_k and E_c are almost entirely stored into or damped by the opponent robot, because in this case we would have $E_{k2} \cong E_k$ and $E_{c2} \cong E_c$. It may seem strange, but the above equations show that the component in a series connection that has the lowest damping or stiffness coefficients is the one that will damp or elastically store the most amount of energy. This is analogous to what we see in electric circuits with resistors or capacitors connected in parallel: the resistor with lowest resistance will dissipate more energy than the others, while the capacitor with lowest elastance (the inverse of capacitance) will store more energy.

6.6.3. Offensive Strategies

From the equations above, we conclude that the strategy for the attacking robot is not only to maximize the impact energies E_c and E_k, but also to concentrate them on the opponent robot's structure (maximizing E_{c2} and E_{k2}) or contact surface (maximizing $E_{ccontact}$ and $E_{kcontact}$). To achieve that, the attacker must have c_1, k_1, c_b and k_b much higher than the opponent's c_2 and k_2 (to maximize E_{c2} and E_{k2}), or try to make the contact values $c_{contact}$ and $k_{contact}$ as low as possible (to maximize $E_{ccontact}$ and $E_{kcontact}$).

This is why the attacking robot must have a very stiff and robust weapon, with very high c_b and k_b. A very flexible weapon would end up vibrating a lot after the impact and dissipating most of its energy, instead of transferring it to the opponent. A horizontal spinner, in special an offset spinner, also needs to have a very stiff and robust connection between its weapon shaft and its center of mass, to maximize c_1 and k_1, as we can see for instance in the rigid trussed weapon support from Last Rites. And a vertical spinner or drumbot needs a very stiff and robust structure linking its weapon shaft with the ground supports, to maximize its c_1 and k_1. Here's the reason why robots with active weapons should not have their structure made out of plastic (such as UHMW): their probably low c_1 and k_1 would make the plastic structure deform and absorb most of the impact energy, instead of delivering it to the opponent.

The contact behavior between the weapon tip and the opponent's armor can be understood using a simplified model (adapted from the Hertz contact theory between two solids). The $c_{contact}$ and $k_{contact}$ of the contact between a sharp blade, with a small tip radius R, and an armor plate (pictured to the right) is proportional to the square root of R. This is why it is good to have a razor-sharp weapon, its tip radius R can reach values below one thousandth of a millimeter, lowering $c_{contact}$ and $k_{contact}$ to concentrate most of the impact energy on $E_{ccontact}$ and $E_{kcontact}$, penetrating the opponent's armor. But, to keep this sharpness, the weapon tip must be made out of a very hard material. Also, the lower the tip radius R, the more often you'll need to resharpen the weapon edge due to chipping and blunting.

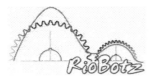 

6.6.4. Defensive Strategies

The values of the contact coefficients $k_{contact}$ and $c_{contact}$ are proportional to, respectively, the stiffness (measured from the Young modulus E, see chapter 3) and the hardness of the armor material. If the weapon tip has very high stiffness and hardness, then $k_{contact}$ and $c_{contact}$ will not depend much on the material properties of the weapon, they will mostly depend on the material properties of the armor.

There is a very old hardness test, using a testing instrument called a Scleroscope, where a diamond tipped hammer (which would be analogous to the weapon tip) is vertically dropped from a 10" height onto the surface of the material under test (analogous to the armor plate). A low hardness material results in a low $c_{contact}$, which causes large indentations that absorb most of the impact energy due to the high $E_{ccontact}$, lowering the height of the rebound of the hammer. So, the higher the material hardness, the higher will be the rebound height, resulting in less damping.

We can conclude then that the attacked robot has three different strategies to defend itself from a sharp blade:

1) absorb the energy at the contact – this strategy involves using an ablative armor, made out of materials with very low hardness (such as aluminum or magnesium alloys, see chapter 3), which will make sure that $c_{contact}$ will be much lower than the c_2 and k_2 coefficients of the attacked robot structure, directing most of the impact energy to $E_{ccontact}$ to be dissipated in the ablation process. With this strategy, the attacked robot structure will only need to deal with relatively small residual energies E_{c2} and E_{k2}. But make sure that your ablative armor is thick enough not to get pierced.

2) absorb the energy at the shock mounts – this strategy involves using a shock-mounted armor. The armor is usually of the traditional type, very hard, but an ablative armor would also work. The shock mounts make the c_2 and k_2 coefficients become very low. The high resulting energies E_{c2} and E_{k2} do not damage the attacked robot structure because they are almost entirely dissipated or stored by the shock mounts. The challenge here is to make sure that the shock mounts won't rupture while absorbing such high amounts of energy.

3) break the weapon – this is the strategy of very aggressive rammers. It involves having a very hard and stiff traditional armor mounted to a very stiff chassis, without any shock mount in between. Shock mounts should only be used for critical internal components. The goal here is to reach high $c_{contact}$ and $k_{contact}$ (due to the traditional armor) as well as high c_2 and k_2 (from the stiff chassis without shock mounts). If these coefficients end up much higher than c_b, k_b, c_1 and k_1, then most of the impact energy will be diverted back to the attacker robot, either breaking its weapon (if E_{cb} and E_{kb} become high) or its structure (if E_{c1} and E_{k1} become high).

In summary, stiffness and damping are key properties for both attacking and attacked robots, so always design your robot keeping this in mind.

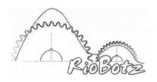

6.6.5. Case Study: Vertical Spinner Stiffness and Damping

The following robots exemplify the application of several of the presented concepts. The middleweight vertical spinner Docinho (pictured to the right) is a high power vertical disk spinner driven by 2 pairs of wheels in an ingenious invertible design. Despite its high power, it seemed to have some trouble launching the opponents. Its disk usually grinds the opponent instead of launching it, mainly because it has 3 teeth spinning at high speeds (n = 3, instead of better values such as 1 or 2), and also because the teeth are not hard enough to keep their sharpness. Another reason for not living up to its potential is the use of compliant wheels supporting the robot under its disk. These compliant wheels act as dampers. They significantly lower the k_1 and c_1 coefficients, ultimately making $K \cong k_1$ and $C \cong c_1$. This makes most of the deformation energy E_d go to E_{k1} and E_{c1}, permanently deforming the wheels, instead of transferring the energy to the opponent. Also, the relatively thin disk (due to its large diameter) and the Lexan armor makes it vulnerable to powerful horizontal bar spinners.

The beetleweight Altitude (pictured on the right) has addressed most of these issues. Its single-piece disk has only 2 teeth, with high hardness to prevent them from getting blunt. It is supported under the disk by skids made out of solid steel bars, which are much stiffer than rubber wheels, resulting in high k_1 and c_1 to deliver much more impact energy and peak forces. It is very important to keep in mind the force path in the robot, as seen in the picture as a dashed line. This path must only have components with high strength and stiffness to be able to guarantee high c_1, k_1, c_b and k_b and thus deliver high energy blows. Also, you must avoid any sharp notches (which are stress raisers), especially along this critical force path. The middleweight Terminal Velocity (pictured below) is another example of

a vertical spinner with a stiff force path. It also has rigid skids to support its vertical spinning bar, using roller bearings to minimize sliding friction, as pictured to the right, without compromising stiffness.

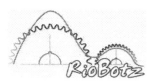

Finally, avoid installing any sensitive components, such as receivers or other electronic parts, close to the force path between the weapon tip and the ground (for drumbots or vertical spinners) or between the weapon tip and the chassis center of mass (for horizontal spinners). The impact vibrations along this path are very high, causing the sensitive components to malfunction if not shock mounted. In most drumbot and vertical spinner designs, the weapon motor ends up very close to this force path. To avoid a broken weapon motor due to impact vibrations, you can either shock mount it to the robot structure or, if possible, move it a little further towards the back of the robot.

6.6.6. Equivalent Electric Circuit

For those of you more electrically inclined, the entire impact problem has exactly the same dynamic equations as the resonator circuit below, if we consider all masses and inertial terms as if they were inductors, the elastic terms as capacitors, the damping terms as resistances, and the electric current as speeds. All blue components would come from the attacking robot, the green ones from the opponent, and the grey ones from the mechanical contact between them. The inductances would have the same numerical values (but with different physical units, of course) as the masses m_b, m_1, m_2, and inertial terms I_b/r^2, I_1/a_1^2 and I_2/a_2^2. The stiffnesses k_1, k_b, $k_{contact}$ and k_2 would be numerically equal to the elastance (the inverse of capacitance) of each capacitor, while the damping coefficients c_1, c_b, $c_{contact}$ and c_2 would be the resistances.

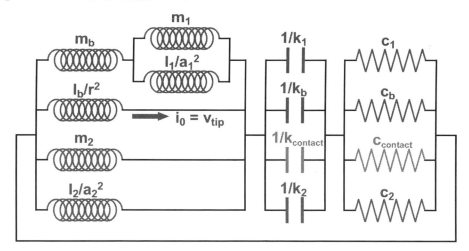

It is easy to see that the equivalent inductance of all blue inductors (attacker inertia) has exactly the same equation as the effective mass M_1, while the equivalent inductance of all green inductors (from the opponent) is M_2. The equivalent inductance of the entire circuit, as expected, would be M. Similarly, the equivalent elastance would have the same equation as the effective stiffness K, while the equivalent resistance would be numerically equal to the effective damping C. So, the circuit behavior would be similar to the one from the equivalent circuit pictured to the right.

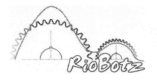 

Before digital calculators and digital computers were available, the first circuit shown above would be useful as an analog computer to calculate all the speeds and energy values of the impact problem.

After building the circuit, the I_b/r^2 inductor, which represents the spinning weapon, would have to be initially energized with an electric current i_0 numerically equal to the speed v_{tip} of the weapon tip, while all other components would be shorted out, with the capacitors discharged. The stored energy in this inductor would be numerically equal to the initial kinetic energy E_b of the bar. If the international system of units (SI) was used, the initial energy of both mechanical and electrical systems would be exactly the same, in Joules.

The circuit would then be connected as it was shown in the figure. During the first part of this simulated impact, the capacitors would be charged, equivalent to the compression between both robots, accumulating the same energy E_k that the equivalent mechanical system would elastically store. During the second part, the capacitors would be discharged, giving back the energy E_k to the system. As soon as the capacitors are first discharged, the electric currents in all inductors need to be immediately measured, and the simulation ends. The resonator circuit will continue to cyclically charge and discharge the capacitors, but only the first cycle is relevant to our simulation, because the impact ends after that. The subsequent cycles would only make sense if both robots would get stuck together after the impact, which is unlikely.

The initially energized inductor would contribute with E_v to the final energy of all inductors, making the total energy of all inductors at the end of the simulation equal to $(E_v + E_k)$. And the dissipated energy in the resistors during the entire cycle, which can also be measured, would be E_c. Needless to say, these energy values E_v, E_k and E_c would be the same as the ones from the mechanical system.

And how about the speeds after the impact? Well, if you measure the electric currents in all inductors at the end of the first resonator cycle, immediately after the capacitors are first discharged, then you'll see that the currents in the inductors m_1 and m_2 would be numerically equal to the attacker and opponent chassis speeds v_1' and v_2', respectively. The speed of the weapon tip after the impact would be the current going through the equivalent inductor M_1, while the current through m_b would give the speed v_b' of the center of the weapon.

6.7. Hammerbot Design

Hammers usually need to be pneumatically powered to be effective. This is because they have to reach their maximum speed in only 180 degrees. Since most pneumatic actuators are linear cylinders, you'll need some type of transmission to convert linear into rotary motion. This can be done in several ways. One of the lightest solutions, adopted by the super heavyweight The Judge, is implemented using a pair of opposing heavy-duty chains, colored in red and blue in the figure in the next page. When the right port of the cylinder in the figure is pressurized, it makes the piston move to the left and pull the red chain, which generates a rotary motion in the hammer.

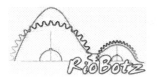

The hammer could have a spring mechanism to move back to its starting position after an attack. But the best solution is to have a double acting cylinder to retract the hammer at high speeds, with the aid of the blue chain shown in the picture. This allows the hammer to get ready in less time for the next attack. Also, and most importantly, it guarantees enough torque to the hammer in both directions to work as a self-righting mechanism in case the robot gets flipped upside down.

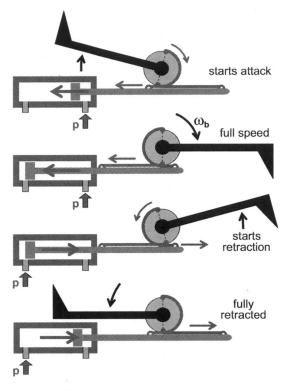

6.7.1. Hammer Energy

No matter which mechanism you use to generate a rotary motion, it is not difficult to estimate the energy and the top angular speed of the hammer in a pneumatic robot. If we assume no energy loss due to friction or pneumatic leaks, then the energy E_b delivered by the cylinder is approximately equal to its operating pressure p times its internal volume V, so $E_b \cong p \cdot V$. If the hammer has much more inertia than the cylinder piston and the transmission mechanisms, then we can say that this energy is entirely converted into kinetic energy of the hammer, $E_b \cong I_b \cdot \omega_b^2/2$, where I_b is the hammer moment of inertia and ω_b is its top angular speed right before hitting the opponent.

For instance, assume your hammerbot uses a pneumatic cylinder pressurized at 1000psi, with a 4" diameter bore and an 8" stroke. The hammer should be able to hit the arena floor before using its entire 8" stroke. So, when hitting a tall opponent, the piston will surely have traveled significantly less than 8". If, for instance, the actual useful stroke during an attack is 6.5", then the useful cylinder internal volume is $V = 6.5" \cdot (\pi \cdot 4"^2/4) \cong 81.68in^3$. Since p = 1000psi = 1000lbf/in^2 (pounds-force per square inch), we get $E_b = p \cdot V = 81,680$lbf·in $\cong 9,229$J. The piston force would be $F = p \cdot A$ = 1000psi·$(\pi \cdot 4"^2/4)$ = 12,566lbf (5,700kgf or 55,896N), where A is the internal cross-section area of the cylinder.

Note that this energy would be obtained while pushing the piston. When it is pulled, the energy is slightly lower, because you have to subtract the piston rod volume when calculating V. If, for instance, the piston rod has a 1.25" diameter, then $V = 6.5" \cdot [\pi \cdot (4"^2 - 1.25"^2)/4] \cong 73.70in^3$, and therefore $E_b = 73,700$lbf·in $\cong 8,327$J. The force would also be slightly smaller, due to the smaller area $A = \pi \cdot (4"^2 - 1.25"^2)/4 \cong 11.34in^2$, resulting in $F = p \cdot A \cong 11,340$lbf (5,144kgf or 50,443N).

So, it is slightly better to design the transmission system such that the hammer hits when the piston is extended. But, depending on the transmission design, this might place the cylinder in the front of the robot, more exposed to attacks, and limiting the reach of the hammer head.

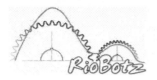

6.7.2. Hammer Impact

If in our example above the hammer handle is 36" (0.91m) long with a mass of 15lb (6.8kg), with a 10lb (4.5kg) hammer head, then its moment of inertia is $I_b \cong 6.8 \cdot 0.91^2/3 + 4.5 \cdot 0.91^2 \cong 5.6$kg·m². If we place the cylinder in the back of the robot, using the mechanism from the previous figure, then the energy $E_b = 8{,}327J \cong I_b \cdot \omega_b^2/2$ from the pulling motion would accelerate the hammer up to $\omega_b = 54.5$rad/s = 521RPM, resulting in a hammer head speed of 54.5rad/s·0.91m = 49.6m/s (179km/h or 111mph).

Note that the robot will tend to move backwards during the acceleration of the hammer, therefore it needs to compensate for that by braking its wheels. The chassis will also tend to tilt backwards, from the reaction force of the hammer accelerating forward. Powerful hammerbots may even see their front wheels lift off the ground because of that, as seen in the middle picture below, which shows The Judge tilting backwards right before it even touches the opponent. Excessive tilting may leave it vulnerable to wedges or launchers that might sneak in underneath (as shown in the picture below to the right, right before The Judge was launched by Ziggy). To avoid that, it is a good idea to move forward the center of mass of the hammerbot.

The picture above to the right shows that the tilting angle of the chassis is increased even more after the hit, due to the reaction impulse J_1 from the impact, pictured to the right. The speeds after the impact and all the involved impact energies can be calculated from the very same equations used for spinners. Since the attacked robot is hammered against the arena floor, it usually does not move its center of mass, it only deforms due to the attack, so the impact problem is similar to an offset bar spinner hitting a flexible but very heavy wall, as shown in the picture to the right.

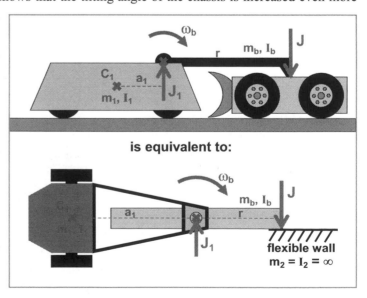

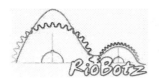

The attacked robot would then have an infinite effective mass M_2, while the hammerbot's M_1 would have the same equation from an offset spinner, where m_1 and I_1 are the chassis mass and moment of inertia in the direction that the hammer rotates, m_b and I_b are the corresponding values for the hammer, and a_1 is the horizontal offset between the chassis center of mass C_1 and the hammer pivot. The speeds of the attacked robot after the hammering are then $v_2' = \omega_2' = 0$, while the hammerbot chassis gains a vertical speed v_1' in the direction of J_1, and it may spin backwards, if $a_1 > 0$, with an angular speed ω_1' calculated from the spinner equations.

Note that, if the back wheels of the hammerbot are still in contact with the ground immediately after the impact against the opponent, then a second impact will probably occur. With the back wheels gaining a downward speed after the impulse J_1, they will press against the arena floor and receive a vertical reaction impulse J'. This back wheel impulse J' is good for the hammerbot, because it avoids its chassis from tilting too much backwards. The final linear and angular speeds v_1'' and ω_2'' of the hammerbot chassis can be calculated using the very same equations from the second impact that happens when a robot is hit by a drumbot or vertical spinner, as studied before.

6.8. Overhead Thwackbot Design

Overhead thwackbots need to be well balanced with respect to the wheel axis, otherwise they won't be able to have enough torque to lift the weapon to strike. This balancing can be done using counterweights opposite to the weapon, to place the center of mass C_1 of the entire robot on the wheel axis.

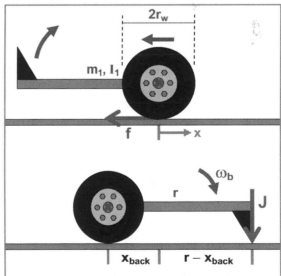

Overhead thwackbots have a few similarities with hammerbots. The main difference is that they use the drivetrain power to accelerate the weapon. This limits the weapon top speed, because a high gearmotor torque would end up making the wheels slip. Both wheels usually bear altogether a ground normal force equal to the robot weight $m_1 \cdot g$, where m_1 is its mass and g is the acceleration of gravity. If μ is the coefficient of friction between the tires and the ground, then the maximum traction force f that the tires can generate together is $f = f_{max} = \mu_t \cdot m_1 \cdot g$, see the picture to the right.

If the wheels have a radius r_w, then the maximum torque τ that both wheels can generate altogether to accelerate the weapon is $\tau_{max} = f_{max} \cdot r_w = \mu_t \cdot m_1 \cdot g \cdot r_w$. The wheel gearmotors need to be able to provide altogether this torque τ_{max}. Less than that would result in a slower weapon impact speed, while more torque would make the wheels slip. If using DC motors, it is a good idea to have a current controller (instead of a voltage controller from most speed control electronics), to

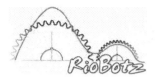 

guarantee a constant current to deliver a constant τ_{max} after the gear reduction, independently of the speed of the motors.

Let's assume now that the robot center of mass C_1 is located along the wheel axis. If the mass and moment of inertia of the entire robot (chassis plus weapon) in the wheel axis direction with respect to C_1 are m_1 and I_1, then the torque τ_{max} would generate an angular acceleration $\alpha = \tau_{max}/I_1$. Note that most of the value of I_1 will come from the weapon, because the rest of the chassis is very close to the wheel axis and do not contribute much to the moment of inertia. For the robot to strike, it needs to turn about 180 degrees (π radians), therefore $\pi = \alpha \cdot t^2/2$, where t is the short time the robot takes to strike.

If the robot is initially at rest, then its chassis will start moving backwards, due to the linear acceleration $f/m_1 = \mu_t \cdot g$ caused by the ground force f. During the strike period t, the distance x_{back} the chassis moves backwards is then $x_{back} = \mu_t \cdot g \cdot t^2/2$. Eliminating t and α from these equations we get $x_{back} = \pi \cdot I_1/(m_1 \cdot r_w)$.

So, if its weapon has a radius r, then it will hit a spot at a distance $(r - x_{back})$ from its position when it started the attack, see the pictures above. The driver has to get a feeling of this distance after some practice, otherwise the weapon will hit short of the opponent's position. Note that x_{back} does not depend on the coefficient of friction μ_t, it is a constant for each robot.

But x_{back} can be compensated for, if the robot starts striking when it is moving with some initial forward speed v_{x1}, not at rest as before. Then $x_{back} = \mu_t \cdot g \cdot t^2/2 - v_{x1} \cdot t$, which would be equal to zero right at the end of the attack if

$$v_{x1,ideal} = \frac{\mu_t \cdot g}{2} \cdot t = \frac{\mu_t \cdot g}{2} \cdot \sqrt{\frac{2\pi}{\alpha}} = \frac{\mu_t \cdot g}{2} \cdot \sqrt{\frac{2\pi \cdot I_1}{\mu_t \cdot m_1 \cdot g \cdot r_w}} = \sqrt{\frac{\mu_t \cdot g}{2} \cdot \frac{\pi \cdot I_1}{m_1 \cdot r_w}}$$

So, if the robot is initially moving towards the opponent at this $v_{x1,ideal}$ speed, it will hit exactly at the distance r from its initial position. If moving slower than that, it will hit short of that distance. If moving faster, it will hit beyond that. This is why overhead thwackbots are so difficult to drive, it is up to the driver to get a feeling of the attack distance as a function of the attack speed. Note that, because this function depends on μ_t, a dirtier arena (leading to a lower μ_t) will result in a lower $v_{x1,ideal}$, which needs to be adaptively controlled by the driver. Poor driver.

Finally, the weapon maximum angular speed ω_b and energy E_b can be calculated from α and t, resulting in

$$\omega_b = \alpha \cdot t = \sqrt{2\pi\alpha} = \sqrt{\frac{2\pi \cdot \mu_t \cdot m_1 \cdot g \cdot r_w}{I_1}} \quad \text{and} \quad E_b = \frac{1}{2} I_1 \omega_b^2 = \pi \cdot \mu_t \cdot m_1 \cdot g \cdot r_w$$

So, since m_1 is basically the mass of the weight class and g is a constant, to maximize the attack energy E_b you must have rubber wheels with large radius r_w, and also with large width to maximize its coefficient of friction μ_t. And, of course, the drivetrain gearmotors should be able to provide altogether at least $\tau_{max} = \mu_t \cdot m_1 \cdot g \cdot r_w$ to reach these maximum ω_b and E_b values.

But be careful, because horizontal spinners love large wheels. It's their favorite breakfast.

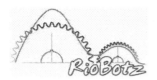 

6.9. Thwackbot Design

A thwackbot can be thought of as a full body spinner. It uses the power of its two (or more) wheels to spin up its entire body. As seen before, its normalized effective mass M_1' can reach very high values, from 1/3 for disk shaped designs tending towards 1/2 for ring shaped designs, being able to store a lot of kinetic energy.

6.9.1. Thwackbot Equations

Let's consider a thwackbot with mass m_1, moment of inertia I_1 in its spin direction with respect to its center of mass, and two wheels with radius r separated by a distance 2·d, as shown in the picture to the right. We'll assume that the robot center of mass is located in the middle of the line between the wheel centers, otherwise the robot would be unbalanced, compromising its

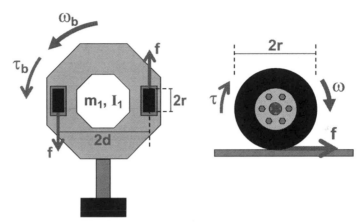

maximum angular speed and drivability. So, if it has some asymmetrical feature such as a single hammer, as pictured above, then the wheel axis location must be carefully calculated to guarantee that it ends up balanced.

Each wheel has a variable angular speed ω, with a maximum value ω_{max}, and it receives a torque τ from the gearmotor output. This wheel torque will cause a traction force f, which is equal to τ/r if the wheel does not slip. Assuming that each wheel bears half of the robot weight, the maximum possible value for f is $\mu_t \cdot g \cdot m/2$, where μ_t is the coefficient of friction between the wheel and the ground, and g the acceleration of gravity. If τ/r is greater than $\mu_t \cdot g \cdot m/2$, then the wheel will slip.

Note that it is not a good idea to increase the width of the tires to increase μ_t, because tires with large width tend to waste a lot of energy while making sharp turns, due to the slipping that always occurs along their width. This slip happens because the inner surface of a wheel with width w is closer to the center of the robot, moving along a circle with radius $(d - w/2)$, while the outer surface moves along a circle with radius $(d + w/2)$. The mid-section of the wheel, which moves along a circle with radius d, makes the inner surface waste some energy by slipping forward to catch up with the mid-section, while the outer surface wastes energy slipping backwards. So, while overhead thwackbots should have wide tires, thwackbots should use instead relatively thin tires to avoid this energy loss.

The wheel torque τ used in the above equations depends on the wheel speed ω, as seen in chapter 5 for DC motors. We can then define an effective wheel torque function $\tau(\omega)$ that includes this dependency, and which is also limited by the value $\mu_t \cdot g \cdot m \cdot r/2$ which would make the wheel slip. This effective torque $\tau(\omega)$ depends not only on ω, but also on the motor and battery properties

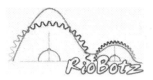 

(such as K_t, K_v, I_{stall}, I_{no_load} and R_{system} for DC motors, see chapter 5), gear ratio n:1, and the value of $\mu_t \cdot g \cdot m \cdot r/2$. For instance, if min(x,y) is the function that returns the minimum value between x and y, then DC motors would result in an effective wheel torque function

$$\tau(\omega) = \min\left\{ n \cdot K_t \cdot (I_{stall} - I_{no\_load} - \frac{\omega \cdot n}{K_v \cdot R_{system}}), \quad \frac{\mu_t \cdot m \cdot g \cdot r}{2} \right\}$$

Let's first calculate the time Δt_{drive} the robot takes to accelerate to, for instance, 95% of its top speed, while driving on a straight line. In this case, the traction forces f of both wheels would be directed towards the same direction, contrary to the figure above. The forward speed of the robot would be $v = \omega \cdot r$, with a maximum value $v_{max} = \omega_{max} \cdot r$. So, the forward acceleration of the robot dv/dt would be equal to the angular acceleration dω/dt of each wheel multiplied by the wheel radius r, thus $dv/dt = r \cdot d\omega/dt$. Both forward forces f cause the robot to accelerate, following the equation $2 \cdot f = m \cdot dv/dt$. We can then use $f = \tau(\omega)/r$ and $dv/dt = r \cdot d\omega/dt$ in this equation, to obtain

$$2 \cdot f = 2 \cdot \frac{\tau(\omega)}{r} = m \cdot r \cdot \frac{d\omega}{dt} \quad \Rightarrow \quad \Delta t_{drive} = \int dt = \frac{m \cdot r^2}{2} \cdot \int_0^{0.95\omega_{max}} \frac{d\omega}{\tau(\omega)}$$

The integral shown above is not difficult to calculate, it was obtained for DC motors in chapter 5.

Let's now calculate the spin up time Δt_{weapon} until the robot reaches 95% of its top weapon speed. The chassis angular speed ω_b is equal to the wheel linear speed $\omega \cdot r$ divided by d, so we have $\omega_b = \omega \cdot r/d$. Thus, their angular accelerations are related by $d\omega_b/dt = (d\omega/dt) \cdot r/d$, and the maximum angular speed of the chassis is $\omega_{b,max} = \omega_{max} \cdot r/d$.

The traction forces f are now in opposite directions, as shown in the figure above, spinning up the chassis with a torque $\tau_b = 2 \cdot f \cdot d = 2 \cdot \tau(\omega) \cdot d/r$, which is equal to the robot moment of inertia I_1 in the spin direction times its angular acceleration $d\omega_b/dt$. We can then use $d\omega_b/dt = (d\omega/dt) \cdot r/d$ to obtain

$$2 \cdot f \cdot d = 2 \cdot \frac{\tau(\omega) \cdot d}{r} = I_1 \cdot \frac{d\omega_b}{dt} = I_1 \cdot \frac{r}{d} \cdot \frac{d\omega}{dt} \quad \Rightarrow \quad \Delta t_{weapon} = \int dt = \frac{I_1 \cdot r^2}{2 \cdot d^2} \cdot \int_0^{0.95\omega_{max}} \frac{d\omega}{\tau(\omega)}$$

So, it is easy to see that decreasing the distance $2 \cdot d$ between the wheels increases the maximum weapon speed $\omega_{b,max}$, but it also increases the weapon spin up time Δt_{weapon}. It is not a good idea to have Δt_{weapon} much higher than 4 to 8 seconds, otherwise the thwackbot won't stand a chance against a very aggressive rammer or wedge, so choose wisely your distance $2 \cdot d$.

Since both Δt_{weapon} and Δt_{drive} depend on the same integral, they are related by

$$\Delta t_{weapon} = \frac{I_1 \cdot r^2}{2 \cdot d^2} \cdot \int_0^{0.95\omega_{max}} \frac{d\omega}{\tau(\omega)} = \frac{I_1 \cdot r^2}{2 \cdot d^2} \cdot \frac{2 \cdot \Delta t_{drive}}{m \cdot r^2} = \Delta t_{drive} \cdot \frac{I_1}{m \cdot d^2}$$

So, if you've already calculated the acceleration time Δt_{drive} of the drive system, as described in chapter 5, then the weapon spin up time is easily obtained from the above equation.

If your thwackbot has its wheels close to its perimeter, it is very likely that I_1 is equal to or a little lower than $m \cdot d^2$, therefore Δt_{weapon} and Δt_{drive} would be approximately equal. This is not good, because the low Δt_{drive} required to make the robot agile would lower the spin up time Δt_{weapon},

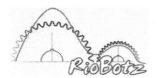 

probably resulting in a low weapon energy. Even if you upgrade your drive motors, they won't be able to deliver a very high power to spin up the weapon in a short Δt_{weapon}, because the wheel forces are limited to $\mu_t \cdot g \cdot m/2$ (for a two wheeled thwackbot), they would slip beyond that. And a larger Δt_{weapon} would naturally result in a large Δt_{drive}, compromising the drivetrain acceleration. So, wheels with a large distance $2 \cdot d$ usually result in a relatively slow robot, or one with low weapon energy.

One way to avoid this is to decrease the wheel distance $2 \cdot d$. But they can't be too close together, otherwise any unbalancing in the robot or any attack from a wedge could make the spinning chassis touch the ground and launch itself. Using casters near the perimeter could help making the thwackbot stable, but there's a good chance they'll be knocked off or broken during an angled impact.

Another way to increase the weapon energy without compromising the drivetrain acceleration is to maximize the value of I_1. This is done by concentrating most of its mass in its outer perimeter, trying to approach the upper limit $1/2$ of its normalized effective mass M_1', as done in the ring-shaped heavyweight Cyclonebot (pictured to the right).

6.9.2. Melty Brain Control

The main drawback of a thwackbot is the required complexity of its drive system to enable it to move around in a controlled way while spinning. The idea, referred to as *translational drift*, is to somehow oscillate the speed or the steering of each wheel with the same frequency that the entire robot spins. The speed oscillation solution is usually called "melty brain" or "tornado drive" control, while the steering oscillation has been called "wobbly drive" or "NavBot steering" control.

Despite its name, the math behind "melty brain" control is not hard enough to melt someone's brain. For instance, when the chassis of a two-wheeled thwackbot is spinning at ω_b, at every time period $T = 2 \cdot \pi/\omega_b$ one wheel will be facing the desired direction to which you want to move, while the other will be facing the opposite direction. If at this moment the first wheel has a slightly larger speed than the other, then the robot will end up moving in that desired direction. The angular speed of each wheel would need to be $\omega = \omega_b \cdot d/r \pm (v/r) \cdot \cos(2 \cdot \pi \cdot t/\omega_b + \varphi)$, with the plus sign for one wheel and the minus sign for the other, where t is time, v is the desired linear speed, r is the wheel radius, and φ is a phase angle that will define the direction of the movement.

"Melty brain" control is not easy to implement, because the time period T is very short. You need a very fast acting control system, powerful drive motors to be able to change the wheel speeds in such high frequency, and some way to measure both the robot angular speed, to obtain ω_b, and its orientation at the end of each time period, to define φ. In theory, wheel encoders or other angular position sensors could be used to estimate ω_b and φ, using dead reckoning, but they do not work in practice because there is wheel slip. Digital compasses do not work well because the high motor currents usually affect the readings of the Earth magnetic field.

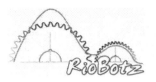 

Most successful implementations of "melty brain" control require that the driver emits some light beam, usually infra-red or laser, in the direction of the thwackbot. This beam is detected by a sensor on the periphery of the robot, allowing it to estimate ω_b from the time period T between two sensor readings, $\omega_b = 2 \cdot \pi / T$. The robot can also estimate its orientation at each reading, which would be the one facing the driver's light source. It is a good idea to use two light sensors close together instead of one, to minimize the chance of both picking up random reflections of the beam from the arena or the opponent robot, which would confuse the control system.

Other successful implementations of "melty brain" control use accelerometers or gyroscopes to measure or infer ω_b, and a led on its periphery that blinks with a period $T = 2 \cdot \pi / \omega_b$, calculated in real time from the current ω_b. If the ω_b measurement is accurate enough, the led will only blink once per revolution, apparently at the same position in space, which would point to a nominal direction. It is then up to the driver to look at the position of the led light and use the radio control to change accordingly the phase angle φ of the robot software, allowing the thwackbot to move around.

A few robots have successfully implemented "melty brain" control through electronics, such as the heavyweight Cyclonebot, the middleweight Blade Runner, the lightweight Herr Gepöunden, the featherweight Scary-Go-Round, and the antweight Melty B, as pointed out by Kevin Berry in his "melty brain" Servo magazine articles from February and March 2008.

6.9.3. NavBot Control

Other robots, such as the middleweight WhyNot (pictured below to the left) and the heavyweight Y-Pout (to the right), have followed a mechanical approach to implement steering control. These robots have three wheels in a 120° configuration, driven by high power motors.

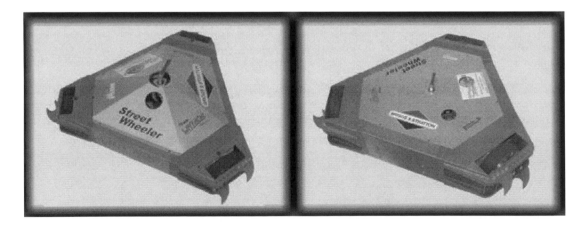

The steering is performed by making each of their three active wheels slightly steer in and out about 3 degrees at every robot revolution. The steering angle of each wheel is approximately given by $\theta_1 = 3° \cdot \cos(2 \cdot \pi \cdot t / \omega_b + \varphi)$, $\theta_2 = 3° \cdot \cos(2 \cdot \pi \cdot t / \omega_b + 2\pi/3 + \varphi)$ and $\theta_3 = 3° \cdot \cos(2 \cdot \pi \cdot t / \omega_b + 4\pi/3 + \varphi)$, mechanically implemented using three steering rods connected to a cam mechanism on a small

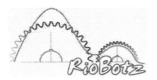

independent radio-controlled two-wheeled robot called NavBot, as seen in the figures below. The phase angle φ that defines the direction of the movement is mechanically controlled by the direction of the NavBot. The NavBot has only one motor, a low power one, which works as a differential drive. When this motor is locked, a worm gear makes sure that both wheels turn together to make the NavBot move straight, pulled by the main robot, keeping constant the phase angle φ. And when this motor is powered, following a radio control signal, one of the small wheels starts spinning with a different speed than the other, making the NavBot turn, which will change the phase angle φ and thus make the entire robot move in the new direction of the NavBot. The NavBot will then be pulled in this new direction that it is facing.

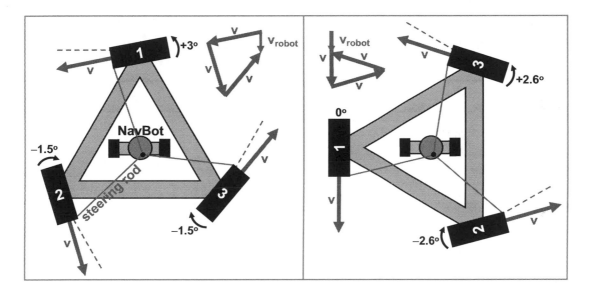

The figures above show that the differential steering of the wheels make the robot move with a constant speed $v_{robot} = v \cdot \sin 3°$, where v is the linear speed of each wheel. The direction of v_{robot} does not change while the robot is spinning, as shown in the figures, as long as the NavBot keeps its orientation. The orientation of the cam mechanism must be setup in such a way that the directions of v_{robot} and of the two small NavBot wheels always coincide, allowing the NavBot to easily follow the robot path while being pulled.

Note that these robots can be categorized as a thwackbot, if we consider that its 3-wheeled drive system makes the entire robot spin, with the NavBot being a secondary robot (almost as a Multibot). Or they can be categorized as ring or shell spinners, if we consider that the NavBot is the main robot, with a two-wheeled drivetrain, and the spinning triangle is the ring or shell, powered by three weapon motors embedded in it. It's just a matter of point of view.

The main advantage of this system is that it is possible to implement a thwackbot (or ring or shell spinner) with a normalized effective mass M_1' close to 1/2, using the same high power motors for both the weapon and drive system. Using high diameter wheels, it might be even possible to make the robot invertible, but the NavBot would be exposed to hammer attacks without a top cover.

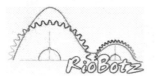

This system, however, has a few disadvantages, because the drive speed of the robot is a function of the weapon speed, due to the relation $v_{robot} = v \cdot \sin 3°$. So, when the weapon is at full speed, with the wheels at full linear speed v_{max}, the robot will have to move around at a full speed $v_{robot,max} = v_{max} \cdot \sin 3°$, all the time. It won't be able to slow down or stop its drive system, it will have to keep moving until it hits something, slowing down the weapon. And, most importantly, after a major hit, the weapon will probably be spinning so slowly that the robot won't be able to move around. An aggressive rammer would only need to survive the first hit, and then keep ramming the thwackbot to prevent its weapon from spinning up. The thwackbot wouldn't be able to run away from the rammer to try to spin up its weapon, because its drive speed would be too slow due to the slow weapon speed.

6.10. Launcher Design

Launchers need to deliver a huge amount of energy during a very brief time. Because of that, they're almost invariably powered by high pressure pneumatic systems. A very simplified estimate can show that a cylinder with piston cross section area A, stroke d, with pressure p, can accelerate a total mass m (including the mass of its piston) with an average power of up to $(p \cdot A)^{1.5} \cdot (0.5 \cdot d/m)^{0.5}$. For instance, a 4" bore cylinder with 8" stroke pressurized at 1000psi would accelerate a 220lb mass with an average power of about 566HP.

Of course, this power is only delivered during a very brief time, but a light weight electric motor or internal combustion engine cannot supply that. Unless the motor is used to store kinetic energy in a flywheel during a few seconds, with an ingenious and very strong mechanism that suddenly transfers this energy to the launcher arm, as done by the Warrior SKF robot (pictured to the right). But such sturdy mechanism is not simple to build.

Hydraulic systems are not good options either for launchers. They can deliver huge forces and accelerations, but their top speed is relatively low.

Most launchers try to either maximize the height or the range of the throw. The "height launchers" try to launch the opponent as high as possible, trying to flip it while causing damage when it hits the ground. The "range launchers" try to launch the opponent as far as possible, not necessarily high, trying to throw it out of bounds to the arena dead zone.

Against horizontal spinners, especially undercutters, there is a strategy used not only by launchers, but also by lifters and wedges, which is to tilt the spinner chassis so that its spinning weapon hits the arena floor, usually launching it. The picture to the right shows the middleweight launcher Sub Zero tilting the chassis of the spinner The Mortician, to launch it with the help of the additional energy from the spinning bar hitting the ground.

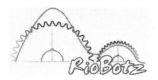

6.10.1. Three-Bar Mechanisms

A very popular launcher design uses a so-called three-bar mechanism, as pictured to the right. The "three bars" are the pneumatic cylinder, with total length varying from d_0 to d_f during the launch, the main structure of the launcher arm, with constant length a, and the part of the robot chassis that connects the arm and cylinder pivots, with constant length x. The launcher arm tip features a wedge-like scoop, at a constant distance r from the arm pivot. The initial and final angles of the

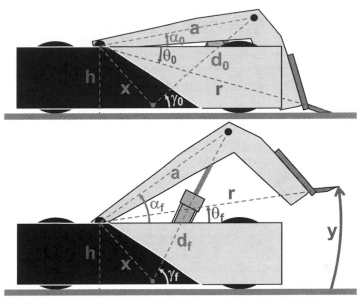

main structure of the arm are defined as α_0 and α_f, while the angles for the arm tip are θ_0 (which is negative for the particular robot shown in the picture) and θ_f. The angle variation during the launch is then $\alpha_f - \alpha_0$, which is equal to $\theta_f - \theta_0$. The initial and final angles of the cylinder are γ_0 and γ_f, as shown in the picture.

During the launch, the arm tip follows the green circular path shown above, with an arc length y that is usually between h and 2·h, where h is the height of the launcher chassis. The direction of this path is, in average, equal to $90°$ plus the average angle θ between θ_0 and θ_f, which is roughly the direction of the launching force. In the picture above, $\theta = (\theta_0 + \theta_f)/2$ is approximately zero, leaving in average an almost vertical ($90°$) force. Note also that a negative θ_0 is a good idea, it makes the arm tip move forward in the beginning of its path, helping to properly scoop the opponent.

The figures in the next page show 4 different launcher configurations. The lightweight Rocket has both α_0 and θ_0 a little above $-45°$, making it a good "range launcher" due to the average $45°$ force it delivers. The only problem with this design is that it requires a high chassis to get a sufficiently long arm with $\theta_0 = -45°$, decreasing its stability and making it more vulnerable to spinners.

The "height launchers" Bounty Hunter and T-Minus were able to lower their height when the arms are retracted, due to their low α_0, about $15°$ and $0°$, respectively. Their average θ during the launch is close to zero, leading to an almost vertical force that allows them to throw the opponents very high in the air. But their low initial cylinder angle γ_0, below $45°$, puts a lot of stress on the back pivot joint of the arm, initially trying to push forward the arm with almost the same force used to launch the opponent. This forward force, which tries to rip off the back pivot, is not necessarily wasted because it does not produce work. But this added force increases the friction losses in the joints. This is the price to pay for a low profile launcher.

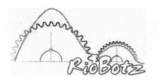

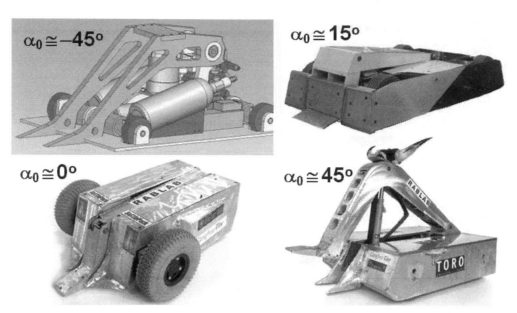

Toro (pictured above), on the other hand, has γ_0 close to $90°$, not stressing too much the arm pivot. It is also a "height launcher" because its average θ is close to zero. But to be able to accommodate its relatively long cylinders with such high γ_0, it needed to increase its α_0 to about $45°$, as shown in the picture, making its tall launcher arm vulnerable to horizontal spinners. Note the curved strap under Toro's arm that limits the stroke of the cylinders, avoiding their self-destruction when reaching their maximum stroke.

The launching calculations are more complicated than in the impact problem, because the arm does not hit the opponent, it shoves it. Therefore, the contact point between the launcher arm and the opponent may move during the launch. The figure below shows an opponent being launched.

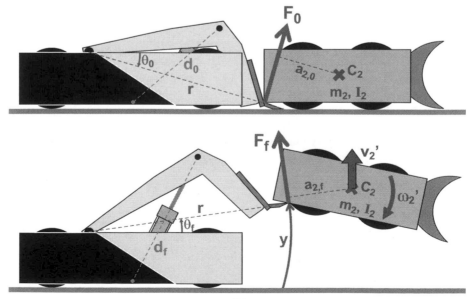

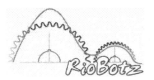 

When the launching starts, the contact point is usually located further in the back of the arm scoop (at a distance smaller than r from the back pivot of the arm), and very far from the opponent's center of mass C_2. The initial launching force F_0 might have a direction θ_0 with respect to the vertical, defining the distance $a_{2,0}$ to C_2. When the contact ends, the contact point of the final force F_f will probably have shifted to the tip of the arm scoop (at a distance r from the back pivot), defining a different distance $a_{2,f}$ to C_2.

The intensities of F_f and F_0 are probably different, even if the force in the cylinder is constant during the entire launch, because of the different mechanism configurations for angles θ_0 and θ_f. Their directions are also different, F_f might have a direction that makes the angle θ_f with respect to the vertical if there's enough contact friction at the arm tip, or it might be perpendicular to the opponent's bottom plate if there's no contact friction, or it may have some direction in between.

Also, finding out at which value of the path length y the contact will end is not simple, it depends a lot on the opponent's mass m_2 and moment of inertia I_2 in the direction it ends up spinning. Only one thing is certain: it is that the kinetic energy you may induce in the opponent cannot be higher than the energy delivered by the pneumatic cylinder, $E_b = p \cdot A \cdot (d_f - d_0)$. In practice, this theoretical value is not reached because of friction and pneumatic losses, gravity effects, and because of the inertia of the launcher arm, which needs to be accelerated as well.

Also, since the opponent usually tends to rotate away from the launcher arm during the launch, it is more efficient to maximize the forces than the path length y, to increase the delivered work. A very large path y is not effective, because the contact between the robots will probably be lost before the end of the stroke of the pneumatic cylinder. It is better to have a cylinder with large diameter and high pressure, to increase the force, than to have a cylinder with very large stroke. The ideal stroke would be the one that ends slightly after the contact between the robots is lost. The straps, or other system that limits the cylinder movement, should also be dimensioned in this way.

However, a short stroke cylinder does not necessarily mean it has a short overall length. For instance, a typical industrial 4" bore hydraulic cylinder, which can be adapted to pneumatic applications, has an overall retracted length of 9.625" plus the stroke length. Even if the cylinder stroke is only 1", its overall retracted and extended lengths 10.625" and 11.625" are relatively high. This short ratio between stroke and total length would not be effective in a three bar mechanism.

6.10.2. Launcher Equations

To properly simulate the launch, it is likely that you'll need some dynamic simulation software. Or you can use the spreadsheet from www.hassockshog.co.uk/flipper_calculator.htm, which has nice launcher models, as pointed out by Kevin Berry in the March 2009 edition of Servo Magazine. But you can also use a simple approximation to get a feeling of what happens during the launch.

Consider that the impulse J that the launcher inflicts on the opponent has an average direction θ with respect to the vertical, as pictured in the next page. The opponent is assumed as a homogeneous rectangular block with width a and height h. This value of a can be either the opponent's length or width, depending on whether you're launching it from its front/back or from its sides.

We also consider that the contact point between the launcher arm tip and the opponent does not shift during the launch, remaining fixed at a point C. This point is located by the distance s shown in the figure to the right, which is related to the length of the arm scoop.

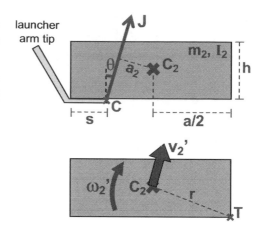

This s value is also increased if the launcher is able to get under the opponent, as shown before in the action shots from the Ziggy vs. The Judge fight.

It is not difficult to calculate the distance a_2 between the impulse vector J and the opponent's center of mass C_2, and the distance r between C_2 and the edge T, they are obtained from

$$a_2 = (\frac{a}{2} - s) \cdot \cos\theta - \frac{h}{2} \cdot \sin\theta \quad \text{and} \quad r = \sqrt{(\frac{a}{2})^2 + (\frac{h}{2})^2}$$

If the opponent's mass is m_2, then its moment of inertia I_2, in the direction it spins due to the launch, can be estimated from the moment of inertia of rectangular bars, $I_2 = m_2 \cdot r^2/3$. If the point T in the figure does not touch the ground during the launch, and if the launching force is much higher than gravity, then we can estimate that the opponent's speed v_2', parallel to J, reaches J/m_2, and the angular speed results in ω_2' = $J \cdot a_2/I_2$, leading to the relation ω_2' = v_2'$\cdot a_2 \cdot m_2/I_2$ (see the figure above).

The vertical component of the speed at point T, equal to v_2'$\cdot\cos\theta - \omega_2$'$\cdot a/2$, must be positive to validate this analysis, making sure that this point does not touch the ground during the launch. So, from the previous equations, the point T does not touch the ground if s is greater or equal than a minimum value s_{min}, where

$$a_2 \leq \frac{2 \cdot I_2 \cdot \cos\theta}{m_2 \cdot a} \quad \Rightarrow \quad s \geq s_{min} = \frac{a}{3} - \frac{h^2}{6a} - \frac{h}{2} \cdot \tan\theta$$

This resulting relationship between s_{min}/a as a function of the opponent's aspect ratio h/a is plotted to the right. Note from the graph that, to launch an opponent without making it touch the ground, a "height launcher" ($\theta \cong 0°$) would need a higher scoop length s than a "range launcher" (which usually has $\theta = 45°$).

In theory, from the gravity potential energy equation, the maximum height an opponent would achieve would be, in theory, $H_{max} = E_b/(m_2 \cdot g)$, where E_b is the launching energy discussed before and g is the acceleration of gravity. But this height could only be reached if the opponent was vertically

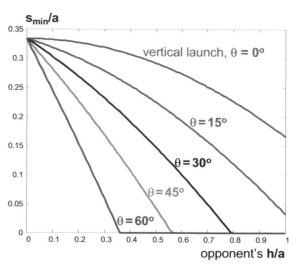

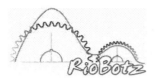

launched (thus $\theta = 0^\circ$) without spinning ($\omega_2' = 0$). If, ideally, the entire energy E_b is transformed into the opponent's kinetic energy (translational and rotational kinetic energies), we can show from the relation $\omega_2' = v_2' \cdot a_2 \cdot m_2 / I_2$, valid if T does not touch the ground, that

$$E_b = \frac{1}{2} m_2 \cdot v_2'^2 + \frac{1}{2} I_2 \cdot \omega_2'^2 = \frac{1}{2} m_2 \cdot \left(\frac{I_2 + m_2 a_2^2}{I_2} \right) \cdot v_2'^2, \quad \text{if } s \geq s_{min}$$

6.10.3. Height Launcher Equations

A "height launcher" with $\theta = 0^\circ$ would have $a_2 = a/2 - s$, while v_2' would be in the vertical direction. Since the maximum height H of a vertical launch is $v_2'^2/(2g)$, the opponent would reach

$$H = \frac{E_b}{m_2 \cdot g} \cdot \left(\frac{I_2}{I_2 + m_2 a_2^2} \right) = H_{max} \cdot \left(\frac{r^2}{r^2 + 3 \cdot (a/2 - s)^2} \right), \quad \text{if } s \geq s_{min}$$

The above expressions are only valid if $s \geq s_{min}$. But if $s < s_{min}$, the point T touches the ground, which makes its vertical speed $v_2' \cdot \cos\theta - \omega_2' \cdot a/2$ equal to zero at the end of the impulse, leading to the relation $\omega_2' = 2 \cdot v_2' \cdot \cos\theta / a$.

In this case, v_2' may not be parallel to J, its direction will depend on the contact friction between point T and the ground. A frictionless contact would keep the horizontal component of v_2' unchanged, while a very high friction could significantly reduce it, making the direction of v_2' becomes steeper than the direction of the impulse J.

A "height launcher" with $s < s_{min}$ would have then $\omega_2' = 2 \cdot v_2' / a$, leading to

$$v_2'^2 = \frac{2 \cdot E_b}{m_2} \cdot \left(\frac{m_2 a^2 / 4}{I_2 + m_2 a^2 / 4} \right) \quad \Rightarrow \quad H = \frac{v_2'^2}{2 \cdot g} = H_{max} \cdot \left(\frac{3}{4 + h^2 / a^2} \right), \quad \text{if } s < s_{min}$$

Note that, for such small s, the launch height H only depends on the opponent aspect ratio h/a.

The graph to the right shows the H/H_{max} ratio as a function of the normalized length s/a, against opponents with aspect ratios h/a. Note that the horizontal lines are the values obtained for $s < s_{min}$, while the curved ones reflect the results from $s \geq s_{min}$.

The horizontal lines suggests that scooping the opponent with $s/a = 0.15$ results in the same height as if s/a was close to zero 0. This is true, at least for the simple model we're using, as long as the contact between the robots is

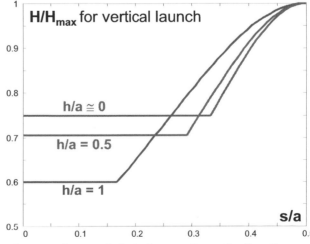

maintained during the launch, which depends not only on s/a but also on the path direction and length y of the arm tip. Obviously, $s/a = 0.15$ would be better to maintain this contact than s/a close to zero.

So, from the graph we conclude that the most effective launch happens when s is close to a/2. One way to do that is to try to launch the opponent from the direction where it is shortest. So, for instance, against a narrow robot, try to launch it from its side, making the distance a become its width instead of length. Getting under the opponent is also one way to increase s, as done by Ziggy with its front wedge.

A long scoop at the end of the launcher arm (as seen in the middleweight Sub Zero, pictured to the right) also helps to increase the distance s. A long scoop will also make sure that the contact between the robots won't be lost during the entire stroke of the launcher arm. But be careful, a very long scoop will be vulnerable to drumbots and undercutters, which may bend it until it loses functionality, not being able to get under robots with low ground clearance. In addition, the previous graph showed that, against very low profile opponents (small h/a), there's no point in having a very long scoop to increase H unless it has s/a greater than 0.35 or if the weapon tip path y is too large. If below 0.35, the value of s/a would only have to be large enough not to lose contact with the opponent during the launch.

Note that the above calculations assumed that the launcher didn't tilt forward too much during the launch, which could make it get unstable. The requirements to guarantee launcher stability are presented in section 6.10.6.

6.10.4. Range Launcher Equations

A similar analysis can be made for a "range launcher" as follows. If the opponent does not touch the ground while it is launched, which happens when $s \geq s_{min}$, then the launch speed v_2' is parallel to the impulse J. The launch angle, with respect to the horizontal, is then $90^\circ - \theta$. The horizontal range R of the launch is then

$$R = \frac{v_2'^2 \cdot \sin[2(90^\circ - \theta)]}{g} = \frac{2E_b \sin 2\theta}{m_2 \cdot g} \cdot \left(\frac{I_2}{I_2 + m_2 a_2^2} \right) = R_{max} \cdot \sin 2\theta \cdot \left(\frac{r^2}{r^2 + 3 \cdot a_2^2} \right), \quad \text{if } s \geq s_{min}$$

where $R_{max} = 2E_b/(m_2 \cdot g)$ is the maximum possible launch range, which only happens when $a_2 = 0$ (the impulse vector passes through C_2) and $\theta = 45^\circ$. The launch angle that maximizes R when a_2 is different than zero must be calculated with a numeric method, because a_2 depends on a, s, h and also θ.

On the other hand, if the opponent touches the ground at point T while it is launched (which happens when $s < s_{min}$), then the equations get much more complicated, because the ground reaction at T will cause a vertical impulse, and also a horizontal one if there's ground friction. This will change not only the magnitude of v_2' but also the launch angle. Even without considering ground friction, the equations are too lengthy to be shown here.

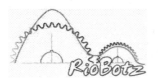

But the results are seen in the graphs to the right, obtained neglecting the effect of ground friction at point T during the launch. The neglected friction effect would only be significant if both θ and a_2 were large and if both s and h were very small, therefore it does not significantly influence the conclusions that are presented next.

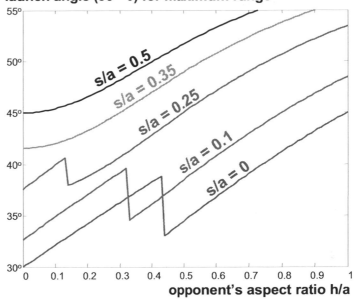

launch angle (90°–θ) for maximum range

The top graph shows the ideal launch angle (90°–θ) of the arm to maximize the range for a given s/a and aspect ratio h/a of the opponent. Against very low profile opponents (h/a close to zero), the best launch angle is 45° if you're able to reach s/a = 0.5, as expected, reaching 100% of the maximum possible range R_{max}, as seen on the bottom graph.

But if you can't get under such very low profile opponent and your arm has a very short scoop (s/a close to 0), then the best launch angle to maximize the reach would be 30° (with respect to the horizontal). This might seem strange but it makes sense: in this case, the ground impulse at point T, which happens due to the small s/a, will add up to the launcher's 30°

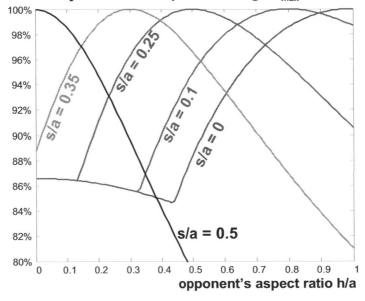

maximized range for a given h/a and s/a divided by the maximum possible range R_{max}

impulse to effectively launch the opponent at 43.7°. As seen in the bottom graph, this best 30° angle will achieve 86.6% of R_{max}, which is still pretty good considering that s/a is so small. It would be impossible to reach 100% of R_{max} with s/a = 0. Unless the opponent was very tall, with h/a = 1, but then the best launch angle would be 45°.

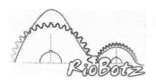

Note that there is a step in the top graph for the curves with s/a < 0.33. This change (or step) is associated with the opponent touching the ground (values to the left of the step) or not touching it (values to the right of the step) during the launch. For instance, if s/a = 0, then an opponent with aspect ratio h/a = 0.4 (a value to the left of the step in the s/a = 0 curve) would be thrown further away if launched at 38°, using the ground to help in its launch.

But if h/a = 0.5 (to the right of the step), then a shallower angle of 34° would decrease the angular speed of the opponent, making it not touch the ground, resulting in the optimal throw in this case. So, depending on h/a and s/a, it may be good or not to use the help of the ground to launch the opponent with maximized range. This conclusion is not trivial at all.

Since most combat robots tend to have a low profile, with 0.2 < h/a < 0.4, to keep low their center of gravity, we can draw several conclusions about "range launchers" as follows. If the launcher can only provide a small s/a, then it is better to set its arm such that its impulse is in average at about $(90° - \theta) = 36°$ from the horizontal, to reach a maximized range of about 85% of R_{max} (in this case, the opponent would touch the ground during the launch, because $s < s_{min}$).

If the arm scoop allows the launcher to reach s/a = 0.25, then a steeper $(90° - \theta) = 41°$ angle would be a better option, typically launching the opponent between 91% and 99% of R_{max} (in this case, the opponent does not touch the ground because $s > s_{min}$).

Finally, if the launcher has a wedge to get under the opponent, or if you decide to use a very long scoop in its arm, then the best choice would be to use s/a = 0.35 and a $(90° - \theta) = 45°$ launch angle. With these values, you can launch typical opponents between 99% and 100% of R_{max}. This s/a and $(90° - \theta)$ combination makes the distance a_2 become very close to zero for most robots, spending most of the launch energy throwing the opponent instead of making it spin.

Note, however, that these ideal launch angles to maximize range are only achievable in practice if they do not cause the launcher to be pushed too much backwards by the reaction force. The requirements to avoid this are presented in section 6.10.6.

Combinations of s/a, h/a and θ that lead to high values of a_2 should be avoided, because they waste too much energy making the opponent spin forward. Also, it is not a good idea to go beyond s/a = 0.35, you'll probably end up with a negative value of a_2, which will waste energy spinning the opponent backwards. The pictures in the next page show three different launch situations. The first one shows the super heavyweight Ziggy launching an opponent with a high average distance a_2 from C_2, resulting in a forward spin. In the second situation, Ziggy is able to launch The Judge with an impulse vector very close to C_2 (therefore a_2 close to zero), resulting in a high range due to the much lower resulting spin. Finally, the lightweight Rocket is able to launch the opponent with a backward spin, because the contact point was beyond the opponent's C_2, making a_2 become negative. Ideally, you should try to keep the opponent's spin as low as possible to maximize the launch range.

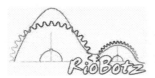

Launching with forward spin, due to the high a_2

High range launch because of the almost neutral spin, due to a_2 close to zero

Launching with backward spin, due to a negative a_2

6.10.5. Four-Bar Mechanisms

As discussed before, the problem with most "range launchers" with three-bar mechanisms is that they usually end up with a tall chassis if they want to throw opponents at the ideal angles $(90° - \theta)$ from $36°$ to $45°$. The tall chassis is a result of their mechanisms, which would need an average θ between $54°$ and $45°$.

One alternative is to use four-bar mechanisms, as pictured to the right. The four bars consist of part of the launch arm (d_1), part of the chassis (d_2) and two auxiliary links (d_3 and d_4). Technically, these launchers have a five-bar mechanism, because the pneumatic cylinder (d) counts as a fifth link.

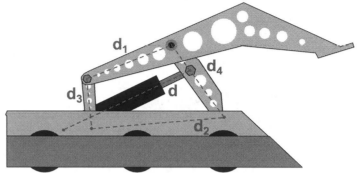

Four-bar mechanisms have two advantages. First, if well designed, they can be completely retracted inside the robot, allowing the use of a low profile chassis. And, if the constant lengths d_1, d_2, d_3 and d_4 are appropriately defined, it is possible to generate optimal trajectories for the arm tip. You can, for instance, make the arm tip trajectory become almost a straight line, with some desired optimal angle (which for "range launchers" would probably be between $36°$ to $45°$ with respect to the horizontal). The four-bar mechanism calculations are too lengthy to be shown here, but you can make them using, for instance, a free static simulation program (screenshot pictured to the right) that can be found on the tutorials in http://www.totalinsanity.net.

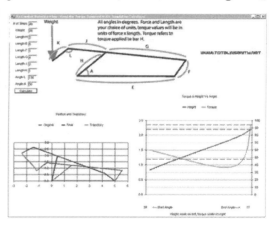

6.10.6. Launcher Stability

During the launch, a launcher should neither tilt forward too much, nor be pushed backwards, otherwise it will lose its effectiveness.

Due to the very high forces involved during the launch, it is very likely that "height launchers" will tilt forward until they touch the ground at their foremost point T, as shown in the figure in the next page as well as in the action shot to the right, featuring Sub Zero launching The Mortician. To avoid tilting forward even more and becoming unstable, it is necessary to locate the launcher's center of gravity C_1 as far back as possible,

to maximize the horizontal distance a_1 to point T, as pictured below. If F is the launcher weight, then the force F_i at any moment during the launch must satisfy $F_i \cdot a_i < F \cdot a_1$, where a_i is the distance between T and the line that contains the vector F_i. It is advisable that this condition is satisfied during the entire launch, for all values of F_i between F_0 and F_f, multiplied by their respective distances to point T.

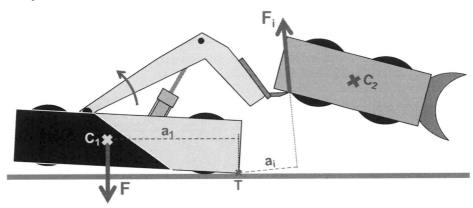

Another concern with "height launchers" is with the stiffness and damping properties of their front wheels. During the launch, these wheels are very much compressed against the ground, storing a great deal of elastic energy. Towards the end of the launch, when the contact with the opponent is almost lost, these wheels will spring back. If their damping is low, as in foam-filled rubber wheels, they may launch the launcher and even make it flip backwards. The action shot to the right shows the launcher Sub Zero off the ground as soon as it loses contact with its opponent The Mortician.

"Range launchers" may tilt forward as well, but it is very unlikely that they lose stability in this way. This is because the line that contains the launch force vector F_i usually meets the ground within the launcher footprint, or very close to is foremost point T, due to the shallower launch angles $(90° - \theta_i)$ involved, see the picture to the right.

But "range launchers" have a problem with very shallow launch angles, because the horizontal component $F_i \cdot \cos(90° - \theta_i)$ may become too large for the wheel friction to bear, pushing it backwards. As

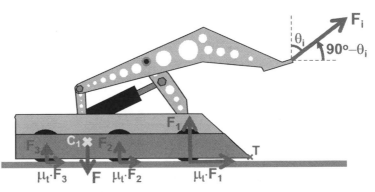

seen in the figure, if F is the launcher weight, F_1, F_2 and F_3 are normal ground forces on each wheel pair, and μ_t is the coefficient of friction between the tires and the ground that cause the maximum friction forces $\mu_t \cdot F_1$, $\mu_t \cdot F_2$ and $\mu_t \cdot F_3$, then

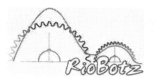 

$$\left. \begin{array}{l} F_1 + F_2 + F_3 = F + F_i \cdot \sin(90^\circ - \theta_i) \\ \mu_t \cdot F_1 + \mu_t \cdot F_2 + \mu_t \cdot F_3 \geq F_i \cdot \cos(90^\circ - \theta_i) \end{array} \right\} \Rightarrow \mu_t \geq \frac{F_i \cdot \cos(90^\circ - \theta_i)}{F + F_i \cdot \sin(90^\circ - \theta_i)}$$

This equation shows that tire friction is very important to avoid "range launchers" from being pushed backwards, probably decreasing the contact time with the opponent and the effectiveness of the launch. A typical high traction tire with $\mu_t = 0.9$ is usually enough for a $(90^\circ - \theta_i) = 45^\circ$ launch angle. But for the 36° or 41° angles, which maximize the launch range for small s/a values, you might need a higher μ_t. If this higher μ_t is not achievable, then the best option is to adopt the lowest launch angle $(90^\circ - \theta_i)$ that satisfies the μ_t condition above.

Finally, note that it is also a good idea for "range launchers" to locate C_1 as far back as possible, to prevent them from even touching their foremost point T on the ground, because this point will probably have a coefficient of friction with the ground lower than μ_t. If some tilting is inevitable, then it might be a good idea to install some anti-sliding material on the bottom of the robot beneath point T, such as a rubber strip, to increase friction.

6.11. Lifter Design

Lifters and launchers have a few common design features. The main difference is that lifters have relatively slow lifting mechanisms, allowing them to use, for instance, highly geared electric motors. The use of electric motors instead of pneumatic systems usually results in less weight to the weapon system, allowing them to improve their armor or drivetrain.

Lifters work better against non-invertible robots, or on arena with hazards where they can shove the opponent to. Clearly, they must have enough wheel traction to be able to push the opponent around the arena while it is lifted.

Similarly to launchers, a four-bar mechanism (as pictured below) is a good option to allow the lifting arm to retract inside the robot chassis, becoming less vulnerable to spinners.

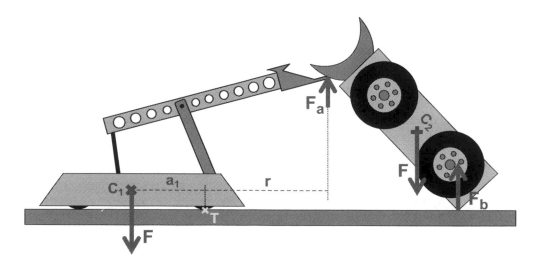

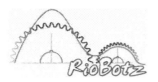 

To be able to lift their opponents without losing stability, it is important to locate their center of gravity C_1 as far back as possible, to maximize the distance a_1 to its foremost ground support (point T in the figure), usually located under the front wheels.

If F is the lifter weight, r is the maximum horizontal distance between point T and the tip of the lifter arm, and if F_a is the lifting force it applies on the opponent, then the gravity torque $F \cdot a_1$ with respect to point T must be higher than $F_a \cdot r$ to prevent it from tilting forward. The force F_a is a function of the weight of the opponent robot (also assumed as F, since both robots should be from the same weight class), of the force F_b from the opponent's ground support, and of the relative horizontal distances among them.

For a symmetric opponent, with a center of mass C_2 at the center of the chassis, it is easy to see that $F_a = F_b = F/2$ when the lifting begins, with the opponent in a horizontal position. As the opponent is lifted, F_b is increased while F_a decreases, as suggested by the picture. This is most noticed on tall opponents, which become easier to lift as they are lifted. So, the worst case scenario would be to consider that F_a is equal to its maximum value F/2, resulting in

$$F \cdot a_1 > \frac{F}{2} \cdot r \implies r < 2 \cdot a_1$$

It is not difficult to satisfy the above condition for the maximum horizontal reach r, so tilting forward is not a major concern for most lifters. Note that a lot of weight will be concentrated on the front wheels, up to 1.5 times the lifter weight in this example. So, make sure that the front wheels have high torque motors, to prevent them from stalling while pushing around the lifted opponent.

6.12. Clamper Design

Clampers are similar to lifters, except that they need to lift the entire weight F of the opponent, instead of just about half of it. To be able to clamp and lift their opponents without losing stability, they should also locate their center of mass C_1 as far back as possible. They usually need an extension on their front to act at point T as their foremost ground support, to increase the distance a_1 shown in the figure to the right.

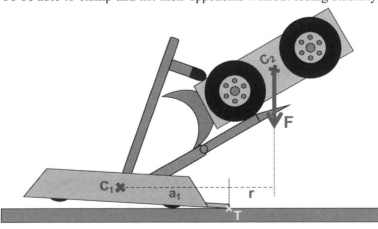

If the lifter and opponent have same weight F, and r is the horizontal distance between point T and the opponent's center of mass C_2, then the gravity torque $F \cdot a_1$ with respect to point T must be higher than $F \cdot r$ to prevent the lifter from tilting forward, resulting in

$$F \cdot a_1 > F \cdot r \implies r < a_1$$

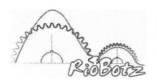 

The above condition is much harder to meet than the stability condition for lifters, because here the distance r is not to the clamper tip, but to C_2. This distance to C_2, which is maximum when the opponent is starting to be lifted, can be very large for a long opponent. In addition, r must be smaller than a_1 for clampers, instead of $2 \cdot a_1$ as found for lifters. This is why tilting forward is a major concern for clampers, usually forcing them to use front extensions to increase a_1.

Similarly to lifters, clampers also need high torque on their front wheels to be able to drive around while carrying the opponent. Their front wheels have to bear up to twice the clamper weight, so make sure that their drive system is very sturdy and powerful.

6.13. Rammer Design

Rammers are usually nicknamed BMW, because they're basically made out of Batteries, Motors and Wheels. They must have a lot of traction to shove other robots around, and high top speeds to be effective as a ram.

Its shield or armor is usually made out of hard materials, used in traditional armors. Ablative materials would also work, however they would have to be changed more often.

There are two design strategies to make them resistant to spinner attacks. Defensive rammers use shock mounts to attach their shield, trying to absorb and dissipate the energy of the attack. They can also use ablative shields for that.

Offensive rammers, on the other hand, have very hard shields rigidly attached to a stiff chassis, trying to divert the impact energy back to the attacker and break its weapon system. But remember to shock mount internal critical components.

Needless to say that taking the hit is not necessarily the best strategy: if you're able to push around a spinner without getting hit by its spinning weapon, it may self-destruct after hitting the arena wall.

Rammers should always be invertible. A few of them, such as the middleweight Ice Cube (pictured to the right), are even capable of righting themselves using only the power of their wheels, similarly to an overhead thwackbot.

One issue with rammers with large shields is to avoid getting "stuck on their nose" (as pictured to the right). Ice Cube faces this problem, even though it is able to rock back and forth using the inertia of its wheels to flip back on its feet after a few seconds (we wish we had known that before the RoboGames 2006 semifinals!).

not good   **better**

But, during the seconds it is rocking, the (orange) chassis gets exposed to attacks. Also, if the rammer is gently pushed while on its nose against the arena walls, it won't be able to get unstuck by rocking. One possible solution to avoid getting stuck is to mount the shield in a slightly asymmetrical position, as pictured above. With the ground projection of the rammer center of gravity closer to (or beyond) the edge of its shield, it becomes much easier to flip back.

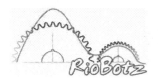

6.14. Wedge Design

Wedges have a very simple and effective design, especially against spinners. They try to use the opponent's energy against it, with the aid of a ground support and an inclined plane (their wedge). The inclined plane idea is so simple and effective that it was used on one of the first combat wedges in history, Leonardo Da Vinci's tank, pictured to the right. Its 35° sloped conical armor, covered with steel armor plates, would work as a

wedge to deflect enemy fire from all sides. It was not built back in 1495 because Italian battlefields were not flat enough to allow it to move without getting stuck. We had to wait 500 years, in 1995, to see a combat wedge on a flat "battlefield", with La Machine (pictured to the right), coincidently with the same slope as Leonardo's tank, taken from his original XV century drawing.

The wedge concept is not limited to wedges. As pictured below, it can be seen in vertical spinners (such as K2), horizontal spinners (Hazard), drumbots (Stewie), lifters (Biohazard), launchers (Ziggy), spearbots (Rammstein), hammerbots (The Judge), overhead thwackbots (Toe Crusher), and even combined with flamethrowers (Alcoholic Stepfather).

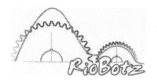

6.14.1. Wedge Types and Shapes

Fixed wedges, without any articulation, can either have some ground clearance, or they can be supported by the ground, scraping it. The first type might work well against most horizontal and vertical spinners, but it is vulnerable to undercutters or to a lower wedge.

The fixed wedges that scrape the floor, on the other hand, do not have this vulnerability, but they need to have a very sharp edge to stay flush to the ground. If they carry a significant portion of the robot's weight, they'll be much more difficult to get under due to the increased downward pressure. But, on the other hand, the robot might lose traction due to the decreased weight under the active wheels. Another way to increase the downward pressure is to decrease the ground contact area of the wedge, as done by the middleweight Emily with its narrow frontal titanium insert (pictured to the right).

Articulated wedges are probably the most popular, resulting in a virtually zero ground clearance if their edge is sharpened. Floppy wedges, which have articulations that are not actively powered, should be heavy enough to increase the downward pressure at their sharp edge. On the other hand, active wedges, which have a powered articulation to make them work as lifters, can have their downward pressure increased by their motors just before they hit the opponent. This strategy has worked very well for the heavyweight lifter Sewer Snake, as pictured to the right.

Downward pressure can also be achieved by mounting springs to the robot's walls, keeping the wedges spring loaded flush to the ground. The picture to the right shows a nice wedge from the spinner Hazard. Note that, besides the spring, there are also two triangular-shaped supports underneath the wedge. These supports work as angle limiters, preventing the wedge from articulating too much and lifting the robot's own wheels off the ground, as well as working as stiffener brackets. Note also the rubber sandwich mounts used as dampers, improving the resistance to spinner impacts.

Wedge angle limiters are very important, especially if the robot has internal wheels. The picture to the right shows that, if the angle limiters from our hobbyweight Puminha are removed, its titanium wedge may get stuck and prevent the internal rear wheels from touching the ground, immobilizing the robot.

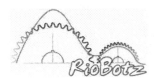

Both pictures to the right show a sturdy Ti-6Al-4V titanium articulated wedge used by our middleweight horizontal bar spinner Titan. Besides the high strength aircraft aluminum angle limiter, which also works as a stiffener, the wedge features Belleville washers on its titanium articulation shaft that work as shock mounts against lateral impacts.

Note that this 32° wedge has a shallower 25° sharp edge, to make sure that only the tip of the edge touches the ground, increasing the downward pressure. Note also that this wedge wouldn't be very effective if upside down, as shown in the bottom picture. This is why we only use it in non-invertible robots such as Titan.

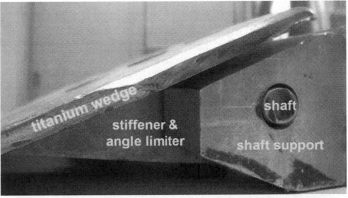

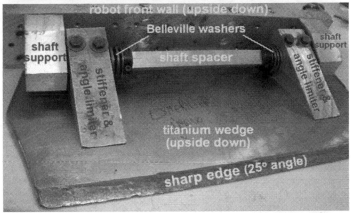

For invertible robots that don't have self-righting mechanisms, it is a good idea to have a symmetric wedge, which has the same effectiveness on either side. For instance, the heavyweight Original Sin has a hollow wedge made out of two rectangular plates separated by triangular spacers/stiffeners (pictured to the right), resulting in a symmetric wedge with a high stiffness to weight ratio.

But, unless your robot has external wheels such as Original Sin, or its wedge is actively powered, avoid using rectangular wedges, they can easily get your robot stuck resting on its side. Instead, use trapezoid-shaped wedges, either with a narrower edge (as pictured to the right) or with a wider edge. It is very unlikely that an internal-wheeled robot gets stuck resting on its side if it has a trapezoid-shaped wedge. A trapezoid-shaped wedge with a narrower edge will probably allow the robot to get unstuck and fall back, as long as there are angle limiters (which were removed before taking the picture to the right). A trapezoid wedge with wider edge will also work, but there's a greater chance that the robot will fall back upside down, which would be a problem if it does not have an invertible design.

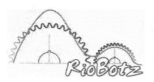

Finally, wedges with blunt edges should be avoided, because they're usually vulnerable to sharp anti-wedge skids, such as the S7 steel ones that support the drum of our hobbyweight Tourinho (pictured to the right). These narrow skids are able to concentrate a significant amount of downward pressure on the ground, in special if they're properly sharpened, easily getting under a blunt wedge.

6.14.2. Wedge Impact

One of the most important features of a wedge is its slope, represented by an angle α with respect to the horizontal. To find optimum α values, it is necessary to study the effect of an impact caused by a weapon that has an effective horizontal linear momentum $p = M_1 \cdot v_{tip}$, where M_1 is the effective mass of the attacker robot and v_{tip} is the speed of its weapon tip. Initially, let's assume that the momentum p is perpendicular to the direction of the wedge's edge, as if facing a horizontal spinner or thwackbot at the side where the weapon tip approaches the wedge, as pictured below, or if frontally attacked by a spearbot.

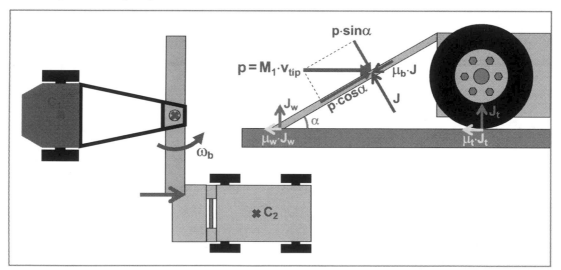

The impact of the horizontal spinner (or thwackbot or spearbot) will cause a reaction impulse J normal to the wedge surface, in response to the $p \cdot \sin\alpha$ component, as shown in the figure. If the coefficient of restitution (COR) of the impact is e, and assuming that v_{tip} is much higher than the speed of the wedge (either before or after the impact), then $J \cong (1+e) \cdot p \cdot \sin\alpha$.

The wedge also responds with a friction impulse $\mu_b \cdot J$ parallel to its surface, decreasing the $p \cdot \cos\alpha$ component as the weapon slides during the impact, where μ_b is the friction coefficient between the weapon tip and the wedge. So, instead of $p \cdot \cos\alpha$, the wedge will only feel $\mu_b \cdot (1+e) \cdot p \cdot \sin\alpha$ parallel to its surface. Because of that, the wedge will effectively receive a

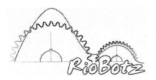

horizontal impulse J_x that is smaller than p. It will also respond with a vertical impulse J_y that will try to launch the horizontal spinner, where

$$\begin{cases} J_x = J \cdot \sin \alpha + \mu_b \cdot J \cdot \cos \alpha = (1+e) \cdot p \cdot \sin \alpha \cdot (\sin \alpha + \mu_b \cdot \cos \alpha) \\ J_y = J \cdot \cos \alpha - \mu_b \cdot J \cdot \sin \alpha = (1+e) \cdot p \cdot \sin \alpha \cdot (\cos \alpha - \mu_b \cdot \sin \alpha) \end{cases}$$

As seen from these equations, it is very important that the wedge is very smooth, to decrease μ_b and thus minimize the backward impulse J_x, while maximizing the "launch impulse" J_y. And the wedge material must also be very hard to avoid dents, which could stick to the weapon tip and make the wedge suffer along its surface the entire component $p \cdot \cos \alpha$ instead of only $\mu_b \cdot (1+e) \cdot p \cdot \sin \alpha$.

Hardened steels would be a good choice to avoid dents, however their stiffness-to-weight and toughness-to-weight ratios are not nearly as good as Ti-6Al-4V titanium, as seen in chapter 3. Since this grade 5 titanium is also relatively resistant to dents, due to its medium-high hardness, it is the material of choice for wedges. A hardened steel weapon tip would have $\mu_b \cong 0.3$ against a very smooth Ti-6Al-4V wedge, or up to $\mu_b \cong 0.5$ against a very rough and battle-battered Ti-6Al-4V wedge.

6.14.3. Defensive Wedges

A defensive wedge has the objective to resist the attack without being thrown backwards, even if it is standing still and not charging the attacker. This can happen if the horizontal impulse J_x that tries to push the wedge backwards is smaller than the vertical impulse J_y multiplied by the coefficient of friction μ with the ground, resulting in

$$J_x < \mu \cdot J_y \quad \Rightarrow \quad (\sin \alpha + \mu_b \cdot \cos \alpha) < \mu \cdot (\cos \alpha - \mu_b \cdot \sin \alpha) \quad \Rightarrow \quad \tan \alpha < \frac{\mu - \mu_b}{1 + \mu \cdot \mu_b}$$

Note that we've neglected above the effect of the robot weight on the friction impulse, because the impact is usually so fast that its forces are much higher than such weight.

If attacked by a "high spinner" as shown in the picture to the right, with the linear momentum p in the upper part of the wedge, then most of the reaction impulse will be provided by the front tires, resulting in $J_x \cong \mu_t \cdot J_t$ and $J_y \cong J_t$, where μ_t is the coefficient of friction between the front tires and the ground, and J_t is the vertical impulse from the ground to both tires altogether.

Assuming a typical high traction tire with $\mu = \mu_t = 0.9$, then a smooth titanium wedge with $\mu_b \cong 0.3$ would need to have about $\alpha < 25°$ to be

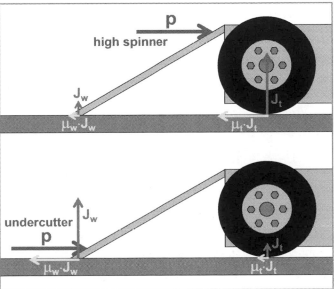

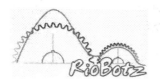 

considered as a defensive wedge, while a rough wedge with $\mu_b \cong 0.5$ would only be defensive if its slope $\alpha < 15°$. Note, however, that it is not a good idea to use a very small α, such as $\alpha < 20°$, because it would lower too much the average thickness (and therefore the strength) of the sharp edge of the wedge.

On the other hand, against an undercutter, the linear momentum p is very close to the wedge's edge, making $J_x \cong \mu_w \cdot J_w$ and $J_y \cong J_w$, where μ_w is the coefficient of friction between the wedge and the ground, and J_w is the vertical impulse from the ground to the wedge's edge, as seen in the figure. Assuming a coefficient of friction $\mu = \mu_w = 0.35$ between the titanium edge and the soft steel arena floor, the equation shows that any $\alpha > 3°$ would allow the robot to be thrown backwards, even if it had a very smooth wedge.

So, in theory, no wedge can be considered defensive against an undercutter: the wedge robot can defend itself from the first attack, but it will be thrown backwards or get spun, making it vulnerable to an immediate second attack. It is up to the wedge driver to keep facing the undercutter at all times.

But it is possible to have a defensive wedge against undercutters. Choosing a very small α is not a good idea, because the resulting low average thickness of the edge would allow the wedge to be easily torn apart by the undercutter, which would hit it at its weakest spot. Perhaps a good idea would be to spread some anti-slip product, such as anti-slip v-belt spray, under the edge where the wedge touches the ground, significantly increasing μ_w. Clearly, the spray should only be applied before matches against undercutters.

Another idea is to use a wedge with variable slope, similar to a scoop, with a lower α near the edge and a higher α near the top, to be effective against both undercutters and high spinners, as pictured below. Due to the variable slope, it would be possible to have an edge with a low α without compromising is thickness and strength.

A defensive wedge may be a good option if the robot also has some active weapon. The wedge could be used to defend the robot from opponents' attacks, or to slow down their spinning weapon, until the active weapon had a chance to get in action against them.

But a defensive wedge by itself would have a hard time winning a fight by knockout. You would need then an offensive wedge.

6.14.4. Offensive Wedges

Offensive wedges have the objective to launch their opponents, as pictured to the right. To most effectively accomplish that, they need to maximize the vertical impulse J_y, which happens for some $\alpha = \alpha_{launch}$ the makes the derivative $dJ_y/d\alpha = 0$, resulting in

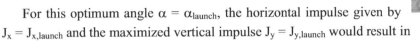

$$\frac{dJ_y}{d\alpha} = (1+e)\cdot p\cdot\frac{d}{d\alpha}[\sin\alpha\cdot(\cos\alpha - \mu_b\cdot\sin\alpha)] = 0 \Rightarrow \tan(2\cdot\alpha_{launch}) = \frac{1}{\mu_b}$$

For this optimum angle $\alpha = \alpha_{launch}$, the horizontal impulse given by $J_x = J_{x,launch}$ and the maximized vertical impulse $J_y = J_{y,launch}$ would result in

$$J_{x,launch} = \frac{(1+e)\cdot p}{2} \quad \text{and} \quad J_{y,launch} = \frac{(1+e)\cdot p}{2}\cdot[\sqrt{1+\mu_b^2} - \mu_b]$$

For a very smooth titanium wedge with $\mu_b \cong 0.3$, the optimum angle to launch the opponent is $\alpha_{launch} \cong 37°$, resulting in a maximum $J_{y,launch} = 0.37\cdot(1+e)\cdot p$, while a battle-battered titanium wedge with $\mu_b \cong 0.5$ would have $\alpha_{launch} \cong 32°$ and $J_{y,launch} = 0.31\cdot(1+e)\cdot p$. Curiously, a hard steel wedge against a hard steel weapon would have $\mu_b \cong 0.4$ and therefore $\alpha_{launch} \cong 34°$, very close to the slope angle from Leonardo's steel-plated tank. This might not be a coincidence: Leonardo Da Vinci was known for performing simple experiments in several areas before proposing a new design.

From the calculations for defensive wedges, a wedge robot with mass m_2 and an optimum angle $\alpha = \alpha_{launch}$, with initial speed equal to zero, would be thrown backwards with a speed v_2' such that

$$J_{x,launch} - \mu\cdot J_{y,launch} = m_2 v_2' \quad \Rightarrow \quad v_2' = \frac{(1+e)\cdot p}{2\cdot m_2}\cdot[1 + \mu\cdot\mu_b - \mu\cdot\sqrt{1+\mu_b^2}]$$

where $\mu = \mu_t$ against high spinners, and $\mu = \mu_w$ against undercutters, as defined before. So, to avoid being thrown backwards, the offensive wedge just needs to charge with a speed of at least v_2' before the impact.

6.14.5. Example: Offensive Wedge vs. Horizontal Spinner

In the calculation example for Last Rites against Sir Loin, the effective linear momentum before the impact was $p = M_1\cdot v_{tip} = 6.21kg\cdot106.4m/s = 661Ns$. If Sir Loin had a smooth $37°$ sloped titanium wedge aligned with J, with $\mu_b = 0.3$, and if the impact had as well a COR $e \cong 0.13$ with the same effective mass $M_1 = 6.21kg$ calculated before (which is not necessarily true, since the calculated M_1 and the measured e were not obtained for an angled impact), then it would only have to take a horizontal impulse $J_x = J_{x,launch} = (1+0.13)\cdot661/2 = 373Ns$, while the arena floor would provide the vertical reaction impulse $J_y = J_{y,launch} = 0.37\cdot(1+0.13)\cdot661 = 276Ns$ to launch Last Rites.

Assuming Last Rites has a mass of 220lb (99.8kg), then this $J_{y,launch}$ would launch its center of mass with a speed $v = 276Ns/99.8kg = 2.77m/s$, reaching a height of $v^2/(2g) = (2.77)^2/(2\cdot9.81) = 0.39m$ (about 15 inches). In addition, this vertical impulse $J_{y,launch}$ would also make Last Rites roll, with an angular speed that would depend on its moment of inertia in the roll direction and on the gyroscopic effect of the weapon. This roll movement could make its spinning bar touch the ground, probably launching it even higher than that.

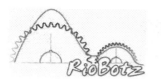 

If the wedge version of Sir Loin had almost no speed before the impact, it would be thrown backwards with a speed v_2', calculated next. If the spinning bar from Last Rites was very low to the ground, then assuming $\mu = \mu_w = 0.35$ and $m_2 = 220\text{lb}$ (99.8kg) we would get

$$v_2' = \frac{(1+0.13) \cdot 661}{2 \cdot 99.8} \cdot [1 + 0.35 \cdot 0.3 - 0.35 \cdot \sqrt{1+0.3^2}] = 2.77\text{m/s}$$

So, to avoid being thrown backwards, the wedge Sir Loin would only need to charge before the impact with a forward speed of at least 2.77m/s (10km/h or 6.2mph), which is not a big deal for most combots. On the other hand, if the bar was spinning higher off the ground (which happens when Last Rites is flipped upside down), hitting near the top of the wedge, then $\mu = \mu_t = 0.9$ would lower v_2' to 1.24m/s (4.5km/h or 2.8mph), which is even easier to reach by Sir Loin.

6.14.6. Angled Impacts

The previous equations assumed that the impact direction was perpendicular to the direction of the wedge's edge, meaning $\theta = 0°$ in the figure to the right.

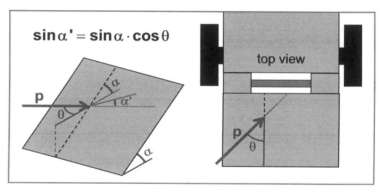

For angled impacts, with θ between $0°$ and $90°$, the wedge works as if it had an effective slope angle α' smaller than α, where $\sin\alpha' = \sin\alpha \cdot \cos\theta$. All previous equations would remain valid, as long as α is replaced by this effective α'.

For instance, in the figure below to the left, the wedge from Pirinah 3 is not able to launch the horizontal bar spinner The Mortician, because an aligned frontal attack with a low forward speed has $\theta \cong 90°$, resulting in $\alpha' \cong 0°$ and therefore $J_y \cong 0$ and $J_x \cong 0$. As long as the wedge is sufficiently smooth, without dents or bolt heads sticking out, it will work as a defensive wedge if properly aligned to the attacking robot.

 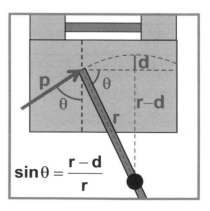

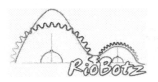

But if the robots are moving forward at high speeds, then the tooth travel distance d (related to a tooth bite d·sinα) can be large enough to significantly lower the attack angle θ, because of the relation sinθ = (r–d)/r, where r is the radius of the spinning weapon, as shown in the previous picture to the right. This lower attack angle θ can then increase α', which can be enough to launch the spinner. So, for an aligned frontal attack of an offensive wedge, forward speed is fundamental.

Another approach is to try to hit the spinner with an offset to the side where the spinning weapon approaches the wedge, as it was assumed in the previous examples, trying to make θ close to zero. But, against a skillful driver, it might be difficult to make the wedge hit the spinner with such offset. The spinner will probably be trying to face the wedge robot at all times.

This is one of the reasons why several wedges have angled side edges, such in the floppy wedge from the middleweight Devil's Plunger (pictured below). Its wedge has angled sides usually fabricated through bending or welding. These angled sides can turn even a perfectly aligned hit at a low forward speed, which would have θ ≅ 90° leading to α' ≅ 0°, into a spinner-launching hit. The effective slope angle α' is a design parameter for the side edges, measured on a vertical section of the

wedge as shown in the picture. In addition, these side edges protect the wedge from being knocked off due to a side hit.

A suggested value for the side angle α' to launch spinners would be, for a smooth titanium wedge, equal to α' = α_launch = 37°. And, for a hardened steel wedge, which has a larger coefficient of friction than smooth titanium, a suggested side angle α' to launch spinners would be Leonardo Da Vinci's α_launch = 34°.

6.14.7. Wedge Design Against Vertical Spinners

One final concern regarding wedge design is to make sure it will be effective against vertical spinners.

As seen in the picture in the next page, a vertical spinner with weapon radius r and weapon shaft height h_1 will most likely hit a wedge with height h_2 and slope angle α at a height $y = h_1 - r \cdot \cos\alpha$. Clearly, the height h_2 must be larger than y, so α cannot be too high to make sure that

$$y = h_1 - r \cdot \cos\alpha < h_2 \quad \Rightarrow \quad \cos\alpha > \frac{h_1 - h_2}{r}$$

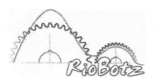 

Note that if the spinning weapon has large tip inserts, or if it is a very wide bar, then the radius r used in the above equations must be calculated as in the picture to the right, considering the bar width and tip insert dimensions in the value of b, as well as the bar length 2·a.

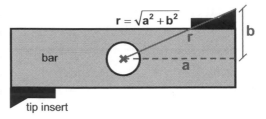

But even if the above condition for α is satisfied, there is no guarantee that the vertical spinner won't bounce off and end up hitting the top plate of the wedge robot. This situation is very likely to happen when low profile wedges charge forward at high speeds against tall vertical spinners. To avoid that, the top of the wedge should have an overhanging

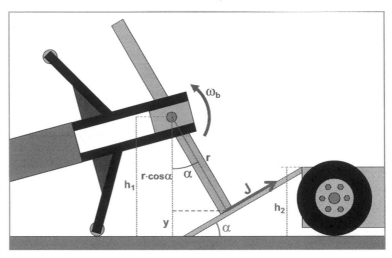

section, such as the small one on Devil's Plunger or the large one on Pipe Wench, pictured below. A large overhung section is very effective against vertical spinners, working as a scoop, as shown in the action shots below featuring Pipe Wench vs. Terminal Velocity.

Note from the action shots that the wedge should be charging forward towards the vertical spinner to most effectively launch it and eventually flip it. This is because the vertical spinner weapon always hits tangentially to the wedge surface. If the robots are not moving forward and the wedge is sufficiently smooth, you should only see sparks and both robots repelling each other, but no major hits.

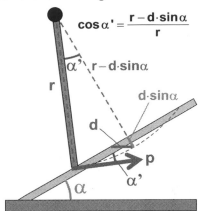

$$\cos\alpha' = \frac{r - d\cdot\sin\alpha}{r}$$

But if the robots are moving forward at high speeds, then the tooth travel distance d will result in a tooth bite d·sinα, due to the slope of the wedge. This tooth bite will result in an effective angle α' ≠ 0 between the speed of the weapon tip and the wedge surface, as pictured to the right, where r is the radius of the spinning weapon and cosα' = (r – d·sinα)/r. So, the higher the forward charge speed, the higher will be the tooth travel distance d and the angle α', resulting in a higher impact that might launch the vertical spinner.

6.15. Gyroscopic Effect

An interesting characteristic of robots with spinning weapons is their gyroscopic effect. Vertical spinners and drums tend to lift off their sides (tilting) when making sharp turns with their weapon turned on. The picture to the right shows our drumbot *Touro*, which is able to make turns with only one wheel touching the ground. This wheel lift-off, besides impressing well judges and audience, works as an excellent "victory dance" at the end of a match.

However, if the robot's wheels lift too much off the ground it can be a disadvantage, because you'll have a hard time making turns, risking being flipped over. But what causes the gyroscopic effect?

The gyroscopic effect comes from the fact that bodies tend to remain in their state of motion, as stated by Newton's first law. In this case, they tend to maintain their angular momentum. When *Touro* tries to turn with the weapon turned on, it is forcing the drum to change its spinning orientation, making it harder to maneuver.

Horizontal spinners don't have this problem, because when turning they don't change the orientation of the weapon axis, which remains vertical. However, spinners have a small difficulty when turning in the opposite direction of the one that its weapon spins, and they turn more easily in the same direction. This doesn't have anything to do with the gyroscopic effect, it is simply caused by the friction between the weapon and the robot structure, which tends to turn the robot in the same sense of the weapon. This effect is usually small. The gyroscopic effect in the horizontal spinners appears when an opponent tries to flip them, because the high speed of the spinning

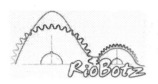

weapon provides them with a certain stability that helps them to remain horizontal. Our spinner *Ciclone* escaped from several potential flips because its weapon was turned on.

There are spinners that have weapon spinning axis that are not vertical, such as the robot Afterthought, pictured to the right. They have a small slope forward, intended to hit lower opponents because the weapon tip gets closer to the ground. Those tilted spinners suffer a little from the gyroscopic effect, which is proportional to the sine of the slope angle between the weapon spinning axis and the vertical. The smaller the angle, the smaller the effect.

This tilted spinner type has a serious problem: there is a chance that the robot gets flipped over during its own attack. If for instance the weapon spins in the sense shown in the picture above, it won't have problems when its opponent is right in front at its left side (the "good" side shown in the figure, which hits like an uppercut). But if the opponent is on its right side (the "bad" side), then the tilted spinner may be flipped over when hitting from top to bottom.

The gyroscopic effect, besides making it harder for vertical spinners and drumbots to make turns, also causes the phenomenon called precession, the same that explains why a spinning top can have its spin axis sloped without falling over. This phenomenon explains the wheel lift off (tilting) during sharp turns. The precession of a robot's weapon can be calculated from the principle of angular momentum (which results in what is known as Euler's equations), which depends on the rotational moment of inertia of the weapon in the spin direction (horizontal, in this case) I_{zz}, and on the moments of inertia in the 2 other directions, I_{xx} and I_{yy}. The weapons of those robots usually have axial symmetry (as in the disks of the vertical spinners or the cylinders of the drums), and therefore $I_{xx} = I_{yy}$.

In the figure to the right, the z-axis was chosen in the direction that the weapon spins, with angular speed ω_z, from ground up in such a way to throw the opponents into the air. Notice that the robot's right and left wheels are represented by the letters R and L respectively. The y-axis was chosen as the vertical one, and the x-axis is directed horizontally towards the front of the robot. When the robot is turning around its y-axis with angular speed ω_y and with its weapon turned on spinning with ω_z, the principle of the angular momentum results in

$$\tau_x \cong I_{zz} \cdot \omega_z \cdot \omega_y$$

216

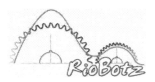 

where τ_x is an external torque applied in the direction of x. This equation is a good approximation if the tilting angle of the robot is small.

If the robot turns left, then $\omega_y > 0$, and therefore from the above equation we get $\tau_x > 0$. This means that the gravity force needs to generate a positive torque in the x direction to keep the system in balance, which happens when the right wheel (R in the figure above) lifts off the ground, see the left figure below. Similarly, if the robot turns right, then $\omega_y < 0$ and therefore $\tau_x < 0$. This negative torque that the gravity force needs to generate is obtained when the left wheel (L in the figure above) lifts off the ground, see the right figure below.

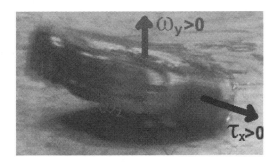 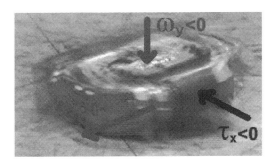

Those results are the same for drumbots as well as for vertical spinners. For instance, during the final match of the RoboCore Winter Challenge 2005 competition, the middleweight vertical spinner *Vingador* was spinning its disk with $\omega_z > 0$. After receiving an impact from our horizontal bar spinner *Ciclone* (left figure below), *Vingador* began to twirl with a clockwise speed $\omega_y < 0$. To balance this movement, a negative torque $\tau_x < 0$ would be necessary. However, even lifting its left wheel, the gravity force wasn't able to generate enough torque. *Vingador* continued tilting (with an angular acceleration $d\omega_x/dt > 0$, see the right figure below) until it ended up capsizing over its right side.

Vertical spinners have more problems with the gyroscopic effect than drumbots. The reason for that is because the gyroscopic effect is proportional to the angular speed of the weapon ω_z, while the kinetic energy depends on ω_z^2. Vertical spinners usually have lower weapon speed ω_z and higher moment of inertia I_{zz} than drums.

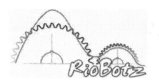 

Therefore, for instance, a vertical spinner that spins a solid disk of mass $m_b = 10kg$ (22lb) and radius $r = 0.3m$ (almost 1ft) at $\omega_z = 1,000RPM = 105rad/s$ has a weapon with moment of inertia $I_{zz} = m_b \cdot r^2/2 = 10 \cdot 0.3^2/2 = 0.45kg \cdot m^2$ and with kinetic energy equal to $E = I_{zz} \cdot \omega_z^2 / 2 = 0.45 \cdot 105^2 / 2 = 2,481J$. A drumbot that spins a cylinder with the same mass $m_b = 10kg$ (22lb) and external and internal radii $r = 0.06m$ and $r_i = 0.04m$ at $\omega_z = 4,173RPM = 437rad/s$ has $I_{zz} = m_b \cdot (r^2 + r_i^2)/2 = 10 \cdot (0.06^2 + 0.04^2)/2 = 0.026kg \cdot m^2$, and the kinetic energy of the weapon is $E = I_{zz} \cdot \omega_z^2 / 2 = 0.026 \cdot 437^2 / 2 = 2,483J$, practically the same energy of the vertical spinner. Therefore, both robots have similar destruction power.

However, the angular momentum of the vertical spinner weapon is $I_{zz} \cdot \omega_z = 0.45 \cdot 105 = 47.25$, much larger than the one from the drum, $I_{zz} \cdot \omega_z = 0.026 \cdot 437 = 11.36$. Because the gyroscopic effect depends on the product $I_{zz} \cdot \omega_z$, a drumbot usually tilts much less than a vertical spinner while making turns.

It is possible to get a better estimate of the gyroscopic effect, explicitly considering the tilt angle α with respect to the horizontal (as pictured to the right), which had been assumed to be very small in the previous calculations. As the robot turns with speed ω_y and with its weapon spinning with ω_z, the robot tilts by the angle α. The projection of the vector ω_z onto the vertical, $\omega_z \cdot \sin\alpha$, doesn't change direction, but the horizontal projection $\omega_z \cdot \cos\alpha$ does, rotating around the y-axis with speed ω_y, which is responsible for the gyroscopic effect.

The gravity torque τ_x is equal to $m \cdot g \cdot d \cdot \cos\alpha$, where m is the mass of the entire robot, g is the acceleration of gravity, and d is the distance between each wheel and the robot's center of mass, as pictured above. Assuming that ω_z is much larger than ω_y (because the weapon spins much faster than the robot turns), the principle of angular momentum states that the tilting movement is in equilibrium if $\omega_y = \omega_{y,critical}$, where

$$\tau_x = m \cdot g \cdot d \cdot \cos\alpha = I_{zz} \cdot (\omega_z \cos\alpha) \cdot \omega_{y,critical}$$

Canceling the $\cos\alpha$ term from the equation, we get

$$\omega_{y,critical} = \frac{m \cdot g \cdot d}{I_{zz} \cdot \omega_z}$$

In other words, if you turn with speed ω_y equal to $\omega_{y,critical}$, the robot will tilt with an arbitrary angle α (the robot stability does not depend on α, at least in the considered model approximation). If ω_y is smaller than $\omega_{y,critical}$, the robot doesn't lift any wheel off the ground. And if ω_y gets larger than $\omega_{y,critical}$, the robot tilting will become unstable, capsizing over its side. At the final match of the RoboCore Winter Challenge 2005, *Ciclone*'s impact made the robot *Vingador* twirl with a speed ω_y larger than its critical value $\omega_{y,critical}$, capsizing it.

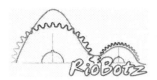 

In the previous example, which compared a vertical spinner and a drumbot with same weapon system energy, assuming that m = 55kg (about 120lb) and d = 0.2m (7.9"), the above equation would result in the conclusion that the vertical spinner would not be able to make turns faster than $\omega_{y,critical}$ = 2.28rad/s = 21.8RPM without lifting its wheel and risking capsizing. On the other hand, the considered drumbot would be able to turn even at $\omega_{y,critical}$ = 9.5rad/s = 90.7RPM without lifting off the ground, a much more reasonable value.

Finally, to avoid that the robot capsizes on its side, it is necessary that $\omega_{y,critical}$ has the largest possible value. Therefore, it is important that the base of a vertical spinner or drumbot is wide, because a larger distance 2·d between the wheels increases $\omega_{y,critical}$, as seen in the above equation. This explains, for instance, the reason for the large distance between the wheels of the fairyweight vertical spinner Nano Nightmare, pictured to the right.

If the vertical spinner from the previous example had a wider base with d = 0.3m (due to a distance 2·d = 600mm between the wheels, about 23.6"), then the calculations would result in allowable turning speeds of up to 33RPM. In other words, it would be able to turn 180 degrees in less than 1 second, a reasonable value to keep facing the opponent.

6.16. Summary

In this chapter, it was shown that weapon design can benefit a lot from basic physics calculations. It was concluded that spinning disks have better inertia and in-plane bending strength than bars with same weight, however they suffer from a lower out-of-plane bending strength. The concepts of tooth bite, as well as effective mass, stiffness and damping, were introduced, showing that the effectiveness of an impact depends on properties of both attacker and attacked robot. It was shown that drumbots have one of the highest effective masses M_1, however they usually suffer limitations regarding the speed v_{tip} of the weapon tip. Impact equations were presented for several robot types, including spinners and hammerbots. The difference between defensive and offensive rammers was discussed, including information on how to setup their shields. It was shown why thwackbots and overhead thwackbots are difficult to steer to try to hit an opponent. Lifter and clamper stability equations were also presented. It was seen that "height launchers" can benefit from a long scoop, while "range launchers" should choose average impulse angles between 36° and 45°, depending on the opponent's aspect ratio, as long as they have enough tire friction not to be pushed backwards. It was found that defensive wedges have a slope angle of at most 25°, while an ideal offensive wedge would be made out of Ti-6Al-4V titanium with a smooth surface and a 37° slope. And, finally, gyroscopic effect equations were presented to help in the design of vertical spinners. In the next chapter, the main electronic concepts to power such weapon systems and the robot drivetrain are introduced.

Chapter

7

Electronics

There are countless electrical and electronic options to use in a combat robot. This subject by itself could result in an entire book. Because of that, in this chapter we try to summarize and limit the discussion to the most used components in combots, with effectiveness verified in practice, in the arena.

Combat robot operation demands a great number of electronic components, among them: radio transmitter, receiver, RC interface, speed controllers, relays/solenoids and on-off power switches, connected by plugs, terminals and wires. These components are described next.

7.1. Radio Transmitter and Receiver

7.1.1. Transmitters

A radio transmitter allows the driver to send commands to a receiver inside the combat robot. There are several radio manufacturers, such as Futaba, Airtronics, JR, Hitec, GWS and Spektrum. There are also other cheaper solutions, such as radios adapted from toys, wireless gamepads, and transceiver circuits, however you must guarantee that these low-cost systems will have enough power to avoid signal loss when the robot is inside the arena, as well as implement failsafe features in all channels, as explained later.

Radio systems are named for their number of channels, which is the number of outputs that a transmitter-receiver set has. For instance, a four-channel set can control four different devices. Most combat robots only use three channels: two for the drive system (forward/backward and left/right) and one to control the weapon, if any. Three and four-channel radios are cheaper and, in general, enough for a combot. However, radios with 6, 7 or more channels usually have more functions, being programmable and including internal memory, such as the Futaba 7CAP 75MHz radio pictured to the right.

Most radios use the 72MHz frequency band, which is reserved for air models only, which makes them prohibited in almost all combot competitions. The usual ground band is 75MHz, but several others are also used: 27MHz, 50MHz, 433MHz (only in Europe, Africa and Middle East), 900MHz (only in the Americas), 2.4GHz and 5.8GHz. All these frequencies are part of the Industrial, Scientific and Medical (ISM) radio bands, originally reserved by an international treaty for the use of radio-frequency in the cited fields other than communications.

The 27 and 50MHz bands are normally employed in radio-controlled toys and old cordless phones, 433MHz is found in European transceiver pairs, while 900MHz and 2.4GHz are a commonplace in modern life, being present in wireless LANs, Bluetooth, cordless phones, and even microwave ovens. The 5.8GHz band is not very common, but it is also present in wireless network systems and phones.

Each band is divided into channels. There are only 30 channels for the 75MHz band, from channel 61 (75.410MHz) to 90 (75.990MHz). The 72MHz band, which is not allowed in combot competitions, has 50 channels, from 11 (72.010MHz) to 60 (72.990MHz). And the 27MHz band has only 6 channels, while the 50MHz has 10.

In a competition, it is forbidden to have two radios using the same channel at the same time. This is a safety measure, since the radio from one team could accidentally activate the robot from another. With this in mind, in the events it is mandatory to only turn your radio on if it has an appropriate frequency clip. Since there is only one clip available per channel at the event, the problem is solved.

However, if a distracted builder forgets to return a clip after using it, the event might suffer delays. As a result, a few events require the use of single bind systems, which only allow a receiver to follow commands from a single radio, normally using the 900MHz or 2.4GHz frequencies.

Those systems rely on Frequency Hopping Spread Spectrum (FHSS) and Direct Sequence Spread Spectrum (DSSS), which use a unique identifier code for the receiver and transmitter. They dynamically change the paired transmitter and receiver channels upon signal quality loss, caused normally by interferences (noise). This bi-directional system is only possible due to the larger bandwidth and higher frequency they use. A few 900MHz and 2.4GHz systems can use the available bandwidth to perform robot telemetry as well, such as in the IFI FRC or Spektrum Telemetry packs.

In 72 and 75MHz systems, channels are defined by a pair of crystals (pictured to the right), usually sold together. The Tx crystal must be installed into the transmitter, while its pair Rx is placed into the receiver. Always have a spare crystal pair from a different channel, in case you have to face an opponent that uses the same channel. Only buy crystals for the same band as your radio - 72MHz  crystals do not work on 75MHz radios and vice-versa. It is possible to convert a radio from 72MHz to 75MHz, a procedure that costs between US$20 and US$50 if done by professionals.

Besides frequency bands and channels, there is also modulation, which is the way that information is encoded into the radio wave. The most commonly used modulations among the countless existing standards are AM, FM, PCM and DSM, described next.

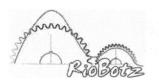

AM stands for Amplitude Modulation, in which information is transmitted varying the radio signal amplitude.

FM means Frequency Modulation, it is a standard less prone to noise than AM. Control information is transmitted adding a variable frequency wave into a carrier wave. The receiver then extracts the information from the carrier and sends the appropriate commands to the devices attached to it.

The third modulation type is PCM, which means Pulse Code Modulation. Technically, this transmission is also FM, the difference is that in PCM the information is digitally transmitted. Instead of an analog transmission, signals are send in digital form, coded, becoming a lot more reliable. PCM provides an even greater noise immunity.

DSM, which stands for Digital Spectrum Modulation, is Spektrum's proprietary modulation for 2.4GHz systems. It divides the 2.4GHz band into 80 channels (slots), using some DSSS and FHSS features, with a unique identifier that only allows communication between a single transmitter and its bound receiver. In the unlikely event of all channels becoming occupied, the link between the Tx and the Rx won't happen. This standard has been updated to DSM2; unfortunately, older radios such as Spektrum DX6 only accept DSM receivers, although DSM2 radios work with DSM receivers.

Besides Spektrum and JR (which also uses DSM), other radio brands have their own 2.4GHz modulation, such as Futaba's FASST, Airtronics' FHSS-2, and Hitec's AFHSS. Their differences are minimal, so choose them keeping in mind your budget and favorite brand.

7.1.2. Receivers

A receiver is the component responsible to demodulate the radio-transmitted signals and direct the commands to servos and other electronic circuits. A typical receiver is pictured to the right, a Futaba 75MHz. They come in several sizes and weights.

For insect robots, such as fairyweights (150g), antweights (1lb, equivalent to 454g) or beetleweights (3lb, equivalent to 1361g), it is a good choice to use the GWS micro receiver (left picture to the right) or the Nano receiver (right picture). They are really small and weigh between 2 and 8 grams without the crystal. Be careful, they do not work with regular crystals, they need special ones.

It is extremely important that the Rx crystal (if needed) is well attached to the receiver. It is usually a good idea to use an adhesive tape around the receiver to prevent the crystal from becoming loose. Many fights have been lost because of a knocked off crystal due to the vibrations from a major hit. Tape as well the connectors, to avoid the cables from becoming loose. It is also a good idea to use some adhesive tape to cover the receiver inputs that are not used, to prevent metal

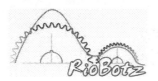 

debris from causing a short-circuit during a round. Also, it is fundamental to shock-mount the receiver, using for instance foam, EVA or rubber.

All modern receivers pick up the modulated signal and then decode it, resulting in a Pulsed Position Modulation (PPM) signal that contains the information about all output channels. This information is framed in an envelope of 20ms (or some other fixed value between 18 and 25ms), consisting of a train pulse with several 5V pulses, one for each output. These 5V pulses have periods that vary between 1ms and 2ms, depending on the command sent by the driver. If a radio stick is completely to the left or down, then the period of the associated pulse from this channel is usually 1.0ms. If the stick is centered in neutral position, then it is 1.5ms. And if it is to the right or up, then the pulse lasts 2ms. For switches, as those that are usually placed on the top portion of a radio transmitter, 1ms would be associated to the off position and 2ms to the on position. Most high-end systems allow these configurations to be changed on the transmitter unit. For instance, a channel can be reversed, or it can be mixed with others. PPM is further explained in section 7.8.2.

Among the radio features, *failsafe* is one that is required by all combot events. A few vendors call it *failproof* or *smartsafe*. It consists of a subsystem that allows to pre-program each receiver channel output in case of a signal loss. Program the failsafe to, in the event of signal loss, send a 1.5ms pulse signal to the drive system channels (associated with a centered radio stick, which would stop the robot's translational movement), as well as send a 1ms pulse signal to a channel from a solenoid or relay from the weapon system (if the 1ms pulse is associated with the off condition). This is an obligatory feature, checked in the safety inspection from any combot event. For hobbyweights (12lb) or heavier robots, all used channels must have a failsafe. Usually, robots in insect classes only need to have failsafe on the weapon channel.

AM, FM and a few other radio systems do not have built-in failsafe. Failsafe must then be implemented between the receiver and the commanded devices, which can be accomplished using an appropriate RC (radio-control) interface board, or through a dedicated module such as the Micro Failsafe Dynamite (pictured to the right), which can be bought in R/C hobby stores.

Listed below is a summary of the main features from radio-receiver systems:

- number of output channels: in general from 4 to 9 in air radios, or 3 in pistol-style ground radios; for a robot with an active weapon, at least three channels are needed;
- reversion: ability to program channel output inversion;
- ATV / EPA: adjustable maximum and minimum values that an output can have;
- dual rate / exponential: output sensibility and linearity adjustment;
- mixing: ability to mix channels (very useful), which sometimes can be programmable;
- multiple models: allows the storage of several distinct programs, one for each robot;
- failsafe: allows to program the channel outputs in case of a signal loss; only digital radios have this feature, but a few do not have it in all channels; analog radios must use an external module;
- frequency channel reassignment: ability to automatically switch the transmitter and receiver channels.

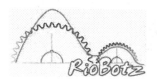

7.1.3. Antennas

A huge problem that strikes combat robots is signal loss. Combots are made out of metal, therefore antennas placed inside them may suffer from the Faraday's Cage effect, where the robot chassis blocks the signal or at least considerably reduces its strength.

Robots with polycarbonate covers don't suffer from this problem, since this material is radio transparent. Sometimes, if a robot is completely shielded by metal, then the solution is to place the antenna outside it. There is a risk that the antenna will get damaged, impairing signal reception, however it is a better choice than to not have a control signal at all.

The good news is that the higher the frequency, the smaller is the wavelength. Therefore, modern systems that use 2.4GHz have waves that can go through the breaches and gaps on the robot covers (such as the holes around the wheels). This allows you to place the antenna inside the robot. All our bots have internal antennas, since we've changed the radio system from 75MHz to 2.4GHz.

Antennas are usually a conductive wire with one fourth of the signal wavelength. Therefore, a 75MHz signal traveling at light speed (about 300,000 km/s) has a wavelength of (3×10^8 m/s) / (75×10^6 Hz) = 4 meters (about 13 feet). Since a 4 meter long wire would be too long, the 75MHz systems use one with 1/4 of this length, 1 meter (a little over 3 feet). In practice, it is a good idea to place the antenna wire in a zig-zag pattern inside the robot. Often, the longer the antenna, the higher is its gain, so if your robot needs a signal boost then try longer antenna lengths.

You can replace the 3-foot wire from the receiver antenna by a mini-antenna with less than 8 inches long. But you need an amplified antenna for a good result, such as the Deans Base-Loaded Whip (pictured to the right). Before switching from 75MHz to 2.4GHz, we used this antenna in our middleweights Touro and Titan without any problems, however it had to be placed outside the robot due to Fadaray's Cage Effect.

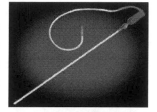

Avoid placing the antenna directly over metal. Ideally, it is good to have at least 6mm (1/4 inch) between any part of the antenna and a metal part. This can be achieved with a large rubber grommet or with some Lexan or Delrin spacer, such as the Delrin Antenna Mount pictured to the right, sold at The Robot MarketPlace.

Since the wavelength in 2.4GHz systems is only (3×10^8 m/s) / (2.4×10^9 Hz) = 0.125m (less than 5 inches), their antennas are really tiny, between 31mm and 62.5mm in length. These receivers (such as the Spektrum 2.4GHz BR6000 model pictured to the right) normally have two antennas, due to wave polarization. It is recommended to place the antennas forming a 90° angle to maximize signal reception.

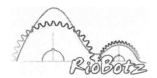

7.1.4. Gyroscopes

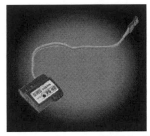

To guarantee that a robot can follow a straight line, you can use gyroscopes (a.k.a. gyros), which are sensors that measure the robot orientation angle. Easy to use, they can be directly connected to most receivers. Almost all R/C hobby stores sell gyros, which are usually used in radio-controlled model helicopters. One of the cheaper models is the GWS PG-03 Micro Gyro (pictured to the right), which costs US$38 at The Robot MarketPlace, which has a great cost-benefit.

The secret to use gyros in combat robots is to adjust the feedback gain to a maximum of 20% of its full scale. This worked very well with our hobbyweight wedge Puminha using a GWS PG-03. Higher gains can make the system become unstable, since the gyroscope will pick up motor vibrations and try to compensate them.

It's not recommended to use gyros on invertible robots, because when the robot is upside down the control gain is inverted, giving positive feedback and making the robot spin out of control (a.k.a. the "Death Spin"). To solve this problem, it would be necessary to have an electronic system to turn the gyro off when the robot is flipped over, or even better, to invert the gain in this case.

Note that gyros must be very well shock-mounted inside the robot not to break. And they must be well secured because, if they get loose or shift their position inside the robot, it will go crazy or go into "Death Spin" mode.

7.1.5. Battery Elimination Circuit

In several model airplanes, there is a small battery pack used exclusively to power the receiver. If your electric motors generate a lot of noise, it can be a good idea to have separate packs for the motors and for the receiver, to avoid interference or signal loss. However, if the motors have noise suppression capacitors or shielded armature, then it is much simpler to use the robot's main batteries to power as well the receiver, without using a separate pack.

This small receiver pack is troublesome: it can become loose inside the robot, it increases the robot weight, and it is another battery pack that needs to be charged, hogging your chargers. In addition, there's a chance you might forget to charge them during a hectic event.

To eliminate the receiver pack, you need a Battery Elimination Circuit (BEC). The BEC is nothing more than a voltage regulator that guarantees a constant supply of, usually, 5V (or other value between 4.8V and 6V).

A few speed controllers, discussed later, already have built-in BECs. But most of them use linear regulators, such as the LM7805, which can overheat if the voltage drop is too high. Our middleweight Touro needs a BEC because the Victor speed controllers that it uses do not have this feature. We once were out of BECs, so we've experimented using BaneBots BB-12-45 speed controllers just to work as BECs. These speed controllers have a built-in BEC, however due to the voltage drop from 27.2V (20 NiCd cells in series fully charged) to 5V they overheated, malfunctioned, and sometimes even got burnt. This speed controller is rated for 24V at most, so it is

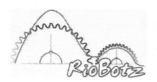 

not recommended to power it beyond this voltage, unless some heatsink is attached to it to dissipate the heat from the linear regulator.

If a speed controller has a built-in BEC but you don't need to use it (because you have a separate stand-alone BEC, for instance), then it is a good idea to remove the red wire (the middle one) from the crimp connector that goes into the receiver. This will avoid it from overheating, especially if it uses a linear regulator. This tip is valid for both brushed and brushless speed controllers.

There are commercially available stand-alone Battery Elimination Circuits, named UBEC (U stands for Universal), such as the S-BEC Super BEC 5V model pictured to the right. They are usually cheap, found in R/C hobby stores. A few UBECs use switching regulators, such as the LM2575, which switches the voltage on and off in order to drop its value, instead of dissipating the power from the excess voltage as heat. In this way, they are less prone to overheat or malfunction.

You can also develop your own BEC. We've used our own BECs for years with great results, as discussed in section 7.8. If you decide to build your own BEC, choose switching regulators, they don't overheat as easily as linear regulators.

7.1.6. Servos

Servos, a.k.a. servo-motors, are motors with embedded position control, such as the Hitec standard model pictured to the right. In model aircrafts, small low-power servos are directly connected to the receiver, and powered by it. Servos are very practical and cheap, a few combat robots use them in the throttle system of internal combustion engines, or to mechanically control some electric switch. The problem with this approach is the great risk of servo failure after an impact, which can break them or let them become loose. In combots, always implement your control system electronically (in solid state), avoiding moving parts or servos.

Servos have been used in insect robots, actuating lifter or clamper/grabber mechanisms that need position control to properly function. With a simple modification, servos can become DC motors with continuous rotation, with an embedded electronic control to convert PPM signals into movement. The modification consists of disassembling the servo, exchanging the internal potentiometer by a two-resistor ladder, removing the plastic stop from the largest gear, and reassembling the unit. A good tutorial on this servo modification can be found at http://www.acroname.com/robotics/info/ideas/continuous/continuous.html. The modified servos can be used in the drivetrain of small robots such as insect combots or light sumo bots, however only the spin direction can be controlled, not its speed, as in the bang-bang control discussed later.

Note that servos included in radio transmitter-receiver packages do not have enough power to actuate most systems in larger bots. High torque servos should be preferred, however most of them will need to be modified if you need continuous rotation. If you need continuous rotation, including speed control, then it is better to stick to DC motors with speed controllers, discussed next.

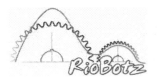

7.2. Controlling Brushed DC Motors

7.2.1. Bang-Bang Control

The most primitive way to drive a DC motor is through a bang-bang controller. Is consists of using relays to create a basic H-Bridge scheme, in such a way that the motor can spin at full power forward, full power backward, or stop. The figures below show how it is possible to use 4 relays to create a bang-bang controller for two opposite wheel motors.

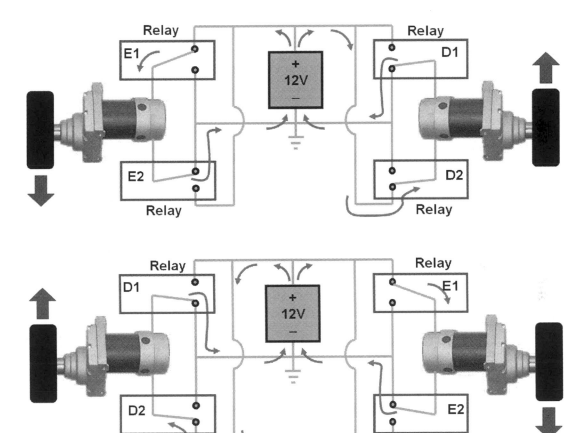

As shown in the upper figure, when the E1 and E2 relays are closed and the D1 and D2 are open, the robot turns to the left. On the other hand, closing relays D1 and D2 while opening E1 and E2 would make the robot turn to the right, as shown in the lower picture. To move forward, you only need to open all four relays. To move backward, you should close all relays. Finally, to stop the robot, you can choose for instance to close E2 and D2, while opening E1 and D1. In the above example, if a 12V battery was used, then each motor would only receive either +12V, 0V or −12V.

Avoid using this type of controller, because it always provides the maximum voltage (in absolute value) to your motors, abruptly reversing their direction. This sudden reversion of the movement causes premature brush wear and can lead to broken gears because of the associated

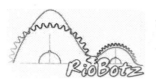

impacts (which are the origin of the "bang-bang" expression). In addition, the high inductance of the motors will create voltaic arches on the relay terminals, shortening their useful life. This kind of controller is only acceptable in low speed motors, which are not very common in a combat robot, except perhaps for a slow lifting or clamping/grabbing mechanism.

A slightly improved version of a bang-bang control would use the 4 relays in series with a single adjustable linear voltage regulator. As seen in chapter 5, the speed of a brushed DC motor is proportional to the applied voltage, so the linear regulator would control speed, while the relays would control direction. Our first combot, the middleweight Lacrainha, used such bang-bang implementation to control the speed of its wheels. The electronic board that we developed also featured an automatic system that would briefly lower the applied voltage down to zero using the linear regulator during the reversion of the relays, to avoid voltaic arches on the relay terminals.

Although very simple, this version of bang-bang control has serious issues, mainly due to its low efficiency at low speeds, since all the energy from the excess voltage that is not utilized by the motor is dissipated as heat on the linear regulator. In addition, the resulting electronic system may still be unreliable since it depends on mechanical moving parts from the relays.

To achieve a higher motor efficiency and more compact and reliable circuits, it is necessary to vary the motor speed in a different way, known as PWM, described next.

7.2.2. Pulse Width Modulation

The Pulse Width Modulation (PWM) method consists of turning the motor voltage on and off, on a fixed frequency basis, through an electronic switch, usually some kind of transistor. The transistor can be, for instance, a Bipolar Junction Transistor (BJT) or a Metal-Oxide Semiconductor Field-Effect Transistor (MOSFET, a.k.a. FET).

The motor speed is then proportional to the ratio between the time interval T_{on} during which the motor is on and the pulse period T. This relation is named Duty Cycle (D) and, if multiplied by the peak voltage supplied to the motor, it results in an average voltage that can be controlled.

The figures below show three PWM signals that use the same frequency. Figure A shows the PWM signal with almost 100% D, because the time T_{on} is almost equal to the period T (the signal is high almost 100% of the time). Therefore, a motor subject to the pulse A would behave as if it was powered by almost the nominal voltage.

The pulse in figure B, on the other hand, is high half of the time, therefore T_{on} is equal to T/2, and the motor would receive about half of the nominal voltage, spinning at half-speed (50% D). Finally, the pulse in figure C would make a motor spin very slowly, since the time T_{off} during which it is off, equal to $T - T_{on}$, is almost equal to T (almost 0% D).

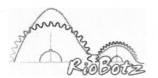 

The efficiency of a PWM circuit is, in an ideal case, 100%, since the switching process that would power the motor wouldn't have any losses. But in practice there are losses in the used transistors, due to their resistance. Despite this, the efficiency of a well-designed PWM controller is usually above 90%.

Be careful not to confuse PWM with PPM (or even PCM), several people mix these concepts, saying that a receiver output signal is PWM. Although these signals are similar, this is not true. Both PWM and PPM are analog pulse trains, yet their functionalities are completely distinct. One of the differences is that the PPM needs to use a train pulse with a precisely defined period. The PWM, on the other hand, can have any pulse period T, what really matters is only the Duty Cycle, the ratio between T_{on} and T. The pulse frequency $1/T$ is an arbitrary value, but it should be large enough to avoid undesired oscillations in the motor. Typical values of $1/T$ are above 4kHz, sometimes higher than 16kHz. Using 20kHz or higher is a good idea to avoid buzz sounds, since it is above the range of human hearing.

7.2.3. H-Bridge

PWM by itself is only able to control the absolute value of the motor speed. There are a few possible ways to control the motor direction. You can, for instance, physically invert the terminals using mechanical switches such as relays and solenoids, however this would result in a bulky and impact-sensitive system, with lower reliability, as discussed before. Another option is to generate a negative voltage, however this can result in a very complex system since the main power supply is a battery, which provides direct current.

The best option is to use an H-Bridge, as mentioned before, named after the disposition of the switches in the circuit. It does not need to generate negative voltages, or to mechanically disconnect the terminals. The H-Bridge can use transistors, which easily stand high currents.

The picture to the right shows a basic H-Bridge, with a motor M and 4 transistors S1, S2, S3 and S4 that work as solid-state switches.

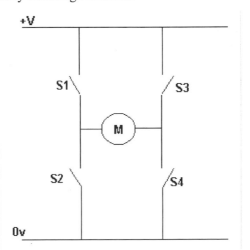

To make the motor spin forward with a voltage V, you just need to activate S1 and S4 and deactivate (open) S2 and S3. To spin backward with voltage V, activate S2 and S3 and deactivate S1 and S4.

To brake the motor, you can either activate only S1 and S3, or activate only S2 and S4, shorting the motor terminals. This braking effect is called motor brake and happens due to the entire energy being dissipated by the motor internal resistance, which is usually very low. This results in quick energy dissipation, which happens until the motor stops. Note that motors with very small internal resistance can generate a lot of heat in this process, so be careful!

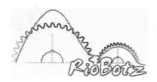

In the figure to the right, the switches are implemented using MOSFETs, resulting in a simple H-Bridge circuit.

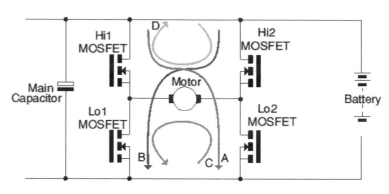

To spin the motor forward, the current from the battery needs to go through the FETs Hi1 and Lo2, following the path A shown in the figure. To spin backward, Hi1 and Lo2 must be deactivated, and immediately after this Hi2 and Lo1 need to be activated, making the current follow path B. Clearly, to brake the motor you can either activate Lo1 and Lo2 (for the current to follow path C, in either sense), or activate Hi1 and Hi2 (to follow path D, in either sense). Note that paths C and D are only possible because these FETs feature integrated free-wheeling diodes, which allow the current to go from source to drain (the upward direction in the figure). Make sure these diodes are present, otherwise the motor inductance will fry the FETs.

PWM is very easy to be implemented into the H-Bridge. When spinning forward, it is enough to keep Hi2 and Lo1 always deactivated and Hi1 always activated, while Lo2 is only active during the T_{on} interval from the PWM, making the current follow path A.

Due to the motor's inductance, the current tends to keep a constant flow even during the T_{off} period. Therefore, during T_{off} from a forward movement, the current will follow the opposite sense of path D, flowing through the motor in the same sense as it did in path A. Note that Hi2 can remain deactivated during T_{off}, because its free-wheeling diode always allows conduction from source to drain, necessary for the opposite sense of path D (Hi2 only controls the flow from drain to source).

This path during T_{off} shorts out both motor's terminals, however with minimal energy losses. It indeed brakes the motor, but this is not a problem, because it only happens during T_{off}. It is the combination of motor acceleration (during T_{on}) and braking (during T_{off}) that allows the resulting average speed to achieve any value between zero and the top speed. As mentioned before, the period T must be small enough so that the acceleration and braking effect cannot be noticed, avoiding oscillations in the motor.

Analogously, to spin backward, you just need to keep Hi1 and Lo2 always deactivated and Hi2 always activated, while Lo1 is only active during the T_{on} interval from the PWM. During T_{on}, the current will follow path B, while during T_{off} the motor inductance will make the current flow through the motor in the same sense as it did in path B, resulting in path D in the sense described in the figure (not in its opposite sense). Note that Hi1 can remain deactivated during T_{off}, because its free-wheeling diode always allows conduction from source to drain, necessary for path D.

A major issue with H-Bridges is an effect called shoot-through, which occurs if two switches from the same side are active at the same time, for instance Hi1 and Lo1. If this happens, the battery is shorted out, generating a very large current that usually destroys the MOSFETs. To address this issue, resistors are installed in series with the MOSFET gates, delaying their activation, while fast diodes are installed in parallel with those resistors, as described in section 7.8.

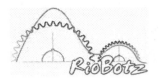

7.3. Electronic Speed Controllers

Electronic Speed Controllers, or simply ESC, are electronic power systems that implement an H-Bridge with PWM, to control both spin direction and speed of a motor. There are several ESCs in the market, so we'll focus on the ones that we have already tested in combat: OSMC, Victor, Scorpion and BaneBots, which are all brushed Permanent Magnet DC (PMDC) motor controllers. All these systems can be purchased, for instance, at The Robot MarketPlace, IFI Robotics, BaneBots, Trossen Robotics or Robot Power.

To control Brushless DC (BLDC) motors, a Brushless Electronic Speed Controller (BESC) is required, explained in section 7.3.6.

7.3.1. OSMC – Open Source Motor Controller

OSMC is a speed controller board capable of powering a single DC motor with nominal

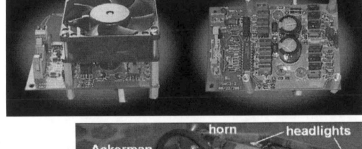

voltages between 13V and 50V, handling continuous currents of up to 160A, with 400A peaks. They are very robust, using 16 MOSFETs on an H-Bridge (4 transistors per leg), cooled by a fan. The pictures to the right show the OSMC board with and without the fan.

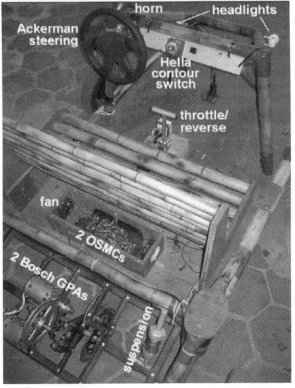

We have used OSMCs in a few of our combots, such as the middleweights Anubis, Ciclone and Titan. We also use 2 OSMCs to power the 2 Bosch GPA motors that drive our bamboo electric vehicle, pictured to the right. This vehicle, built by the students, is useful to carry two people and a middleweight from our lab to the "weapon testing field" from our University (a.k.a. the soccer field).

The OSMCs are a little bulky if compared with other ESCs available in the market, but they are very reliable. The middleweight wedge Max Wedge, honorable mention in the Robot Hall of Fame, used OSMCs to control the speeds of its power hungry D-Pack motors.

The Open Source Motor Controller is the collaborative result of several combot and electric vehicle builders. The OSMC

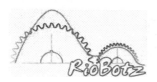 

diagrams can be found at http://robotpower.com/osmc_info. In this very site it is possible to buy an assembled and tested OSMC board for US$219, or a full kit including a bare-board and all components needed for the assembly at US$169.

To save some money, our team bought several bare-boards for US$29 each at Robot Power, and looked for information about the needed components and their suppliers in the OSMC Discussion Group (http://groups.yahoo.com/group/osmc) to buy them individually. The savings were not significant if compared to the Robot Power full kit, our cost per board was slightly below US$159. But at least we've learned a lot about each individual component and its vendors.

One disadvantage of the OSMC board for use in combots is the need for a separate electronic RC interface between the OSMC and the receiver, called MOB (Modular OSMC Brain), or μMOB for a smaller version. Unfortunately, the MOB/μMOB interfaces have been discontinued, so we've developed our own RC interface board, which is able to control 2 OSMCs. Our RC interface, which can also trigger a solenoid to be used in the weapon system, is detailed in section 7.8.

7.3.2. IFI Victor

Victor is a family of speed controllers from IFI. These ESCs are extremely compact, almost as robust as the OSMC. We use Victors in our middleweight Touro to save precious space for other components (Touro ended up so compact that OSMCs wouldn't fit inside it). Each Victor only controls one motor, so you'll need at least two in the robot drivetrain.

There are several Victor models, however externally they have the same look, as pictured to the right. A fan is used to cool the MOSFETs from the H-Bridge. There are several different models and prices, listed next.

- Victor 884 – US$ 114.95: single channel, forward and reverse, from 6 to 15V, handling up to 40A continuously; it uses 12 MOSFETs, 6 for each direction;
- Victor 883 – US$ 149.95: single channel, forward and reverse, from 6 to 30V, up to 60A continuously, with a surge current capability of 100A for less than 2s, and 200A for less than 1s; it uses 12 MOSFETs, 6 for each direction;
- Victor 885 – US$ 199.95: single channel, forward and reverse, from 6 to 30V, up to 120A continuously, surge currents of 200A for less than 2s, and 300A for less than 1s; it uses 12 MOSFETs, 6 for each direction;
- Victor HV-36 (the model shown in the picture above) – US$199.95: single channel, forward and reverse, from 12 to 42V, up to 120A continuously, surge currents of 250A for less than 2s, and 275A for less than 1s; it uses 16 MOSFETs, 8 for each direction;
- Victor HV-48 – US$199.95: single channel, forward and reverse, from 12 to 60V, up to 90A continuously, surge currents of 200A for less than 2s, and 225A for less than 1s; it uses 16 MOSFETs, 8 for each direction;

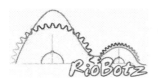 

- Victor 883SC – US$169.95: single channel, forward only (therefore it is usually used to power a spinning weapon), from 6 to 30V, up to 90A continuously, surge current capability of 100A for less than 2s, and 200A for less than 1s; it uses 12 MOSFETs, all of them for a single direction.
- Thor 885SC – US$219.95: single channel, forward only, from 6 to 30V, up to 150A continuously, surge current capability of 200A for less than 2s, and 300A for less than 1s; it uses 12 MOSFETs, all of them for a single direction.

Victors can be connected directly to the receiver, without the need for an external RC interface (such as the MOB needed by the OSMC). However, a signal booster cable (US$ 15, pictured to the right) is highly recommended between the receiver and the Victor, improving the quality of the signal. The booster cable is also interesting for robots that suffer from radio signal noise or loss, especially in combots with internal combustion engines.

IFI also sells speed controllers specifically developed to drive spinning weapons, denoted by the SC (Spin Controller) suffix, such as the Victor 883SC and Thor 885SC discussed above. They only work in one direction, but they feature softer acceleration and deceleration ramps that minimize current peaks, saving battery capacity. With them it is possible to control the speed of a spinning weapon, which can be especially useful in three situations:

1. batteries not lasting the entire match: you could spin the weapon at an intermediate speed, saving the batteries, and accelerate to full speed a few moments before striking your opponent;
2. damaged or cracked weapon: if, after a tough match, some structural damage or crack is found on a weapon component that cannot be replaced, then it is possible to avoid a broken weapon in the following fights adopting a lower speed;
3. robot vibration when the weapon is at full speed: slightly slowing down the weapon can minimize this problem, allowing you to control the weapon speed to avoid natural frequencies.

Despite the acceleration and deceleration ramp features from the SC types, most builders prefer to use in their weapon systems ESCs that are capable of forward and reverse spin, resulting in a fully reversible weapon system (which is very useful, for instance, for invertible drumbots). If you don't need a fully reversible weapon, such as in most horizontal bar spinners, and if you don't care about controlling the weapon speed, then it is a good idea to use solenoids, due their lower price tag and higher current limits. Solenoids are studied in section 7.4.

We have been using Victors HV-36 to control the drive system of all our combots from 30lb to 120lb. Our 4-wheel drive hobbyweight wedge Puminha also uses Victors HV-36, each one is used to power both drive motors from each side of the robot. But, since Victors in hobbyweights are usually overkill, we've removed their fans in Puminha without needing to worry about overheating problems, saving precious space.

The Victor HV-36 is a better option than the Victor 885 even for 24V combots, because it uses 16 MOSFETs instead of only 12, better handling high peak currents from high power motors. And the use of the Victor HV-36 also allows future robot upgrades up to 36V.

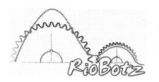

The Victor HV-36 is also a good option for the weapon system. After RoboGames 2008, we've replaced the weapon solenoids from our middleweights Touro and Titan, lightweight Touro Light, and hobbyweight Tourinho, with Victors HV-36, resulting in fully reversible weapons with speed control. One Victor HV-36 is needed for each Magmotor S28-150 or DeWalt 18V motor from the

weapon systems of Touro Light and Tourinho. But we use 2 Victors HV-36 for each Magmotor S28-400 from the weapon system of our middleweights, as explained in section 7.7.3, for redundancy and to avoid burning the ESCs due to the high stall current of this motor.

Since the fans are really important to cool down the MOSFETs, it is recommended to protect them. Debris or even loose wires from the electric system can touch the fan blades, making it stop. To avoid this, you can use a fan grill (pictured to the right), which can be made out of steel wire or from an aluminum sheet. Another good idea is to use the IFI Stainless Steel PWM Clip, which locks the signal cables onto the Victors, preventing the connectors from popping loose.

7.3.3. Robot Power Scorpion

Victors and OSMCs are suited for larger robots, typically featherweights (30lb) or heavier. They are expensive and relatively large to be used in robots that weigh 15lb or less. Robot Power sells four ESCs that are a good option for those lighter combots.

Scorpion XL and XXL

The most famous product from the Scorpion line is the Scorpion XL (US$119.99, pictured to the right showing its front and back sides), a controller that offers two PWM output channels, normally used to control both drivetrain sides. It can handle from 4.8 to 28V, up to 12.5A continuous, and 45A peaks per output, which can be combined into a single channel to provide higher current limits.

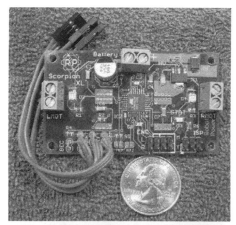

In 2008, a beefier version was introduced, the Scorpion XXL (US$159.99). It uses the same board from the XL version but, instead of one BTS7960B power IC per H-bridge leg, this version has a pair of them, one on top of the other. This allows a higher continuous current of 20A. The channels can also be combined, resulting in a single PWM output that could handle continuous 40A. To withstand the higher currents, the battery screw terminals were replaced with wire pig-tails.

Since both XL and XXL versions use the same board,

most features are the same, such as current and temperature limiting, flip input to allow a radio channel to invert the robot drive system commands (to be used if an invertible robot is flipped over), five signal mixing options, exponential output, built-in BEC, and failsafe. They have a downloadable quick start guide, however they lack a full user manual.

These controllers are a good choice for combots up to 15lb, especially for their size and built-in features. As a matter of fact, until 2007 our hobbyweights Tourinho and Puminha used Scorpions XL. However, in 2008 we've reduced the size of Tourinho's chassis to save weight, which required us to switch the Scorpion XL board to two BaneBots BB-12-45 ESCs, which are smaller in size. In addition, in 2008 we've replaced Puminha's drive motors to higher power models that would greatly exceed the 45A peaks, so its Scorpion XL was switched to a pair of Victors.

Scorpion XL is a very good board, however there are a few tips to make it bulletproof, as we've learned after two years using and abusing its 2006 version in our hobbyweights:

- the large SMD capacitor (silver cylinder with black notch under the battery writing in the top picture) can become loose or even break during an impact, unless the board is very well shock-mounted; to avoid this, you can use epoxy resin without metal additives, hot glue, or even tape, to better secure the capacitor; if the capacitor gets knocked off, the board will present an inconsistent behavior; you can replace the capacitor by carefully soldering it onto the surface contacts, bending it towards the board over the Scorpion writing, and gluing it to the PCB;
- the green screw terminals are prone to become unfastened during combat, enabling the wires to fall off; we've removed these terminals and soldered the wires directly to the board;
- the flip feature doesn't work in the 2006 versions that we've bought; the solution, if available, is to implement this function using mixes in the radio transmitter;
- to improve heat dissipation and stretch the current limits a little bit, it is possible to attach heatsinks onto the BTS chips; old CPUs are a good source for small aluminum heatsinks.

Scorpion HX

The Scorpion HX (pictured to the right) is a great option for lighter combots such as beetleweights and antweights. The board weighs only 0.78oz (22g), with a compact 1.6" × 1.6" × 0.5" size.

Unlike the XL/XXL controllers, this board features 3 channels. Two of them are PWM outputs, usually used to control drive motors, and one on-off switch (Aux/Weapon channel) that can be used to operate a brushed DC motor in a single direction, usually in the weapon system.

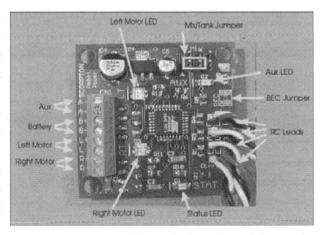

The PWM channels accept motors between 4.8V and 22V, delivering each 2.5A continuously with 6A peaks. If you need a few extra amps, it is possible to attach a heatsink onto the drive chips. Regarding the Aux/Weapon channel, it can handle constant 12A and as much as 35A peaks.

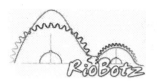

Other characteristics include flip (which works very well), current and temperature protection, channel mixing, BEC (which must be deactivated for input voltages above 12V), and safe weapon channel start (where the robot won't start if the weapon switch is activated during power-up), besides failsafe in all channels.

Mini-Touro used a Scorpion HX ESC during RoboGames 2006, with great results. However, in 2007 we've decided to use a brushless motor to power its drum, instead of a brushed one, so we had to replace the HX board with two BaneBots BB-3-9 ESCs (for the drive system), to give room for the BESC needed by the weapon motor.

Scorpion Mini

The Scorpion Mini (pictured to the right) is the smallest ESC from Robot Power, weighing only 0.21oz (6g) and measuring 0.625" × 1.6" × 0.4". It is a single channel PWM H-Bridge, which takes input voltages from 4.8 to 18V, or up to 34V if an external 5V supply is used. Current limits are 2.5A continuous without a heatsink, up to 4A with one, and 6A peaks.

Its version 2.4 includes BEC, which must be disabled if using voltages above 16V, and failsafe. It also features limit switch inputs that stop motion when pressed, which makes it an interesting choice to actuate clamper or lifter weapon systems in insect combats.

7.3.4. BaneBots

BaneBots is another manufacturer with ESCs aimed for light robot classes, from fairyweights to hobbyweights. There are three versions, all featuring a single fully reversible PWM output that can handle inputs from 6 to 24V: the BB-12-45, the BB-5-18, and the BB-3-9, pictured below from left to right, respectively.

The larger unit, the BB-12-45 (US$57), can withstand 12A continuously and 45A peaks (thus its name 12-45), presenting current limiting and thermal protection. It also counts with an integrated

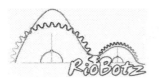

BEC, failsafe, and neutral start. It also includes receiver cables and leads to the motor and power supply. It weighs 0.98oz (26.3g) and measures 1.7" × 1.1" × 0.5", including the orange shrink wrap. It is a good option for most hobbyweights.

The BB-5-18 (US$ 46.50) is the intermediate model, it can withstand 5A continuously and 18A surges (thus its name 5-18). Similarly to the BB-12-45, it includes all wires, it has an integrated BEC, failsafe, and neutral start. It measures 1.45" × 0.825" × 0.375" and weighs only 0.60oz (17g).

The tiniest model is the BB-3-9 (US$28.75), it only weighs 0.33oz (9.4g). Its dimensions are 1.2" × 0.52" × 0.29", allowing it to fit in small spaces inside the robot. As its 3-9 name suggests, it can handle 3A continuously and 9A peaks.

One great feature that is common to all BB speed controllers is their modularity. Single channel units are easier to fit inside the robot, because several different orientation combinations can be experimented. In addition, if for instance a wheel locks up and damages your ESC, there is no need to replace the entire electronic system, only replacing the ESC associated with that channel.

We use BaneBots speed controllers in our lighter combots: the hobbyweight Tourinho uses a pair of BB-12-45 for its drive system, while the beetleweight Mini-Touro, the antweight Micro-Touro, and the fairyweight Pocket, use each a pair of BB-3-9 ESCs.

Despite their advantages, BaneBots ESCs aren't fail-proof. The BEC circuit heats up very easily, leading to unexpected failures when used at 24V. One of our BB-12-45 caught fire while tested at 24V drawing only 500mA, due to the BEC issue. It's probably a good idea to disable their BEC if you're using input voltages near 24V, which might require you to install a UBEC, discussed in section 7.1.5. Another issue is that the names from all of its electronic components have been sanded out by the manufacturer, making any homemade repair impossible.

7.3.5. Other Brushed Motor Speed Controllers

There are several other speed controllers in the market. A few of them are very sophisticated, including several programmable features, input voltages of up to 48V with peak currents exceeding 100A, tachometer and potentiometer inputs for closed-loop speed and position control, regenerative braking, temperature sensors, RC and microcomputer interfaces, and much more.

A few examples of sophisticated controllers are the RoboteQ, AmpFlow and Sidewinder, pictured below. They can independently control 2 brushed DC motors at the same time, which is perfect for the drivetrain of most combat robots. They also have mounting brackets for easy installation. They are good options for middleweights, heavyweights, and super heavyweights.

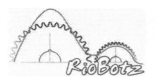

7.3.6. Brushless Electronic Speed Controllers

As seen in chapter 5, brushless DC (BLDC) motors demand a specific controller to be powered. This kind of motor resembles an alternate current (AC) synchronous three-phase motor, in which three waves with 120° shifted phases and same frequency actuate the rotation. Similarly to AC motors, brushless motors also spin due to three 120° phase-shifted waves but, instead of sinusoidal, these waves are three trapezoidal PWM signals, as pictured below. Each PWM signal acts over each one of the three motor windings from the brushless motor.

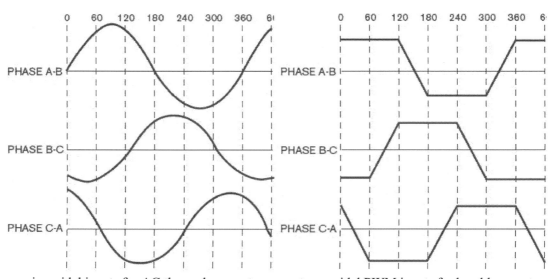

sinusoidal inputs for AC three-phase motors   trapezoidal PWM inputs for brushless motors

The wiring scheme to control brushless motors using a BESC is a little different from the one from brushed DC motors controlled by an ESC, as pictured to the right. Never connect the BESC to the battery with reverse polarity. Note that

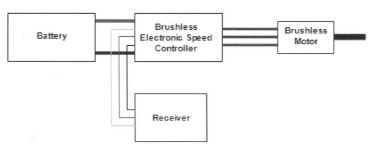

the BESC can be connected to the brushless motor using any combination of the motor's 3 wires. However, if the brushless motor spins in the wrong direction, then simply switch connection between any 2 wires. If, when the motor is unloaded, it takes too long to start spinning, you could try other wiring combinations between the BESC and the motor, this might solve the problem.

In order to start the motor, the controller must know which of the three windings should be first triggered. There are two methods to do that. The first is to have position sensors on the motor to measure the angle of the rotor, however brushless motors with sensors usually have high cost and complexity.

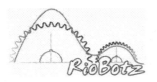

The other method is sensorless, where the controller is able to figure out the angle of the rotor by measuring the differences between the windings inductances. This is why you should never shorten brushless motor wires, this would change their inductances and confuse a sensorless system. Sometimes the motor wires are the coil wires extended out, so if you don't cut them at exactly the same length, getting them cleaned up nicely, removing all the varnish, and evenly tinning them, then your motor will single phase and not work, but just chatter. So, leave the motor wires alone.

On the other hand, long wires between the battery and the BESC should be avoided. The inductance of such long wires would cause voltage spikes on the BESC's input power lines due to its voltage switching. This is why most BESCs (and ESCs as well, for brushed DC motors) have a capacitor between their input power lines: its role is to suppress voltage spikes that could fry the speed controller. So, it is a good idea to shorten the wires between the battery and the BESC, placing these components close together inside the robot, and leaving the brushless motor wires with their original length. Long motor wires only lose a bit of power, they don't damage the BESC.

There are several sensorless Brushless Electronic Speed Controllers (BESC) in the market, from various manufacturers, such as Castle Creations, Hextronik, and Dynamite. The picture to the right shows the Castle Creations Phoenix 25 BESC, used in the weapon system of our beetleweight Mini-Touro.

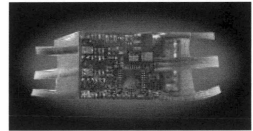

The downside of most BESCs is that they can only spin the motor in one direction. To reverse rotation during combat, you can commute any two of the motor wires, which can be done, for instance, using traditional or solid-state relays. Or you can use a reversible BESC, usually developed for R/C model cars, which can reverse the spin direction of brushless motors using a three phase H-Bridge circuit.

A few important aspects when choosing a BESC are its maximum voltage capacity, maximum continuous and peak currents, and motor reversion ability. Brushless motor reversion is still uncommon in combats, however it is very useful to reverse the direction of a weapon, allowing for instance an invertible drumbot that was flipped over to reverse its drum to continue launching the opponent. Brushless motor reversion has also potential applications in the robot drivetrain, resulting in a very efficient drive system with a high power-to-weight ratio.

A usual feature found in most BESCs is a BEC to power the receiver. But most BECs found in BESCs use linear regulators, which may overheat at high voltages. You may need to disable the BEC if the battery voltage is high. For instance, the HXT120 BESC from Hextronik (pictured to the right) can handle 120A continuously if properly ventilated, at input voltages of up to 24V. The BEC can handle up to 2A, but only if the input voltage is below 12V, otherwise it will overheat. For voltages higher than 12V, the BEC needs to be disabled, which is done by removing the

middle red wire from the receiver connector. We use this BESC to power the weapon motors from our featherweight Touro Feather and hobbyweight Touro Jr.

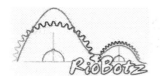 

High quality BESCs usually feature soft start, to prevent damages to fragile gearboxes, along with several programmable features such as:

- low-voltage cut-off: if the battery voltage per cell drops below a certain threshold, the motor speed is reduced or even brought to zero, to prevent damaging lithium batteries; the problem with this feature is that combat robot weapons usually require high current bursts for short periods, which can momentarily lower the battery voltage below the threshold, turning your weapon off; the default threshold settings are usually too conservative to use in combat, so program this feature to the lowest allowed value;

- over-current protection: if the current gets higher than a programmed threshold, the motor is turned off; so, choose the highest possible value for this threshold, if your motor can take it;

- brake type: option to brake the motor when the radio transmitter stick is in neutral position; this can be useful to stop spinning weapons in less than 60 seconds, as required in most competitions; note that the entire energy is dissipated in the motor, which might overheat;

- throttle range: option to limit the maximum power output, which is an interesting option in case the motor is overheating and/or drawing too much current from the batteries during combat;

- timing advance: ability to advance motor timing, resulting in a higher top speed in one direction;

- reverse delay: if a BESC is capable of reversing the motor by itself, without the aid of any external system, then it will probably feature this option; it defines a time delay between the moment when the radio control stick is moved to a reverse position and the moment when the spin reversion is fully commanded; this feature is used to prevent damages to gearboxes from the impact caused by a sudden reversion; if you need a very fast reversion, then set this delay to the lowest possible value, however keep in mind that the entire braking energy will need to be dissipated by the motor, which can end up overheating.

BESC programming methods vary among manufacturers and models. High-end models can be programmed via USB (using an appropriate connector, such as the one pictured to the right) in a computer. This is a really nice feature, making it possible to tweak the performance on the fly using computer software with intuitive user interfaces.

Cheaper models are usually programmed using the radio transmitter throttle stick, with the aid of feedback beep sounds emitted by the BESC, which can be quite confusing (and very annoying).

Other BESCs allow the use of a special programming card, such as the one pictured to the right, which significantly eases the task. It is plugged between the receiver and the BESC, allowing you to program it according to a series of LED indicators. The card usually has buttons at the bottom to navigate

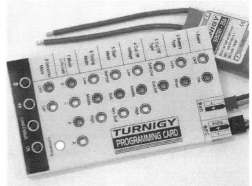

through the menu, select the desired feature, and then effectively change it in the BESC.

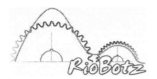

7.4. Solenoids

High power motors will probably require expensive speed controllers to be activated. A few motors, such as the D-Pack, are so powerful that they can even blow tough ESCs such as OSMCs, if care is not taken. Using your ESC in combat near its operational limit is risky, not to mention that its life will be significantly shortened.

So, if you don't really need speed or direction control, such as in the spinning weapon motor of most combots that have self-righting mechanisms, then solenoids might be a good option. Solenoids, a.k.a. contactors, are basically relays on steroids, capable of handling high currents to operate powerful weapons. Two of the most famous solenoids used in combots are the White-Rodgers 586 SPDT, and Team Whyachi's TW-C1, discussed next.

7.4.1. White-Rodgers 586 SPDT

The White-Rodgers 586 SPDT (pictured to the right, sold at The Robot MarketPlace for US$96) handles 200A continuous and withstands peaks higher than 600A, being, therefore, appropriate for almost any combot weapon even in super heavyweights. It is the solenoid used to power the Etek motor from our middleweight horizontal bar spinner Ciclone.

The SPDT in its name stands for Single Pull Double Throw, in other words, a single signal can switch between the Normally Open (NO) and Normally Closed (NC) terminals, necessary to activate or brake the weapon.

Most competitions require a robot's weapon to stop in less than 1 minute, therefore spinners with large weapon inertia might need some sort of braking system - it is not enough to just turn the motor off. A very simple braking system can be implemented in a brushed DC motor by shorting its leads. A spinning motor that is shorted out will become a generator, the same principle used in power plants, producing a current that is dissipated by the internal resistance of the motor, converting its kinetic energy into heat. This will effectively brake the system, as it can be easily verified by turning by hand the shaft of a motor with or without its leads shorted out. It will be much harder to turn the shaft when the leads are shorted.

Therefore, to implement a braking system, it is enough to connect the motor to the solenoid NO terminals, while shorting the terminals from the NC pair passing through the motor. When the solenoid is activated, current will flow through the NO terminals and the motor will accelerate. When deactivated, the motor leads will be shorted out at the NC terminals, braking the motor.

However, the kinetic energy of a few weapons, such as spinning bars or disks, is so high that shorting out the motor would simply fry it, due to the entire energy being dissipated as heat by the motor's usually small internal resistance. To prevent this, a power resistor must be placed between the NC leads, instead of just shorting them. Its value can be, for instance, 9 times the internal

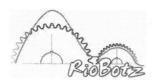

resistance of the motor, which would result in 90% of the weapon's energy being dissipated by this resistor, and only 10% by the motor. Note, however, that higher resistance values mean a longer braking time, because the dissipated power is inversely proportional to the resistance of the circuit. Therefore, choose a power resistor with a high enough resistance to avoid frying the motor, but not too high to guarantee that the weapon will stop in significantly less than 60 seconds.

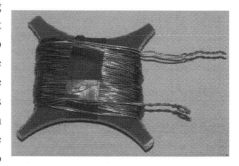

Instead of a power resistor, it is possible to use a long coated copper wire that is wound up on some heat resistant material (as pictured to the right, wound up around a blue piece of garolite). Knowing the wire resistivity, you can calculate the needed length for the desired resistance. Then, wind it up as a reel to keep its size very compact. The advantages of this braking system are its low cost, its low weight, and the efficiency of the heat dissipation provided by the long wire length. It is also easy to change its resistance: if the weapon is taking too long to stop, then just cut a few feet of wire from the reel to lower its resistance. It is important to use a heat resistant material such as garolite, because the temperatures while braking a powerful weapon can reach very high values. The blue garolite from our braking system usually comes out with a burnt-black color after a competition. In addition, never mount this braking reel close to the electronic system or to any flammable material.

We have used this wire braking solution in our middleweight horizontal bar spinners Ciclone and Titan. It works really well. Touro's drum doesn't need a brake, despite the weapon's high kinetic energy, because the moment of inertia of the drum is not as high as the one from the spinning bars. Even if two spinning weapons have the same kinetic energy, the one with lower moment of inertia will most likely stop faster, because it will have a much higher spin speed (to result in the same energy) that will result in larger bearing friction. Touro's drum stops in a few seconds, without the need for a braking system. Drumbots in general do not need a braking system to stop under 60 seconds. But horizontal and vertical spinners, which have weapons with very large moments of inertia, might need it. A few powerful spinners hit the arena walls after the end of a match to slow down and stop their weapon in much less than 60 seconds, to avoid delaying the event - but you must check if this is allowed, especially with the arena's owner!

7.4.2. Team Whyachi TW-C1

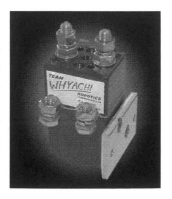

Team Whyachi's TW-C1 contactor (pictured to the right) is also SPDT, however it is smaller, lighter and cheaper than White-Rodger's 586. It tolerates, at 48V, currents of 80A continuous, 240A for 3 minutes, and 500A for 25 seconds.

Until 2008, both Touro and Titan had TW-C1s powering their weapons. The only issue with both presented solenoids is their plastic casing, which can break due to the high impact accelerations. Therefore, always shock-mount solenoids inside the robot structure.

7.5. Wiring

To connect the previously presented components, high quality wires and connectors are needed. The wires must bear high currents, while the connectors can't become loose during impacts. These components are presented next.

7.5.1. Wires

Wires must be very flexible, making it easy to route them through the robot's inside without rupturing the solders during impacts. Therefore, never use cables with a solid metallic core, use instead cables with multiple wires. A good example is the Deans Wet Noodle (pictured to the right), formed by over a thousand extremely thin  wires. Also, it is important to leave a little slack in all wires, in order to avoid them from getting stretched, ruptured or disconnected during combat, especially if the robot suffers structural deformations or if internal parts slightly move.

It is important to keep in mind the current ratings, which depends on the wire diameter (gauge, usually measured in AWG, which stands for American Wire Gauge) and the isolation material. The higher the AWG, the smaller the wire diameter is. When it comes to isolation, there are two usual types: PVC, which withstands temperatures up to 221°F (105°C), and silicone, withstanding up to 392°F (200°C). The highest current ratings for typical wire gauges are the following:

- 8 AWG: 70A to 80A continuous (PVC); 100 to 110A continuous and 500A peaks (silicone);
- 10 AWG: 50 to 60A continuous (PVC); 75 to 85A continuous and 350A peaks (silicone);
- 12 AWG: 35 to 45A continuous (PVC); 55 to 65A continuous and 200A peaks (silicone);
- 14 AWG: 30 to 35A continuous (PVC); 45 to 50A continuous (silicone).

The picture to the right shows a typical device used to measure wire gauges, ranging from the 0.3249" (8.252mm) diameter 0 AWG to the 0.005" (0.127mm) diameter 36 AWG.

Note that most wires can withstand without problems very brief and sporadic current peaks that are 4 times higher than the continuous limits.

A good tip is to use zip-ties to organize the wiring inside the robot. This can save a lot of time during a pitstop.

Also, always use rubber grommets (pictured to the right) to protect wires that need to go through metallic plates: if not protected, the friction between the wires and a hole in a metallic part can cut the isolation layer and cause short-circuits. Smoothing out with a metal file the borders of the hole is also a good idea.

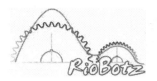 

7.5.2. Terminals, Plugs and Connectors

Terminals, plugs and connectors are critical in combat, because they must withstand impact accelerations and high currents. Avoid using fork, slide or quick connection terminals, since they are prone to disconnect upon impacts. Always use ring terminals (pictured below to the left). Fasten

tightly the connectors with nuts, along with a pressure washer. Never place a washer between the contacts, because it usually has a large electric resistance. Also, apply liquid electrical tape (as pictured in white to the right, applied to a White-Rodgers 586 SPDT solenoid) to avoid shorted contacts due to metal debris.

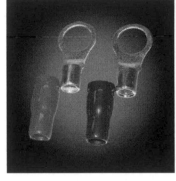 

It is a good idea to stick several pieces of double face adhesive tape inside the robot, especially on the inner part of the bottom cover, near the electronics. If the tape is very sticky, such as the VHB4910 tape, it works as a "flypaper" to pick up any debris that enters the robot, which could cause problems such as shorting out the electronics or getting stuck in the clearances between the wheels and the structure. We always use this technique: at the end of a competition, our VHB4910 tapes are filled with metal chips, small bolts and dirt. The tapes are replaced before every event.

Connectors must have a very low resistance and lock in well. An excellent connector is the Deans Ultra (pictured to the right), which withstands continuous 80A. Their maximum peak current is much higher: Touro's weapon motor draws almost 300A for a couple of seconds in the beginning of the drum acceleration, as we've measured, which goes through a single Deans Ultra without problems. We use these connectors on the batteries, motors and ESCs from all our hobby, feather, light and middleweight combots. An extra protection is, after connecting them, to duct tape them together to make sure they won't get disconnected. Be careful with knock-offs with cheap plastic (not nylon) housings that easily melt during soldering, letting their contacts come loose.

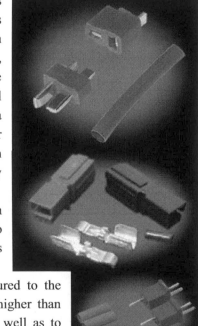

Another high power connector we've used is the Anderson PowerPole. The most common are the 45A version (pictured to the right) and the 75A. The 75A tolerates higher current peaks than the Deans Ultra, however it is a little bulky.

For lower currents, we use the Deans Micro Plug (pictured to the right), which is much smaller but even so tolerate currents higher than 20A. We use them as fan connectors in the bigger bots, as well as to connect the brushed DC motors from our beetleweight Mini-Touro.

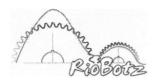

7.6. Power Switches

Power switches, or other on/off mechanisms, are mandatory in every combot (except, in a few cases, in insect class robots without active weapons). The switch must interrupt any current flux inside the robot, even low power ones. It is not enough to be able to switch off the RC interface or the receiver, even if this is enough to stop your robot, because any malfunction could still activate your weapon or drive system. It is necessary to be able to completely disconnect all batteries from every system.

For this reason, such switches must handle the entire amperage that goes through the robot, including the current required by the drivetrain and weapon systems. In the case of our middleweight Touro, current peaks can reach up to 400A.

There are automotive switches that can handle this level of current, however they are usually bulky and heavy. Two small and light weight switches, popular with combot builders, are the Hella Master Power Switch and the Team Whyachi MS-2.

The Hella switch (US$18, pictured to the right) can take continuous 100A, 500A for 10 seconds, and 1,000A current peaks. It is turned on or off by a red key that is not very convenient to use in combat. Most builders cut off the head of the red key, and file a notch on the remaining stub (as pictured to the right), allowing it to be switched on or off using a flathead screwdriver. We have used this switch in our middleweights Ciclone and Titan.

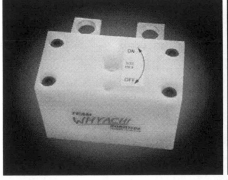

Team Whyachi's MS-2 switch (pictured to the right) is better and smaller than the Hella switch, however it is more expensive (US$65). It withstands continuous 175A, 500A for 3 minutes, and 1,000A for 25 seconds. To turn the robot on or off, you need to insert an Allen wrench (also pictured to the right) and turn it four times in the appropriate direction. These four turns make it almost impossible to have the switch disconnected due to vibrations during combat.

There is a smaller version that is rated for a lower current, the MS-05, good for hobbyweights (12lb), however it isn't much cheaper, at US$45. It takes continuous 40A, 140A for 3 minutes, and 250A for 25 seconds. In high power featherweights or heavier combots, it is better to use the MS-2, due its higher current rating. Our featherweight Touro Feather, which has a PolyQuest lithium-polymer battery capable of delivering up to 225A, had a few problems with the MS-05; since we've replaced it with the MS-2 switch, we haven't had any more problems.

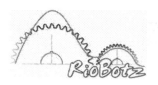

The MS-2 and MS-05 switches seem a little pricey, but they have an excellent cost-benefit relation considering that they are a vital part of the robot. And their manufacturing is not so simple, it involves the milling of two Delrin blocks, one for the body of the switch and the other for its cover, as pictured below.

milled Delrin body of the switch (inner and outer views)	milled Delrin cover of the switch (inner and outer views)
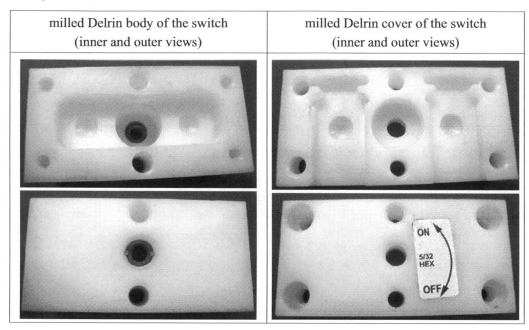	

The Delrin body has an embedded threaded nut (in black at its center), to which a long Allen screw is attached, along with a copper bar with gold contacts, two plastic non-conductive washers, and a spring, pictured below to the left. After the switch is assembled, the long Allen screw will have the function to position the copper bar to close or open the circuit. As it is screwed, it will drive the copper bar away from the two copper terminals (pictured below in the center), opening the circuit and turning the robot off. When the long Allen screw is unscrewed, the spring-return will make the copper bar touch the two copper terminals, turning the robot on.

The picture below to the right shows the assembled body and cover, before they are attached together using 4 small Allen screws. The switch itself is attached to the robot with the aid of two threaded holes in its Delrin body.

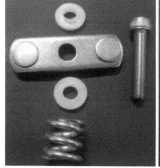

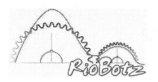

Note that the non-conductive washers are important to electrically isolate the long screw from the copper bar, besides reducing friction between them.

Note also that the spring is important to guarantee contact between the copper bar and the terminals. It also works as a spring-lock to avoid the screw from turning due to vibration. And, if the switch breaks due to an impact, it is likely that it will remain in the "on" position due to the spring, keeping the robot alive during a violent match. This feature has saved us during RoboGames 2006: the violent impacts during the match against the undercutter The Mortician managed to detach the long Allen screw from the threaded nut in the Delrin body of Touro's MS-2 switch; however, Touro continued to fully function because the spring was able to guarantee contact between the terminals.

There are also even simpler and cheaper switches that can be made. One of them is the one used in the drive system of our middleweight Ciclone. A Hella switch controls the weapon, at 24V, but the drive motors use an 18V cordless drill battery pack, which needs a second switch. The adopted solution costs only a few bucks: a pair of Deans Ultra connectors wired as a jumper, as shown to the

right while turned off (left picture) or on (right picture). The wire that is connected to the battery positive was cut in two, opening the circuit. Then, the two cut pieces were soldered to a female Deans Ultra plug. Next, a short wire was soldered to both terminals of a male Deans Ultra plug. Don't forget to isolate well both

plugs (the Deans jumper in the pictures is shown without its isolation tape). Connecting the male into the female plug closes the circuit and turns the robot on. It's important to insert the connected plugs well inside the robot, to avoid them from being knocked-off by an opponent. The bar spinner The Mortician was able to win a RoboGames 2006 match by knock out after knocking off a jumper switch from the launcher Sub Zero (as pictured to the right), even though the jumper was almost completely inserted into the robot.

An even simpler solution is to not use a switch. The robot must have an opening that allows the driver to directly connect or disconnect the battery (or batteries). This is the solution used in our hobbyweights (as pictured to the right) and smaller robots. A Deans Ultra female connector is

soldered to the battery leads (never solder the male connector to a battery, to avoid accidental shorts). The robot electronics uses a male Deans Ultra. To turn the robot on, just connect the plugs, insert them into the robot opening, and cover it to protect against debris. Since this must be done by the driver in the arena, make sure that the cover can be easily attached to the bot.

7.7. Connection Schemes

In this section, it is shown how to connect the presented components in a combat robot. A typical configuration is pictured to the right.

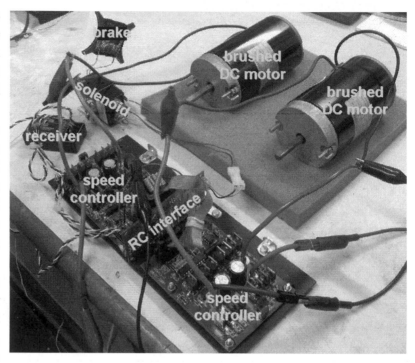

In this picture, the receiver converts the radio signals into a PPM signal, which is then interpreted by the RC interface (which is described in details in section 7.8.2).

This RC interface then generates low power PWM signals that are sent to the speed controllers, such as OSMCs (as shown in the picture), which amplify them to a high power PWM output to control the speed of the brushed DC motors used in the robot drivetrain.

The RC interface is also able to activate the solenoid that triggers the weapon motor (not shown in this picture), using a copper wire reel connected to the Normally Closed (NC) terminals to brake the weapon.

Three connection schemes are presented next. The first is a classic scheme, often used by beginners, which will surely not pass safety inspection. The second one is an improved version, which addresses all the issues from the first one. Finally, a third connection scheme is shown, better than the second one if you need a fully reversible weapon with speed control.

7.7.1 Classic Connection Scheme

The figure in the next page shows a classic connection scheme, including components that are sized to a middleweight combat.

It uses one Nickel-Cadmium (NiCd) battery pack to power the weapon, which is a good option due to its ability to provide very high peak currents (as it will be studied in chapter 8).

And it uses one Nickel-Metal Hydride (NiMH) battery pack for the drivetrain, which is also a good option because it has more capacity than NiCd, lasting longer (the current peaks from the drivetrain are usually much lower than the ones from the weapon system, because wheel slip acts as a torque limiter).

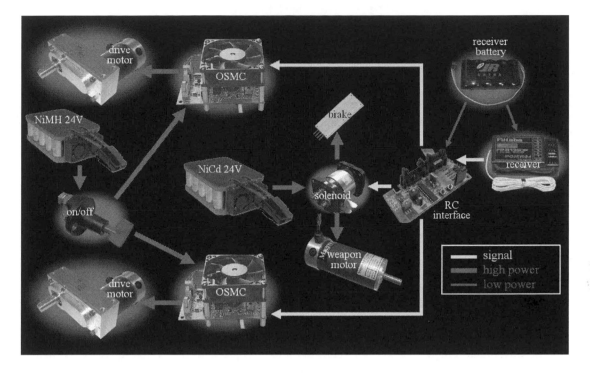

In the configuration shown in the scheme, the NiMH pack is connected to a Hella switch, which powers the OSMC speed controllers. The NiCd pack is connected to a White-Rodgers 586 SPDT solenoid that activates the weapon, while a power resistor is used for braking. A small battery pack powers both the receiver and the RC interface. The RC interface interprets the PPM signal from the receiver, sending a low power PWM signal to each OSMC. Each OSMC amplifies the received signal, sending a high power PWM output to the drive motor it is connected to. The RC interface is used as well to trigger the solenoid that powers the weapon motor.

Note that this RC interface only works if powered by both 5V from the small battery pack and 12V from the OSMCs. Therefore, if the Hella power switch is off, the NiMH pack won't provide 24V to the OSMCs, which in turn won't provide 12V to the RC interface, which in turn won't be able to keep the solenoid active, turning the weapon off. In theory it would work, but it is unsafe.

The above scheme looks good, including the battery optimization feature: NiCd for the weapon, to deliver high peak currents, and NiMH for the drivetrain, for improved capacity. However, it has serious flaws:

- if the single NiMH pack breaks, the robot will stop working and, therefore, lose the match;
- if the small battery pack voltage is too low, the robot will become unresponsive;
- there isn't a power (on/off) switch between the NiCd pack and the weapon solenoid, thus if due to a surge current the solenoid terminals get soldered, the weapon won't stop, even with the main switch off – therefore this scheme won't pass safety inspection; you would need to include another switch between the NiCd pack and the solenoid;
- there isn't an on/off switch between the small pack and the receiver and RC interface, which is

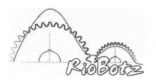

required to pass safety inspection; note that even the fans used in the robot must be turned off by the robot switch(es); therefore, the robot would need 3 on/off switches, because you can't connect the three battery packs in parallel due to their different types or voltages; the driver would need then to remember to turn on all three switches in the beginning of every match;

- the White-Rodgers 586 SPDT Solenoid is relatively large and heavy;
- OSMC speed controllers are not very compact, occupying a lot of the internal space;
- a resistor with both low-resistance and high-power, typically with less than 1Ω and more than 1kW for middleweights, needed to brake the weapon, isn't cheap and it can burn.

7.7.2. Improved Connection Scheme

To solve the problems presented above, you should use an improved scheme, such as the one pictured below. It includes 2 (or more) identical battery packs in parallel (NiCd in this example), connected to a single MS-2 power switch. A second power switch could be used in parallel to both packs, as a redundancy measure in case one of the switches breaks in the off position. Both packs need to be exactly the same, with same type, voltage and capacity, to be connected in parallel without any problems. This is why we use 2 identical NiCd packs. This switch powers the Victors, the TW-C1 solenoid, and the RC interface. This interface, needed to activate the solenoid, has a built-in BEC to power the receiver. The Victors can be directly connected to the receiver without an RC interface. A copper wire reel is connected to the solenoid to act as a weapon brake. A Deans Base-Loaded Whip antenna is attached to the 75MHz receiver, enhancing reception quality.

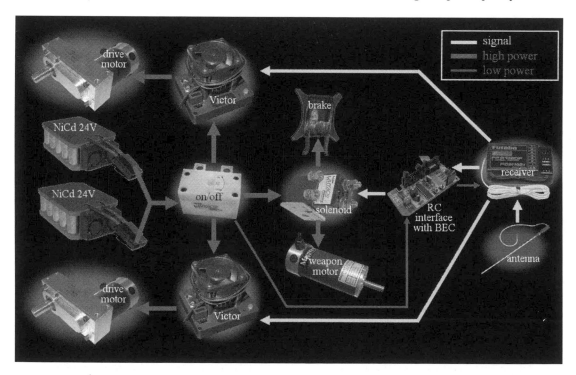

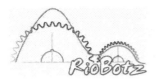

This improved scheme addresses all the issues from the classic scheme, because:

- if one of the batteries fails, due to a broken solder or connector malfunction, the robot will continue to fully function using the remaining pack(s), however with lower total capacity;
- the weapon will accelerate faster, since two packs in parallel can supply twice the current, assuming that the drivetrain isn't demanding too much from them during this acceleration;
- there is no need to have a small battery pack for the receiver, due to the RC interface BEC;
- a single on/off switch can power down the entire robot, including drivetrain and weapon motors, as well as the receiver and RC interface, which is required by safety inspections in the events;
- the TW-C1 solenoid is smaller, lighter and cheaper than the White-Rodgers 586 SPDT;
- Victor speed controllers, besides smaller than the OSMC, can be directly connected to the receiver, without needing the RC interface (which is only used above to power the solenoid and to work as a BEC to power the receiver);
- Victors have a brake/coast jumper, used to set its action during a neutral condition from the radio; the brake setting sets the output to a short-circuit during neutral, while the coast setting sets an open circuit; when used in the drive system, the brake setting will stop your robot when you release the radio control stick, while the coast setting will let your robot continue moving due to its inertia; the brake setting is a good option for sharp turns in agile robots, while the coast setting is good to prevent the drive motor from overheating due to the short-circuits;
- the RC interface can become smaller, because it only needs to actuate the solenoid and to work as a BEC, without any need for PWM outputs for the drive system;
- the copper wire reel is cheaper and it dissipates heat better than the power resistor.

The presented scheme is also pictured below, showing a close up of the electronic system of the 2006 version of our middleweight Touro. Note that all components from this improved scheme are included below, except for the weapon brake, usually not needed in drumbots.

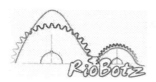

7.7.3. Connection Scheme for Reversible Weapons

After RoboGames 2008, we've decided to try a new connection configuration, one capable to reverse Touro's drum rotation and to solve minor issues from the previous schemes. To improve the battery capacity and voltage, the NiCd packs in parallel were replaced with A123 packs. Instead of using an RC interface and solenoid, two Victors are used to control the weapon motor, one for each pair of brushes. Using 2 ESCs to power the same DC motor is only possible if it has independent circuits for each pair of brushes, such as the Magmotors (which have 4 brushes).

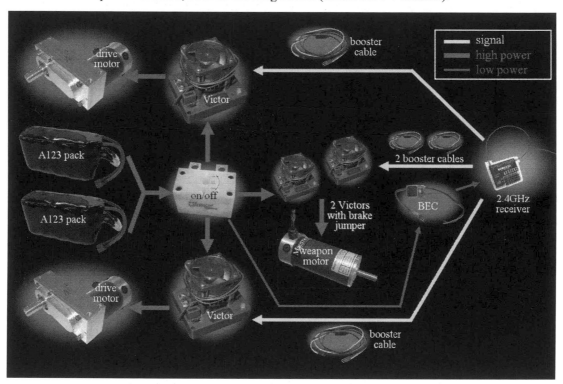

This scheme has improvements that are important for fully reversible spinners:

- the two Victors HV-36 powering the weapon allow its spin sense to be inverted if the robot is flipped, maximizing weapon effectiveness;
- one Victor for each pair of brushes from the weapon motor improves reliability, since if one of them fails the other will be able to spin the motor, although with less power;
- Victors weigh less than the TW-C1 solenoid, therefore their inertia is lower, reducing the risk of mechanical damage due to high impact accelerations;
- if the weapon is taking too long to brake, then simply set the jumpers from its Victors to the brake setting; otherwise, choose the coast setting to prevent the motor from overheating;
- Victors are less prone to lock-up than solenoids, therefore safety is also improved;
- a dedicated Universal BEC is used to power the receiver, instead of one integrated in an RC interface that also controls other devices, improving reliability; note that this BEC features a switching voltage regulator instead of a linear regulator, avoiding overheating problems.

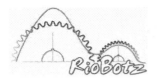

7.8. Developing your Own Electronics

The speed controllers presented in this chapter and their radio-control interface (RC interface, a.k.a. logic interface, which interfaces the controllers with the receiver) are not cheap. But even so they are an off-the-shelf solution with very good cost-benefit, considering their complexity. Since

most good quality electronic components needed by these systems are expensive, developing your own speed controller or assembling yourself an existing one doesn't save you too much money. For instance, a fully assembled and tested OSMC (pictured to the right) costs US$219 at www.robotpower.com, while its bare board (US$29) and components (about US$140) will set you back about US$169. You're basically paying US$50 for assembly and testing, which is quite reasonable, in special considering that you could burn out the entire controller if it's not carefully soldered.

But developing your own speed controller or RC interface, or even just assembling an OSMC (as pictured to the right), is a great learning experience. In addition, you'll be able to tailor the RC interface to your needs. For instance, Robot Power sells a very high-performance microcontroller-based closed-loop interface that controls two OSMCs, named Dalf. It is great for an autonomous robot, but it would be overkill for a radio-controlled combot, in special considering its US$250 price tag. The simpler MOB (Modular OSMC Brain) interface, or its more compact version µMOB,

would be a much better option for robot combat, however both have been discontinued. This was the motivation for us to create a compact 3-channel RC interface to control two OSMCs (for the drivetrain motors) and one heavy duty relay or solenoid (such as the White-Rodgers 586 solenoid, for the weapon system), including a BEC to power the receiver, as explained in section 7.8.2.

Note, however, that bulletproof speed controllers and RC interfaces are not trivial to build. So, if you're planning to use your own system in combat, it is fundamental to perform several benchmark tests to avoid any surprises. There are a lot of things that can go wrong with these systems.

7.8.1. Speed Controller Development

Before we discuss the RC interface, it is important to understand the speed controller it will interface with. In this section, we'll introduce the main features of a typical high power speed controller, based on the OSMC design.

A typical speed controller is basically an H-Bridge (introduced in section 7.2.3) used to power DC motors with a controllable voltage. To do that, the H-Bridge uses one or more transistors in parallel at each of its four legs.

To activate the H-Bridge, do not use a Bipolar Junction Transistor (BJT), it is not efficient when dealing with the high electric currents needed in combat. Instead, use a MOSFET (Metal-Oxide Semiconductor Field-Effect Transistor, a.k.a. FET), such as the IRF1405 used in our OSMC boards. It has several advantages, despite its relatively high cost.

The first advantage of FETs is that they are voltage-activated (instead of current-activated such as in a BJT), making it easy to activate them. It is enough to guarantee that its input voltage (at the gate) is higher than its threshold voltage V_{th}, to allow the high currents to go through the drain and source.

When the FET is activated, in saturation mode, it behaves as a resistor, with resistance R_{on}. Very good quality FETs can have R_{on} as low as 5mΩ. To continuously supply, for instance, 160A to a motor, you will need more than one FET at each leg of the H-Bridge. If 4 FETs are used in parallel at each leg (totaling $4 \times 4 = 16$ FETs for all 4 legs from the H-Bridge), then the power dissipated by each FET is about

$$P = I^2 \cdot R_{on} = \left(\frac{160A}{4}\right)^2 \cdot 5m\Omega = 8W$$

which is an acceptable value for use with small heatsinks coupled with a fan to actively cool down the FETs. If the heatsinks were not used, then the maximum continuous current acceptable going through a system with 4 FETs in parallel would be approximately 100A. Another great advantage of the FETs is that they don't have any current limitation, as long as their maximum temperature is not exceeded. Therefore, FETs can easily take very high current peaks, as long as they are brief enough not to overheat them. And the FET commutation usually takes only a few dozen nanoseconds, keeping low the energy losses from this process.

The fact that FETs are activated by voltage helps a lot in the development of an activation circuit. But there's a catch that can cause a few problems. For the FET to conduct (in saturation mode), an electric charge must be injected at the gate of the FET to make the voltage between the gate and the source reach approximately 10V. This 10V is, in general, the voltage required to completely enter saturation mode, minimizing the value of the resistance R_{on}. Such need to charge the FET, called parasitic capacitance effect, can be modeled as a capacitor in parallel with the gate of the FET.

To charge this large capacitance, the integrated circuit HIP4081A can be used. It is a high frequency H-Bridge driver, capable of supplying up to 2A for the four FETs connected in parallel to each output. To avoid shoot-through, which could happen for instance if an upper FET turns on while a lower FET from the same side of the H-Bridge is still conducting, a resistor is connected in series with the gate of each FET, limiting the total current and making the FETs take longer to be activated. In this way, the resistors help to balance the T_{on} and T_{off} times from all FETs in parallel, by equalizing their resistor-capacitor constants.

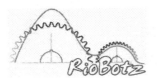

Despite the presence of the resistors, there would still be a chance of happening a shoot-through. Two protection measures exist to avoid this condition. The first is a programmable time in the HIP4081A when both FETs get turned off (in cut-off mode). The second is the addition of extremely fast diodes in parallel with the resistors, so that during the T_{off} time of the PWM the entire current is drained by them, eliminating any chance of happening a shoot-through in the circuit.

The HIP4081A has, therefore, the function to activate the upper and lower FETs, including a circuit to increase the voltage to the levels required by the FETs. In the application example pictured to the right, the HIP4081A is powered by 12V (allowable values are between 9.5V and 15V), while a battery voltage of 80V (or any other value between 12V and 80V) is applied to the load.

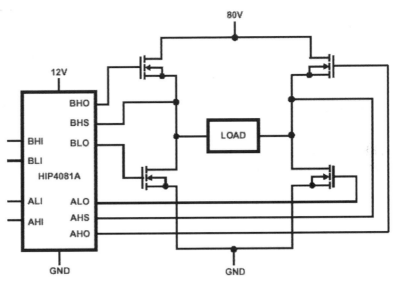

The load in this example can be, for instance, a brushed DC motor. If the voltage supplied to the HIP4081A is lower than 9.5V, then an internal protection turns off the upper FETs. On the other hand, if such voltage is higher than 16V, the HIP4081A can be damaged. In addition, to protect the FETs against voltage peaks, two Zener diodes are used to limit their voltage to 15V.

The HIP4081A has four digital inputs, AHI, ALI, BHI and BLI, each one corresponding to the outputs that power each group of FETs from a leg, respectively AHO, ALO, BHO and BLO, as pictured above. In other words, when an input is enabled, the FETs connected to the corresponding output are activated. The RC interface, which in the case of the OSMCs is an external electronic system (not an integrated one as in Victors), needs to send PWM and direction signals to the HIP4081A to control the H-Bridge. These digital signals are compatible with the TTL logic, but any input voltage above 3V, such as 5V or 12V, is recognized as a high ("1") logic state.

Also, the HIP4081A has a protection in its internal logic against shoot-through, which shuts down the upper FETs connected to AHO (or BHO) when the lower FETs from the same side of the bridge, connected to ALO (or BLO), are activated, independently of the state of the upper inputs AHI and BHI. This protection is implemented using AND logic gates in all the HIP4081A inputs, as seen in the picture on the next page, which shows the functional diagram of half of a HIP4081A driver. The AND gates have as input the values of AHI, ALI, BHI and BLI, in addition to the complement from the DIS (Disable) pin, deactivating all FETs if DIS has a low logic level.

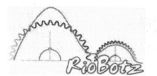

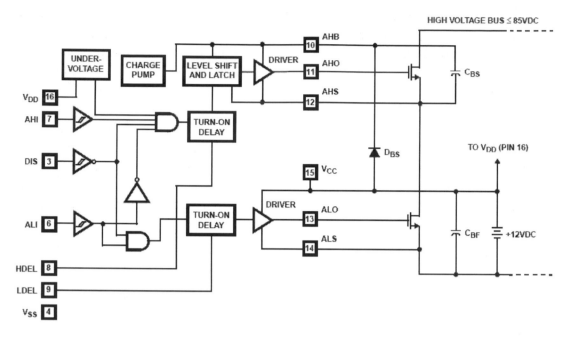

Each upper gate also features two other inputs. One of them is used for protection against low supply voltages, while the other is the complement of the lower gates, to guarantee that the upper gate output AHO (or BHO) will be turned off if the lower gate inputs ALI (or BLI), from the same side, are activated.

Resistors are also connected to the HIP4081A, between the input and ground pins. These resistors guarantee that all FETs will be turned off if no RC interface is connected to the power board. This is an additional protection to make sure that the motors will be turned off if the connection with the RC interface is lost.

Due to the nature of the used FETs, the gate voltage should be approximately 10V higher than the battery voltage to activate the upper FETs. To generate such higher voltage, the HIP4081A has a charge-pump system that, with the aid of a diode and an external capacitor, generates the necessary voltage at the outputs AHO and BHO, making it possible to activate the FETs connected to them.

To protect the circuit from voltage peaks caused by the DC motor brushes and commutators, a Transient Voltage Suppressor (TVS) is used. It works exactly as a Zener diode, in other words, when the voltage on the TVS is above a specified value, it starts conducting, "absorbing" the excess voltage. The TVS is optimized to tolerate voltage peaks with high currents. It is used to absorb the voltage peaks between the battery terminals, and to protect the FETs.

In addition to the TVS, resistor-capacitor circuits between the motor terminals provide an additional protection against high frequency peaks generated by the brushes. Also, large electrolytic capacitors are placed as close as possible to the H-Bridge to reduce the effects caused by the inductances of the wires that connect the battery to the circuit.

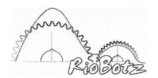

The last part of the circuit comprises the switched-mode power supply, which converts the battery voltage (between 12V and 80V) down to 12V, using a high efficiency regulator that does not need a heatsink. These 12V will also be used to power the RC interface, described next.

7.8.2. RC Interface Development

The power circuit discussed above cannot receive directly the signals from a radio-control (RC) receiver. These signals need to be treated first. This signal conditioning is made through a RC interface circuit, which is an interface between a receiver and a speed controller (or solenoid).

There are several off-the-shelf RC interfaces that you can buy, for instance, at www.robotmarketplace.com. But if you want to build one yourself, then you'll probably need to use a micro-controller, such as a PIC, dsPIC or AVR, capable of executing several million instructions per second. The RC interface that we've developed uses a PIC, capable of decoding the signals from up to four receiver channels, to use them to command power circuits, solenoids or relays.

The input signal in the RC interface comes from the receiver, it is a pulse train following the PPM standard, as pictured to the right. This pulse train has a period that 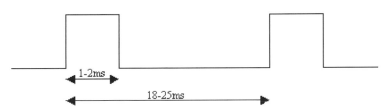 can vary between 18 and 25ms, with each pulse lasting between 1ms (low) and 2ms (high).

As mentioned before, PPM and PWM are two completely different signals, even though they are both pulses. In our application, PPM is a low power pulse train from the receiver that carries the commands from several channels through a code that is based on the absolute width of each pulse, bringing not only the information about the desired speed of several motors, but also their direction. PWM, on the other hand, as explained in section 7.2.2, is a pulsed signal that carries, in our application, the information about the absolute speed (but not the direction) of a single motor, determined from the *ratio* between the periods T_{on} and T (and not from the absolute value of T_{on}). So, the job of our RC interface is to take the single PPM signal from the receiver, decode it, and send one PWM and one directionality signal to the power board of each motor.

When, for instance, a stick of the radio control is completely to the left, the PPM pulse from the associated channel has a 1ms width; if the stick is in the middle, then the width is 1.5ms; and if the stick is moved completely to the right, the pulse will take 2ms. All other stick positions will translate to a pulse width between 1ms and 2ms. The pulse width is either directly or exponentially proportional to the stick position, depending on the radio settings.

Therefore, to control a bi-directional motor, the 1ms pulse is usually associated with a command to move back at full speed, while a 2ms pulse would mean move forward at full speed, and a 1.5ms pulse means that the motor should stop. Also, for instance, a 1.9ms pulse would mean that we want to go forward with $2 \times (1.9 - 1.5) = 0.80 = 80\%$ of the top speed, while a 1.2ms pulse would mean that we want to go back at 60% of the top speed, because $2 \times (1.2 - 1.5) = -0.60 = -60\%$.

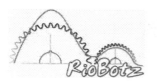 

We've programmed our PIC to validate the PPM signal from the receiver, and then to count the width (time interval) of each pulse. Clearly, each pulse is associated with one receiver channel.

There are five output signals that need to be sent to the HIP4081A driver: AHI, ALI, BHI, BLI, and the DIS (Disable) signal. The DIS signal is only used in case you want to turn off the H-Bridge. Due to the protection against shoot-through in the HIP4081A, it is possible to simplify the involved logic, keeping both AHI and BHI signals in the high logic level, all the time. The suggested signals for the correct operation of the power circuit are shown in the table below.

AHI	BHI	ALI	BLI	DIS	Function
1	1	0	PWM	0	Forward
1	1	PWM	0	0	Back
1	1	0	0	0	Brake
1	1	1	1	0	Brake
×	×	×	×	1	Off

×: the state doesn't matter;   1: means 5V (or 12V);   0: means 0V;

So, we only need to deal with two signals, ALI and BLI, because AHI and BHI are always kept at the high ("1") logic level. But both ALI and BLI deal with the PWM signal, as seen in the table above, which is not good because you would need to use two PWM output pins from the PIC to control a single motor. We need to modify the table above such that only one signal takes care of the PWM (for instance, the BLI, which can be used to define the absolute speed of the controlled motor) while the other defines the direction of the movement (the ALI signal, in this example). In this way, only one PWM output pin from the PIC will need to be used per motor (to carry the BLI signal, in this example). The modification is shown in the table below.

AHI	BHI	ALI	BLI	DIS	Function
1	1	0	PWM	0	Forward
1	1	1	$\overline{PWM}$	0	Back
1	1	0	0	0	Brake
1	1	1	1	0	Brake
×	×	×	×	1	Off

×: the state doesn't matter;   1: means 5V (or 12V);   0: means 0V;

With this new table, the BLI will be the PWM signal, while the ALI will control the direction in such a way that a low ("0") logic level means forward, and a high ("1") logic level means backward. So, when moving forward, the current goes through the FETs connected to the HIP4081A outputs AHO and BLO, and when moving backward the current goes through the FETs at BHO and ALO. However, when moving back, the PWM that goes to BLI must be inverted, either through software or hardware, resulting in the $\overline{PWM}$ signal as shown in the table. The reason for that can be understood through the example in the next page.

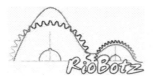

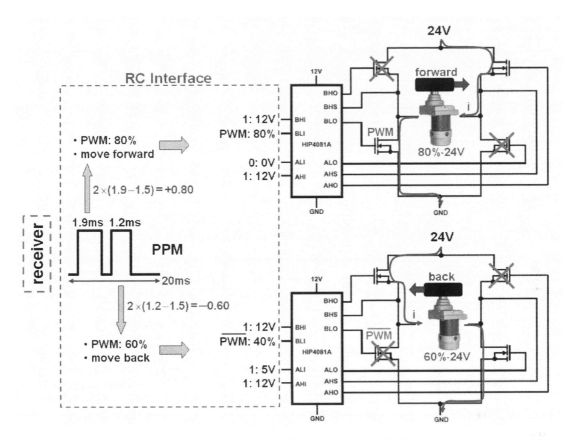

The figure above shows an example of a PPM pulse train with 20ms period, generated by a receiver, used to control two 24V permanent magnet brushed DC motors. The RC interface uses its PIC controller to measure the pulse widths, resulting in 1.9ms for the first motor (the top motor in the figure) and 1.2ms for the second. Then, the RC interface figures out, from the calculations shown in the figure, that the first motor has been commanded to move forward with 80% of the provided 24V, while the second motor needs to move back with 60% of 24V.

For the first motor, the RC interface sets the ALI voltage to 0V (low level "0") to indicate that the motor should move forward, and the BLI receives a PWM signal that is high ("1") during 80% of the time. Note that the BHI and AHI are always at a high ("1") logic level, by default, which can be obtained using 12V hard-wired from the HIP4081A 12V pin (as discussed before, any voltage above 3V translates to "1" in the HIP4081A). With ALI at 0V, the FETs connected to ALO do not conduct. Even though BHI is at the high logic level, the FETs at BHO do not conduct because of the shoot-through protection that prevents a short-circuit of the 24V battery through BHO and BLO.

So, the current from the 24V battery has to flow to the first motor through AHO (which conducts because AHI is always set to "1") and BLO. But BLO only conducts 80% of the time, because of the 80% PWM signal at BLI, so the resulting motor voltage will be, in average, about 80% of the 24V input voltage, as desired.

For the second motor, ALI is set to 5V (at the high logic level "1") to indicate that the motor should move backward, while BHI and AHI are always at 12V (also at the high logic level "1"), by

259

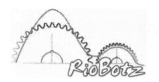

default. But the RC interface, instead of sending the desired 60% PWM signal to BLI, sends the inverse signal $\overline{\text{PWM}}$, which is low (0V, instead of high) during 60% of the time. Therefore, this inverted signal is high (logic level "1") during only 40% of the time. With ALI set to the high logic level, the FETs at ALO will always conduct. Even though AHI is at the high logic level, the FETs at AHO do not conduct because of the shoot-through protection that prevents a short-circuit of the 24V battery through AHO and ALO.

During 60% of the time, when BLI is low ("0"), the FETs at BLO will not conduct, and the current from the 24V battery will flow to the second motor through BHO (which conducts because BHI is always set to "1") and ALO. But during the remaining 40% of the time, when BLI is high ("1") and hence the BLO FETs conduct, the BHO FETs will stop conducting due to the shoot-through protection that prevents a short-circuit through BHO and BLO. Without the BHO FETs to conduct, the second motor won't be powered during 40% of the time. So, the resulting motor voltage will be, in average, about 60% of the 24V input voltage, while moving back, as desired.

In summary, the PWM signal must be inverted at BLI to move back because we make use of the shoot-through protection, which only allows the motor to be powered when BLO is not conducting.

The hardware of the developed RC interface, pictured to the right, is relatively simple and compact in size, measuring 4" × 1.75". It includes a micro-controller PIC16F876A, and a buffer to isolate the signals generated by the PIC from the power board signals, to avoid any problems.

The interface board also features an independent circuit used to activate a high power relay or solenoid, which is completely isolated with the aid of an optocoupler.

In addition, the developed board includes a BEC that takes the 12V from the power board (used in the HIP4081A driver) and converts it into 5V to power both the receiver and the RC interface itself, using a linear regulator. The developed board also includes two buttons, one to reset the PIC and the other to enter into calibration mode.

The micro-controller PIC16F876A (pictured to the right) features in-circuit serial programming, which allows it to be programmed without the need to remove it from the board. The input signals from the receiver are connected through resistors to the pins RB4 through RB7 from the PIC.

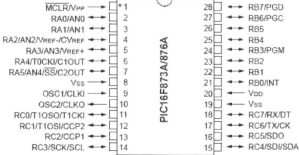

Those pins generate an interruption when the input changes its state, which is perfect to read PPM receiver signals.

The buffer used in both PWM outputs from the RC interface is the chip 74HCT244, consisting of two sets of four buffers each. It is possible to use as well other chips equivalent to 74HCT244, such as the 74HC244, as long as their output voltages are high enough to be used with the HIP4081A from the power board. For instance, the 74LS244 chip is not recommended in this case, because it associates any voltage beyond 2V to a high logic level, while the HIP4081A requires a minimum value of 2.5V.

The figure to the right shows two connectors, named OSMC1 and OSMC2, which are used to connect the developed RC interface to two OSMC speed controllers. The output signals to AHI and BHI (pins 5 and 7, respectively) need to be permanently set at the high logic level "1" as discussed before. This is accomplished with any voltage higher than 3V, not necessarily 5V, so in our case we've connected these pins 5 and 7 to the pins 1 and 2, which provide the 12V supplied by the HIP4081A.

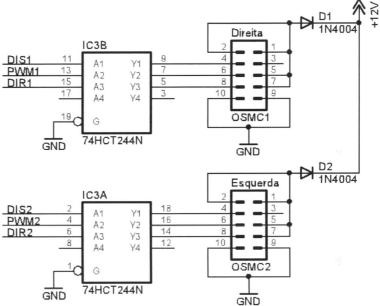

Note also in the figure above that there are two diodes D1 and D2 between the RC interface and both power boards. These diodes are a redundancy measure to ensure that the RC interface, which is powered by the 12V lines from both power boards, will still be functional even if one of the power boards burns out.

The developed RC interface is also able to activate a relay (or a solenoid), usually used in the weapon system, including as well a status LED, as pictured to the right. This circuit uses an optocoupler

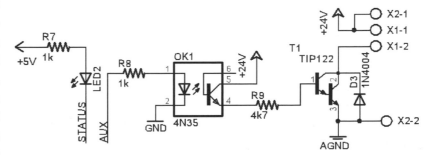

that, when enabled, makes the T1 transistor conduct, activating the relay. To do that, the relay terminals must be positioned at X1-1 and X1-2, while 24V should be applied to X2-1 and X2-2 (assuming a 24V relay). The transistor T1 can handle up to 3.5A with a heatsink, or 1.0A without

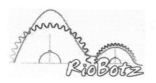 

one. The transistors TIP120, TIP121 or TIP122 can be used in this circuit. The figure also shows the status LED, which is used to inform the state of the program running in the PIC. In this case, the LED is used to indicate whether the circuit is under normal mode or calibration mode.

The software used by the PIC16F876A from the RC interface was written in the programming language C. The entire program, together with more details about our RC interface board, can be found in the undergraduate thesis (in Portuguese) from the RioBotz team member and former advisee *Felipe Maimon*, which can be downloaded at www.riobotz.com.br/en/tutorial.html. I've tried to summarize and translate to English the main points from this thesis here, in section 7.8. Note that the program, which is relatively lengthy, is very specific to the hardware of the developed RC interface. Our RC interface is nicknamed MOB, in honor of the discontinued "Modular OSMC Brain" interface, however here it stands for "*Maimon*'s OSMC Board."

Our RC interface board was successfully used in all our OSMC-powered middleweight combots: the overhead thwackbot Anubis (controlling the speed of both NPC T74 drive motors), as well as the horizontal bar spinners Ciclone (controlling the speed of two DeWalt gearmotors and activating its Etek weapon motor through a White-Rodgers 586 SPDT solenoid) and Titan (controlling 4 Magmotors S28-150, two of them for the drive system using PWM, and the other two for the weapon through a single TW-C1 solenoid). The board withstood well the rigors of combat.

We haven't been using our RC interface board described above since we migrated from OSMCs to Victors in all our middleweights. Victors don't handle as much current as the OSMCs do, but they're more compact and they can be connected directly to the receiver without an RC interface.

However, we still needed an RC interface to activate the solenoid from the weapon system of our middleweights. We've then designed another more compact RC interface, measuring 2" × 1.25" (pictured to the right), without the PWM outputs, featuring 1 output for a high power relay or solenoid and a BEC to power the receiver. It was used until 2008 by Touro and Titan to power the TW-C1 solenoid from their weapon systems.

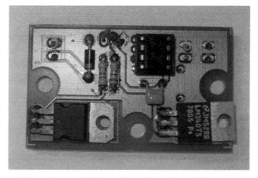

Finally, we've considered designing a third RC interface board, featuring 2 solenoid outputs controlled by independent channels. Having two solenoid outputs, instead of one, would be useful to power a single weapon (such as a drum) in both directions using two SPDT solenoids arranged in the bang-bang configuration shown in section 7.2.1 (or it could be used in combots with more than one weapon). However, the added weight and volume of 2 solenoids, in addition to the risk of shorting out the battery if both of them are accidentally switched to a "shoot-through" configuration, made us choose instead to use Victors to power the weapon in both directions, as explained in section 7.7.3.

To power the presented electronic systems, you'll need batteries that are capable to deliver high currents. The main battery types, along with their advantages and disadvantages, are studied in the next chapter.

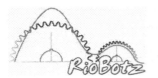 

Chapter

8

Batteries

Batteries are components that usually limit a lot the autonomy of a mobile robot, besides representing a significant part of its weight. Usually, batteries are the heaviest component of a mobile robot. Humanoid robots, for instance, have reached an impressive level of sophistication in the last 20 years. Powerful motors were miniaturized, high performance computational systems became even more compact, however the components that less evolved until now were batteries. In 2000, the most sophisticated humanoid robots needed to be recharged every 30 minutes, even though their batteries accounted for a significant portion of their weight, about 15%. But recent advances in lithium battery technologies, such as the development of A123 batteries, are starting to change this.

Fortunately, combat robots only need an autonomy of about 3 minutes. Combots still need about 15% of their weight in batteries, similar to several humanoid robots, however it is possible to extract from them a much higher power during this short period. But most batteries were designed to be slowly discharged, in 20 hours, in 1 hour, not in 3 minutes. Therefore, it is necessary to know the advantages and disadvantages of each type.

The main battery types are: lead-acid (Sealed Lead Acid, SLA), nickel-cadmium (NiCd), nickel-metal hydride (NiMH), alkaline, and lithium, presented next.

8.1. Battery Types

8.1.1. Sealed Lead Acid (SLA)

SLA batteries have lead-based electrodes, and electrolyte composed of sulfuric acid. Each electrode inside the battery contributes with about 2V, therefore a typical 12V battery has 6 cells connected in series. The SLA types usually used in automobiles cannot be used in combat, because the acid can spill if they are flipped over

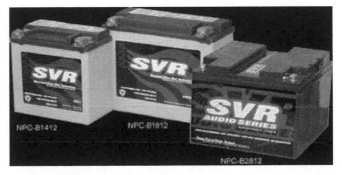

or perforated by an opponent's weapon.

Competitions only allow SLA batteries in which the electrolyte is immobilized, which could work upside down without risk of spilling. The most common technologies to immobilize the electrolyte are gel, where silica is added to generate a semi-solid gel, and AGM (Absorbed Glass Matte), where a fibrous and porous material absorbs the acid and keeps it suspended.

SLAs are usually available in up to 12V, therefore it is necessary to use at least 2 of them connected in series to reach usual combot voltages of 24V or more. They are the cheapest type of battery, however they are the heaviest ones, therefore it is usually better to replace them for NiCd, NiMH, or lithium batteries, which will be discussed next. Another disadvantage is that most of them take several hours to charge.

8.1.2. Nickel-Cadmium (NiCd)

NiCd batteries use nickel as cathode, and cadmium as anode. They supply high currents without significant voltage drops, and because of that they are an excellent choice to power the robots'

weapons. They are more expensive than SLAs, however they can last several years if properly handled, returning their investment. Each cell (pictured to the right) provides about 1.2V. The cells are usually soldered in series to form battery packs (also pictured to the right), with voltages that are a multiple of 1.2V. The packs used in combat usually have 12V, 18V, 24V and 36V, with respectively 10, 15, 20 and 30 cells.

8.1.3. Nickel-Metal Hydride (NiMH)

NiMH batteries also use nickel as cathode, however the anode is composed by a metallic alloy capable to absorb hydrates, replacing cadmium, which is poisonous. NiMH batteries store 30% more energy per weight than NiCd, however they can consistently supply about half the peak currents of a NiCd with same capacity. They are a good choice for the robot's drive system, resulting in a high capacity to avoid having a slow robot at the end of a match (drive systems usually don't require very high current peaks due to wheel slip). A significant problem is that

these batteries lose naturally about 30% of their charge every month (self-discharge), therefore they are not appropriate for applications with sporadic use, such as TV remote controls. Even if the remote is not used, in about 2 months the battery would probably need to be recharged again.

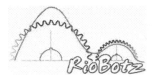

8.1.4. Alkaline

Alkaline batteries are the most common, storing a great amount of energy. They don't suffer as much from the self-discharge problem as NiMH batteries, therefore they are the best option for sporadic use (although they are prone to suffer long term corrosion, which may cause cell rupture and electrolyte leakage – this is why they should be removed if not used for several months). The problem with alkaline batteries is that they are not able to supply high currents, and because of that

they are not used in combat. Besides, they are not rechargeable, and therefore it would be very expensive to use new batteries in every match. There is a rechargeable version, called RAM (Rechargeable Alkaline Manganese, pictured to the right), however it doesn't supply high currents as well, and the number of recharge cycles is relatively low.

8.1.5. Lithium

Very used in cellular phones, portable computers and several other gadgets, lithium batteries (pictured to the right) currently are the ones with the highest charge capacity with lowest weight. However, they are more expensive and, sometimes, dangerous.

The lithium-ion type is the oldest one, and it suffers risk of explosion if perforated and exposed to oxygen, shorted out, or improperly charged, hence it is not recommended for combat robots. This risk is reduced in the lithium-ion-polymer type (a.k.a. LiPo or lithium-polymer), due to its polymeric layer, but it still exists. Newer lithium battery chemistries, such as lithium-manganese and lithium-iron-phosphate, are much safer, although great care and attention is needed when handling this kind of battery, as discussed in section 8.3.

In addition to safety issues, the models that are capable to supply high currents are still expensive, and they require some electronic system in the robot to guarantee that they won't be discharged below a critical voltage, to avoid permanent damage. But the cost-benefit is still very good. The main lithium technologies for use in combat are described next.

Lithium-Ion-Polymer

Most lithium-ion-polymer batteries have discharge rates higher than 20C, in other words, it is possible to completely discharge them in less than 1/20 of an hour, which is exactly the 3 minutes that we need during a combot match. For 2 minute matches, common in insect classes, a discharge rate of 30C or higher would be better.

The nominal lithium-polymer cell voltage is 3.7V, but when fully charged it provides up to 4.2V per cell. It is not recommended to let the battery voltage drop bellow 3.0V per cell. If this happens,

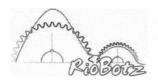

the pack can swell and become permanently damaged, which is also known as puffing or ballooning. This is why it is a good idea to use LiPo battery sets with at least twice the capacity you might think your robot will need during a match, making sure they'll not be completely drained.

Similarly to other battery types, more than one cell is usually needed to power a combat robot. Most manufacturers use the number of cells connected in series or in parallel to describe their products. For instance, if each cell has a nominal voltage of 3.7V and a 500mAh capacity, then a 3S LiPo pack would stand for three cells in series, resulting in 11.1V and 500mAh, and a 3S2P pack would stand for two parallel arrays of three cells in series, resulting in 11.1V and 1,000mAh.

An inconvenience of LiPo batteries is that, despite their short discharge time, usually the charge time is much longer, up to 2 hours in the oldest models, which can be critical between combats. However, newer models can be safely charged at a 1C rate (in other words, in 1 hour), while a few vendors state that their packs can handle a 2C charge rate (charged in 30 minutes). The use of an adequate charger is mandatory, never charge lithium-polymer batteries on lead-acid, NiCd or NiMH chargers, otherwise it will ignite on a strong fire, releasing toxic fumes.

As it can be seen in the picture to the right, LiPo batteries usually have two sets of wires. The twisted pair cable, with black and red wires, is the main power cable. The other cable, with five wires and a white connector in this case, is used for cell-balancing. This process consists of equalizing the cells after charging. There's some controversy on that, because a few manufacturers claim that their battery packs don't need to be balanced, while others recommend to balance the cells regularly. As a rule of thumb, cell balancing is only needed 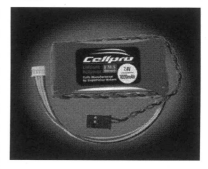 when a fully charged pack presents significant disparities between cell voltages. This can be checked using a voltmeter between the black (negative) and the other wires of the balancing connector. Most vendors recommend balancing if there's a difference higher than 0.1V.

LiPo batteries are mostly used in combat in the lightest (up to 3lb) weight classes, however new technologies are emerging to allow their widespread use. Lithium batteries have a great potential to become the best choice even for the heaviest classes. Nowadays, there are quite a few heavier robots that use this kind of battery, such as all our hobbyweights and featherweight combats, as well as Kevin Barker's lightweight vertical spinner K2, which uses two LiPo 6S1P 5,000mAh packs.

Lithium-Manganese

Lithium-manganese batteries (LiMn, pictured to the right), developed in 2005 by Apogee High Performance Lithium Polymer Technology, use a safer chemistry that can sustain perforation without exploding.

Since the voltage of each cell is 3.7V, these batteries are entirely compatible with lithium-polymer chargers. They can provide the same peak currents of NiCd and the same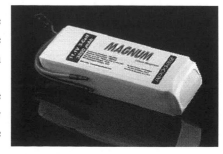

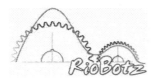

capacity of NiMH with about half the weight. In addition, they can be charged at a 2C rate. Apogee states that their batteries don't need balancing, due to their cell matching process.

An interesting feature of these batteries is their polycarbonate shielding, which can minimize cell damage during rough handling in combat.

Super Charge ion Battery

Toshiba started shipping in 2008 its Super Charge ion Battery (SCiB, pictured to the right), which can be recharged to 90% of its capacity in only 5 minutes, with a life span of over 10 years.

Charging can be performed with currents as high as 50A, which is a real breakthrough. This battery can sustain more than 3,000 rapid charge cycles, with less than 10% capacity loss. It adopts a new negative-electrode material technology that is safer and more stable, being virtually resistant to punctures and short-circuits.

SCiB Cell

Unfortunately, SCiB batteries are currently only available to industrial markets, in either 2.4V/4.2Ah/0.150kg or 24V/4.2Ah/2.0kg versions.

Lithium-Iron-Phosphate

One of the most promising battery technologies is the lithium-iron-phosphate (LiFePO$_4$ or LFP), discovered in 1996 at the University of Texas. In addition to its high peak currents (over 100C pulsed discharge rates) and high capacity, it is a safe technology: it will not catch fire or explode with overcharge. Its charge time is very low compared to other lithium battery types, sometimes as low as 15 minutes. It is also environmentally friendly.

The most famous brand of LiFePO$_4$ battery is A123. Originally, only the M1 cell model (pictured to the right) was available, but now A123 is also producing other models with higher capacity, such as the M1HD and M1Ultra cells. A123 cells are also sold assembled in battle-ready packs, as pictured to the right, available in several configurations at www.battlepack.com.

Nominal voltage varies amoung manufactures, from 3.0V (K2) to 3.3V (A123) per cell, but when fully charged the cells can provide up to 3.6V. To avoid damaging them, these batteries must not be discharged below 2.8V per cell.

Note that LiFePO$_4$ batteries must be charged with a specific charger, such as the Astroflight 109 A123, iCharger 1010B+, or Robbe Power / MegaPower Infinity SR. There are also adapters, such as the Dapter123, which allows the use of most NiCd chargers.

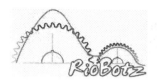 

8.2. Battery Properties

Several battery characteristics need to be considered: price, weight, voltage, shelf life, number of recharge cycles, charge time, self-discharge, discharge curve, internal resistance (which determines the peak current and voltage drop), capacity, de-rating factor and discharge rate, described next.

8.2.1. Price

Price is the first factor in the choice of batteries. SLAs are the cheapest, followed by alkaline, NiCd, NiMH, and finally the lithium batteries. Prices vary a lot depending on the technology, manufacturer, quality and capacity.

8.2.2. Weight

The weight of the battery is crucial in robot combat. More specifically, it is important to know the power-to-weight, energy-to-weight, and capacity-to-weight ratios of each type, the higher the better. SLAs are the worst ones in this requirement, they store less energy per pound than any other type. NiCd and NiMH are much better, while lithium is the best, see sections 8.2.10 and 8.2.12.

8.2.3. Voltage

Battery voltage depends on the number of cells and the electrode chemistry. SLA electrodes nominally supply 2V, usually combined to provide 12V. Alkaline electrodes supply 1.5V, while each NiCd or NiMH cell provides 1.2V. The nominal voltage of lithium batteries depends on their type: lithium-ion-polymer (LiPo) and lithium-manganese (LiMn) provide 3.7V per cell, lithium-ion (Li-Ion) 3.6V, and lithium-iron-phosphate (LiFePO$_4$ / A123) between 3.0 and 3.3V.

Note that the values above are nominal voltages. In practice, the voltage is usually higher than this value, when the battery fully charged, or it can be lower, if it is supplying very high currents, which lead to significant voltage drops due to their internal resistance.

8.2.4. Shelf Life

Shelf life depends a lot on the use and mainly on the storage temperature. In a few cases the batteries can last more than 20 years without significant capacity loss, such as in the case of NiCd stored at 40°F (about 5°C) in a refrigerator. If stored at 100°F (about 38°C), these same batteries would last less than 2 years.

8.2.5. Number of Recharge Cycles

The number of recharge cycles during the useful life of a battery goes from zero (alkaline), up to 300-800 (SLA and NiMH), 500-1200 (lithium-ion), 1500-2000 (NiCd), 5000 (SCiB), and even up to 10,000 recharge cycles in a few special lithium batteries. Note that, as the technology develops, these numbers can be outdated, however they are a good reference for comparison purposes. Forum posts and manufacturer websites are a good source of information to find out more accurate values.

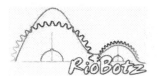

8.2.6. Charge Time

Charge time is another important factor, it determines the minimum time necessary to fully charge a battery without damaging it. The longer it is, the higher will be the number of spare battery sets you will need in a competition. SLAs are the worst ones in that sense, they usually need several hours to fully charge. The Li-Ion and LiPo types usually need at least 1 hour (1C charge rate), however a few of the newer technologies may charge must faster than this, such as A123 (in 15 minutes, with a 4C charge rate) and SCiB (in 5 minutes, 12C). The NiCd is one of the best types, it takes much less than 1 hour to fully charge, in a few cases in only 15 minutes, without permanent damage. Some newer NiMH batteries are reaching similar charge times as NiCd.

8.2.7. Self-Discharge

Self-discharge quantifies which percentage of its capacity a battery naturally loses per month (or per day). Lithium batteries lose about 5% of their capacity per month. NiCd and SLA batteries, if stored at room temperature, may lose about 10% of their capacity per month, while NiMH may lose about 30% per month. Therefore, if you use NiMH batteries in a combat robot, always recharge them again at the day of the competition, to compensate for this loss. This is also a good idea for the other types.

8.2.8. Discharge Curve

The discharge curve of a battery shows its voltage level as it drops off during use. For instance, the graph to the right shows that each electrode (cell) of a SLA battery supplies about 2.1V (up to 2.2V) when fully charged, a value that is gradually reduced until reaching about 1.7V. Therefore, a SLA battery with nominal voltage 12V (with 6 lead electrodes) would have up to 6 × 2.2V = 13.2V when fully charged, and 6 × 1.7V = 10.2V when discharged. This noticeable

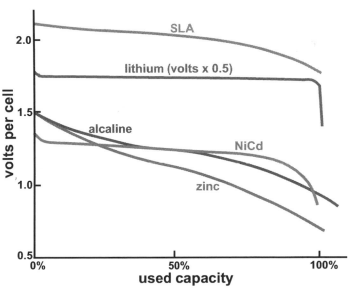

drop has only one advantage, it could be used to indirectly measure the remaining capacity of the battery. But this voltage drop, on the other hand, will make the system lose power and become slower. Also, robot combat judges would be able to tell that the batteries where dying from this sluggishness, awarding damage points to the opponent. This significant voltage drop happens not only with SLA, but also with (disposable) zinc and alkaline batteries, as seen in the graph above.

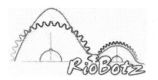 

Lithium and NiCd cells have an almost horizontal discharge curve, keeping constant their voltage level during the entire combat (except during voltage drops due to high currents). The (rather abrupt) voltage drop is only noticeable towards the end of the battery capacity. NiMH curves are not as horizontal as in NiCd, they are slightly sloped, however not nearly as much as in SLA.

8.2.9. Internal Resistance

The internal resistance of a battery is added to the total resistance of your electronic system. Therefore, the smaller the resistance, the larger will be the current peaks that the battery can deliver. SLA and NiCd batteries have very low internal resistance, allowing them to generate very high currents. The problem with SLA is that those current peaks reduce a lot the battery capacity, due to the de-rating factor, which will be discussed later. NiMH batteries have larger resistance than NiCd, and therefore they are not able to deliver such high current peaks (if compared to NiCd batteries with same capacity, of course). The first lithium batteries had high internal resistance, however in the most recent versions, such as the A123, this value is much lower.

The internal resistance is also related with the voltage drop in the battery caused by very high currents. This is simply due to the energy loss caused by the resistance, which is significant under high currents. This energy is converted into heat, which can also cause thermal failure of the battery due to overheating. A123 batteries, due to their very low internal resistance, can deliver very high currents without significant increase in their temperature.

8.2.10. Capacity

Capacity measures the total amount of current that a battery can deliver until it is fully discharged. It is measured in A·h, calculated from the product between the total discharge time and the average delivered current (if the current isn't constant, then it is calculated integrating it along the discharge time). For instance, a 3.6A·h battery would theoretically supply a current of 3.6A, continually, during 1 hour, or 36A for 1/10 hour (6 minutes). Note that if two identical batteries are connected in parallel, the total capacity is doubled.

In theory, the capacity of a 24V SLA battery would be about 1.25A·h per kilogram (about 0.57A·h per pound). This is a relatively low capacity, leading to a low energy density of about 24V × 1.25A·h = 30V·A·h/kg = 30W·h/kg. A 24V NiCd pack would have from 1.7 to 2.5A·h/kg (0.77 to 1.13A·h/lb, with energy density between 40 and 60W·h/kg), a good quality NiMH would have 2.5 to 3.3A·h/kg (1.13 to 1.5A·h/lb, with energy density between 60 and 80W·h/kg), and finally lithium batteries would go beyond 4.2A·h/kg (1.9A·h/lb, with energy densities between 100 and 200W·h/kg).

Regarding energy per volume, which is also relevant if you want to build a compact robot, then SLAs only have between 60 and 75 W·h per liter, while NiCd between 50 and 150, NiMH between 140 and 300, lithium-ion about 270, and LiPo around 300W·h/liter.

But those capacity and energy numbers are theoretical, because in practice it is not so simple, the effect of the de-rating factor must be considered, as discussed next.

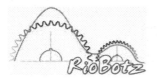

8.2.11. De-Rating Factor

The total capacity of a battery depends on the discharge time. The nominal capacity that is written on SLA batteries, for instance, is related to a discharge time of 20 hours. Therefore, if you discharge a 17.5A·h SLA with a constant current of only 17.5/20 = 0.875A, it will really last 20 hours. But if you discharge it at 17.5A, it won't last 1 hour. This is because the real capacity of this battery in 1 hour would be only about 10A·h, see the graph below. Therefore, the correct value for a 1 hour discharge would be 10A instead of 17.5A. As it is discharged faster, its capacity decreases. That same battery would only supply 5.8A·h if totally discharged in 6 minutes (0.1 hours), and less than 5A·h during a 3 minute combat (0.05 hours). Those values, obtained experimentally, are represented in the graph below.

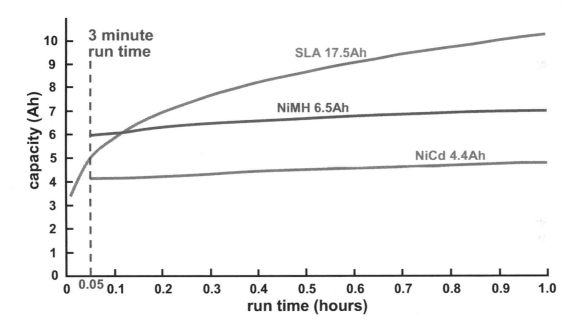

Note in the graph that the capacity of the SLA battery depends a lot on the total discharge time (run time). The value that must be multiplied to the nominal capacity to generate the actual battery capacity is called de-rating factor, a number that is usually between 0 and 1. For instance, the de-rating factor of a SLA battery that is required to be discharged in only 6 minutes (0.1h) is worth 0.33, which would give 0.33 × 17.5 = 5.8A·h for the 17.5A·h SLA, which agrees with the graph above, delivering continuous 5.8A·h / 0.1h = 58A.

If you still need more current than that, to the point of fully discharging the battery during a 3 minute combat (0.05h), the de-rating factor will be even lower, about 0.28. In this case, the capacity would be 0.28 × 17.5 = 4.9A·h for the 17.5A·h SLA, which also agrees with the graph above, delivering continuous 4.9A·h / 0.05h = 98A.

The special SLA Hawker-Odyssey (also known as Hawker-Genesis, pictured to the right) has higher de-rating factors than regular SLA batteries, reaching values between 0.4 and 0.5 for the 6 minute run time (instead of 0.33).

One of the greatest advantages of NiCd and NiMH batteries is that their capacity is almost insensitive to the total discharge time (run time). This can be seen in the previous graph, which shows NiCd and NiMH capacity curves that are almost horizontal. Note that their nominal capacity is measured in a 1 hour discharge time, instead of 20 hours as with SLA (lithium batteries are also measured in 1 hour). Even so, there is a de-rating factor for NiCd and NiMH, which is about 0.9 for run times between 3 and 6 minutes. That de-rating factor is 3 times better than the one from regular SLA, and almost 2 times better than in Hawker-Odyssey.

Therefore, for instance, 2 regular SLA batteries with 12V and 18A·h each, when in series, are able to supply 24V with a combined weight of 6.2kg × 2 = 12.4kg (27.3lb). If their desired run time is 3 minutes (in practice it is safer to design your robot with run times of at least 4 minutes, so it can safely endure a 3 minute combat), their actual capacity is 0.28 × 18 = 5A·h. On the other hand, two 24V NiCd packs with 3A·h each, when in parallel, can supply the same 24V with a nominal capacity of 3A·h × 2 = 6A·h. Their actual capacity in 3 minutes is 0.9 × 6 = 5.4A·h, larger than the one from the SLA, and with a total weight of only 1.8kg × 2 = 3.6kg (7.9lb). This is less than a third of the weight of the SLA set, with an equivalent capacity!

The performance of SLA batteries only approaches NiCd when at least 10A·h is needed by the robot, such as in heavyweights and super heavyweights. Even so, this only happens for special batteries such as Hawker-Odyssey. For instance, two 12V Hawker-Odyssey batteries with 26A·h each, when in series, supply 24V, with a combined weight of 6.1kg × 2 = 12.2kg (26.9lb), with an actual capacity of 0.42 × 26 = 10.9A·h (the 0.42 de-rating factor was experimentally measured). It would be necessary to use four 24V NiCd packs, with 3A·h each, in parallel, to achieve those values, resulting in a combined weight of 7.2kg (15.9lb). The NiCd packs would still be lighter, however the weight difference decreased to 5kg (11lb), a small value if compared to the total weight of a heavyweight or super heavyweight. The advantages of using special SLA batteries are their price (about one third the price of equivalent NiCd) and the achievable peak currents, which would be about 800A for this NiCd arrangement (with high discharge cells) but almost 2400A for the Hawker-Odyssey (watch out not to burn your motors and electronics!).

The previous calculation always assumed that the discharge current was constant, which is certainly not true during combat. To estimate with better accuracy the capacity of a SLA battery, you need to use different values of the de-rating factor. For instance, consider the 17.5A·h SLA

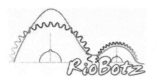 

battery from the previous graph, and assume that your robot needs about 15A to drive around with the weapon turned off, and 100A when it is on. How many minutes would it last with that battery, assuming that it spends 80% of its time with the weapon powered? The answer is obtained calculating the capacity considering the different values of the de-rating factor. From the previous graph, a run time of 0.6 hours (36 minutes) would result in 9A·h, with a continuous current of 9A·h/0.6h = 15A. But a run time of 0.05 hours (3 minutes) would result in only 5A·h, with a continuous current of 5A·h/0.05h = 100A. If the number of minutes to be calculated is t, then the robot spends 0.8·t minutes drawing 100A (3 minute run time), and 0.2·t minutes drawing 15A (36 minute run time), so to completely discharge the battery we would have (0.2·t)/36 + (0.8·t)/3 = 1, thus t = 3.67 minutes.

Let's check the calculations: during the 0.2·t = 0.734 minutes at 15A the robot drains 0.734min/36min = 2% of the battery capacity, and during the remaining 0.8·t = 2.936 minutes at 100A it drains the other 2.936min/3min = 98%. These more sophisticated calculations are not necessary for nickel or lithium batteries, because their de-rating factor varies very little, between 0.9 and 1.0.

8.2.12. Discharge Rate

Finally, the last relevant battery property is the discharge rate, which measures how much current can be continually drawn from the battery without letting it become significantly hot. It is represented by a number followed by the letter C. For instance, 8C means that the battery tolerates without problems a current of 8 times its measured capacity (C, hence the name 8C) in A·h. For instance, a 3.6A·h battery with 8C tolerates continuous $8 \times 3.6 = 28.8A$ without overheating. This is the same as to say that it can be fully discharged in 1/8 of an hour (because 28.8A × 1/8 h = 3.6A·h), which is equivalent to 7.5 minutes.

In practice, most NiCd batteries can withstand more than twice the calculated current from their discharge rate, however they will significantly warm up (but not necessarily overheat, in special if the cells are spaced inside the pack to keep down their temperature). In other words, a NiCd 8C battery could be continuously discharged in only 3.75 minutes, compatible with the duration of a typical combat. Avoid using NiCd batteries rated below 8C, they will very likely overheat during combat.

A few lithium batteries, such as the Polyquest (LiPo) and A123 (LiFePO$_4$), can continuously deliver over 30C and sustain 50C (or higher) peaks without overheating. But, different from NiCd packs, lithium batteries usually do not tolerate current peaks that are more than twice the calculated value from the discharge rate.

Another way to evaluate discharge capacity is through the power-to-weight ratio of a battery, in W/kg. It evaluates the power that a battery can continuously deliver divided by its weight. SLAs can only deliver about 180W/kg, while nickel batteries between 150 and 1,000W/kg, lithium-ion about 1,800W/kg, and LiPo beyond 2,800W/kg.

8.3. Battery Care and Tips

To make your batteries last longer, it is important to follow several procedures, described next.

8.3.1. Shock Mounting

Make sure that the batteries are very well mounted inside your robot, with some cushioning to avoid impact damages. For instance, the instantaneous accelerations that a robot might suffer during an impact from a violent spinner can reach up to 800G, in other words, 800 times the acceleration of gravity. Only for reference, at 10G a person would faint, and at 100G one's brain would detach from the skull, causing instantaneous death. Therefore, 800G is somewhat frightening, even considering that this acceleration lasts only a small fraction of a second. Quick calculations show that a 4.4lb (2kg) battery would suffer an equivalent inertial force of 2kg × 800 = 3,520lb in its support. Of course this would be an extreme case, but even for much smaller impacts it is evident that zip ties are not appropriate (unless it is a very light pack such as the ones used in receivers). Besides, zip ties might also melt due to the high temperatures that the batteries can reach.

Good materials to shock mount your batteries are hook-and-loops and neoprene. Corrugated plastic, cut from file cases or other office supplies, is also an inexpensive and effective shock mounting material, as pictured to the right.

Be careful not to cause short-circuits, you must isolate very well any metal parts that get in touch with the battery. And you must guarantee in your robot design that the batteries can be quickly replaced, to speed up pitstops.

Note that LiPo batteries expand almost 10% in size during use, so make sure that there's extra room inside your robot not to let them get squeezed too much. Using very compliant shock mounts is a good way to accomplish that.

8.3.2. Recharging

To recharge batteries, especially the nickel and lithium types, you must use an electronic charger. They are an indispensable investment. Without them, the chances of damaging batteries are very high. Triton 2 (pictured to the right) is one of the best and easiest chargers to use. It automatically charges or discharges most battery types, with several programming options. It eliminates the infamous "memory effect" that happens when NiCd batteries are not properly charged. It costs a little over US$100 in the US. It is

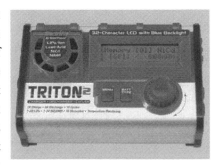

really worth investing in an electronic charger such as Triton 2, or several famous others such as Astroflight, Thunderpower or Dynamite models. It only takes one damaged 24V NiCd pack to set you back more than the price of the charger.

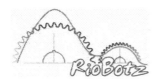

Due to lithium batteries being prone to ignite when mishandled, they need a lot of attention when charged. They need an intelligent charger to charge them following the correct algorithm. This kind of battery cannot be overcharged under any circumstances. In addition, to prevent damages to your robot (and people), always remove the batteries from the robot and charge them inside a fireproof container, such as LP-Guard or LipoSack. The picture to the right shows a LipoSack and the ignited battery that it withstood inside.

If your robot uses more than one pack, you might want to have more than one charger, to be able to charge the entire set of batteries at the same time, in time for the next match. We've built a wooden box, which is a good electric insulator, to mount 4 Triton chargers (pictured to the right), together with 12V power supplies taken from old personal computers. The box also carries our battery packs, which is a practical solution for ground transport and use in Brazilian competitions. For international competitions, which require air travel, we have a more compact version of the charger box, shock-mounted inside a suitcase.

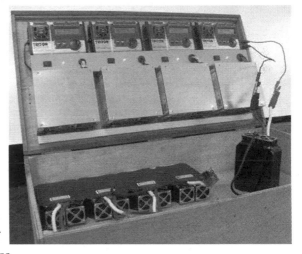

Always have at least 2 sets of batteries for your robot. Nickel and lithium batteries can take up to 1 hour to fully charge, however the pit time between rounds can be as low as 20 minutes. If you can afford it, get 3 or more battery sets. Besides solving the charging time problem between rounds, having a third (or fourth) set is an additional insurance in case a pack gets damaged or shorted out during combat. If you use regular SLA batteries, you might need up to 6 or 7 sets during a competition, because they charge very slowly, sometimes taking several hours.

Another important tip is: never charge hot packs. Its useful life would be very much reduced if charged while still hot after a match. As soon as a combat ends, immediately remove the battery packs and put them over a large fan to cool down (as pictured to the right). Only begin to charge them after they get close to room temperature.

8.3.3. Battery Storage

Always store SLA batteries fully charged. If they are stored discharged for a long time, they can be damaged. Completely recharge SLA batteries every 6 months if they're stored.

Unlike SLA, NiCd and NiMH batteries should be stored fully discharged. Be careful not to discharge them too much, nickel batteries should never get below 0.9V per cell. Lithium-polymer and lithium-manganese batteries must never get below 3.0V per cell, while LiFePO$_4$ should never be allowed to drop below 2.8V per cell.

Also, have in mind that heat kills: even if properly discharged, nickel batteries can last less than 2 years if stored at 100°F (38°C). At 77°F (25°C) they usually last 5 years, at 59°F (15°C) they can reach 10 years, and at 41°F (5°C) up to 20 years. Therefore, store NiCd and NiMH batteries inside a refrigerator, fully discharged. Put them inside a sealed plastic bag to protect them from moisture. Never freeze the batteries. Every 6 months or less, make sure to fully charge and discharge them, and put them back in the refrigerator.

To store lithium batteries, first place the packs into a LiPo sack or equivalent, then charge (or discharge) them to 40%~60% of their capacity (LiPo: 3.8~3.9V per cell; LiFePO$_4$: 3.2~3.3V per cell). If your charger does not support terminal voltage configuration, then fully charge the battery and then discharge it monitoring the voltage value with a voltmeter. Finally, isolate the connectors with tape, place the batteries into separate sealed plastic bags, and store them in the refrigerator at about 41°F (5°C). Once a month, check the voltage of your lithium batteries and, if needed, recharge to keep them between 40% and 60% of their capacity.

Be very careful not to short-circuit your batteries, especially the lithium ones. Be attentive when handling screws near the pack, they can fall inside and cause a short-circuit that can permanently damage the battery. There is also the risk of metal debris entering your pack during combat. This may result in the famous "magic smoke" (pictured to the right), which will either disable your robot or result in damage points to your opponent. Some people say that magic smoke is the robot's soul leaving its metal body.

To avoid this problem, you can wrap up each battery in the pack with Kapton tape, a polymer that resists high temperatures, up to 750°F (400°C), besides being a good electric insulator.

If your pack is getting too hot, an option is to install one or two fans to blow air inside it, helping it cool down. There are a few ready solutions in the market. One of the best NiCd and NiMH packs in the market are the ones from Robotic Power Solutions (www.battlepack.com), they sell both the traditional battlepacks and the intercooled ones (pictured to the right).

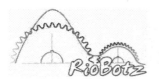 

8.3.4. Assembling Your Own Pack

NiCd and NiMH packs are not cheap, so it is possible to save some money if you assemble them yourself. First, buy individual cells, making sure that their discharge capacity is at least 8C, and wrap each of them with Kapton to avoid shorts.

To assemble the pack, weld the cells using flexible copper braids. Rigid connections can break during combat. At the RoboGames 2006 semi-finals, one of the two 24V battery packs from our middleweight drumbot Touro stopped working right after the first impact against the tough rammer Ice Cube. Touro had to fight the entire 3 minutes with only one pack, which made it slow down near the end of the match. This counted as damage, which was decisive in our split decision loss by 17-16. Back in the pits, we've realized that a rigid connection had broken off inside the battery. The pack was cold, indicating that it had broken in the beginning of the match, with almost full charge. This was confirmed from the 2 minutes it later took to be recharged, after the solder was fixed. Since then, we've only used copper braid connections in our batteries.

Use very fine sandpaper on the battery contacts to remove oxidation, which would compromise the mechanical and electrical resistance of the solder. Use a high power solder iron, with at least 100W. Tin as much as possible the battery contacts and the wire, and weld them quickly to avoid

heating up and damaging the cell. An important tip is, before welding, to put o-ring spacers among the cells, as pictured to the right. These spacers can be, for instance, cardboard rings held together with shoemaker's glue. By doing so, you will leave gaps among the cells that will allow air to flow, making the heat exchange much more efficient, avoiding overheating them.

After having soldered all the cells using copper braid, weld the connector and its wires. The connector can be, for instance, a Deans Ultra or the Anderson Powerpole. Remember that the battery should always use the female connector, never the male, to avoid any chances of short-circuit if the connector accidentally touches some metallic part.

Optionally, you can use nylon plates (or some other insulating and resistant material) covering the top and bottom of the pack, as pictured to the right. It will be necessary to use a mill to carve slots in the nylon to accommodate each of the cells. You can then secure the nylon plates with long screws, as shown in the picture.

With or without nylon plates, it is advisable to protect the pack with shrink-wrap. If you don't have

shrink-wrap, a very cheap alternative is to fit the pack inside a cut Coke bottle, and use a heat blower to shrink it to hold the pack with a snug fit. Using a hot soldering iron, you can make a few openings in the shrink wrap (or Coke bottle) at the spaces between the cells, to improve cooling.

A very similar process can be used to assemble packs made out of A123 packs, which also have a cylindrical shape. But don't forget to solder separate cables and connectors for cell-balancing.

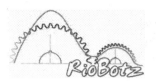 

The pictures to the right show a 20-cell 10S2P A123 pack including cell-balancing connectors. The 10S2P configuration features 2 parallel arrays of 10 A123 cells each in series, resulting in a nominal capacity of 2×2.3Ah = 4.6Ah and a nominal voltage of 10×3.3V = 33V.

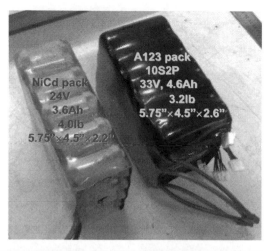

This pack resulted in the exact same width 4.5" and length 5.75" of a 24V 3.6Ah NiCd battlepack (including o-ring spacers, as pictured to the right), which allows both to be interchanged without having to modify existing robots to fit them.

The height of the A123 pack, however, is a little higher: 2.6" instead of 2.2" from the NiCd pack. But the A123 pack is actually lighter, because its 20 cells weigh 20×70g = 1.4kg, instead of 20×88g = 1.76kg from the NiCd CP-3600CR cells. In addition, this A123 pack has higher nominal voltage and capacity than the equivalent sized NiCd battlepack, 33V and 4.6Ah instead of 24V and 3.6Ah. Not to mention the improved properties of A123 cells over NiCd.

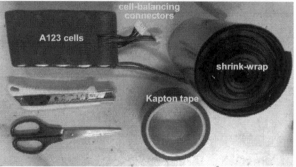

An important advice: you should only assemble your own pack if you know what you're doing. It is not difficult to damage cells by overheating them while they're soldered. It pays off to have them professionally assembled, for instance, at www.battlepack.com.

8.3.5. Billy Moon's Rules for LiPo

Finally, a battery care section would not be complete without the rules from famous builder Billy Moon for handling LiPo batteries:

1. NEVER, NEVER, NEVER charge in your robot;
2. NEVER, NEVER, NEVER charge them hot;
3. always charge them in a LiPo sack or steel tool box;
4. check balance on your batteries before each charge;
5. charge them as slow as you can afford to;
6. never short their leads (if you do, toss them out);
7. never remove them from your bot by the leads;
8. allow extra room for them to expand by 10% in all dimensions while in use;
9. never fully discharge them: plan for at least 50% more capacity than you need;
10. bring or arrange to have a 'class D' fire extinguisher on hand.

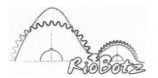 

Chapter

9

Combot Events

This chapter presents several tips related to combat robot events, and how to get ready for them.

9.1. Before the Event

The first step is to find out an event. In the www.buildersdb.com website you'll find all the information on most of the incoming as well as past events. There you'll also be able to register your team, builders and robots, as well as search for other teams and robots. In addition, the Robot Fighting League events can also be checked at http://botleague.net. Don't forget about the registration deadlines, the organizers need to know as soon as possible how many teams will attend to plan accordingly. It's important to register in advance for the events, because there might be a limited number of robots in each weight class. Read carefully all the event rules, to make sure that there are no problems with your robot.

9.1.1. Test and Drive Your Robot

Finish your robot before you travel to the event. There's nothing more stressful than going for an all-nighter on the eve of the event. Especially if you'll be waking up all other hotel guests with grinding noises and the smell of burnt rubber mixed with Dremel disks, as we unfortunately did during our first event back in 2003. Guarantee that your robot will pass safety inspection.

Train driving your robot. A lot. Several matches are won or lost because of the driver's ability. Train slalom using traffic cones. Wendy Maxham suggests a practice technique from Grant Imahara's book [10], in which you mark out a square on the floor and then drive the robot as close to the edges as possible. You'll learn how to drive straight, and how to make sharp turns. Start out slow, then go faster and faster until you reach full speed. Don't forget to train in both directions, to practice both left and right sharp turns.

Another great practice move, suggested by Matt Maxham, is the "James Bond turn." While driving forward on a straight line, quickly spin your bot 180° and reverse the wheels to keep driving in the same sense (you'll be driving backwards, but in the same original sense). Then spin again, to make the bot face forward, always moving in the same sense. It is a good maneuver to make your

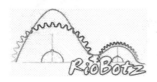

weapon face the opponent while you're escaping from it. It is also a great maneuver to shoot your pursuer during a car chase, if you're James Bond.

During a combat, you can't waste time thinking about which way to steer, left or right, which can be tricky if your robot is moving towards you. As Matt Maxham says, you (the driver) need to imagine that you're sitting on top of the bot, then you'll naturally steer in the correct direction.

Buy a cheap remote control car to play cat and mouse. Actually, buy more than one, they usually don't survive when you catch them. In early 2003 we created a toy overhead thwackbot out of a plastic remote control car (pictured to the right). The robot itself, while driven, was useful to improve our skills controlling overhead thwackbots in general, which are very tricky to handle. And this toy robot also doubled as a very fast and efficient "mouse" when chased by our first middleweight combot *Lacrainha*.

Always test your robot. Test it under real conditions, drive it against a wall, several times. Use its weapon (if any) to hit on junk parts with up to the same weight as your robot. Try to hit solid one-piece blocks, to avoid having small parts flying all over. A low hardness sparring is a good idea to avoid blunting your weapon, such as the 7" diameter solid aluminum block that our combots like to play with, pictured to the right. Use the hit-break-fix it technique, until your robot does not break anymore. It is important to test the robot well in advance, to make sure that there will be time to fix it before the event.

Drop your robot from 3 feet (about 1 meter) in the air over a rigid floor. It needs to resist this fall, no matter which weight class it belongs to. From fairyweights to super heavyweights, all of them are usually thrown higher than that during battle. Drop it several times and always verify if something broke or got loose. If you really trust in your robot's resistance, try dropping it from 6 feet (about 2 meters). Most well designed combat robots can survive such 6 foot fall. During RoboGames 2006 we were able to verify that: the heavyweight Sewer Snake (pictured to the right) was still functional even after it was launched several feet into the air by the super heavyweight Ziggy.

Tests will expose the robot's weak points. It is usually possible to correct them without changing too much the original design. Even a tiny flaw can sometimes be enough to make you lose a match. Therefore, it is very important to have redundancy. For instance, if your robot uses 2 or more batteries in parallel, guarantee that it will keep moving if one of them fails. If your robot has 4 active wheels, guarantee that if one of them is destroyed all the other three will still be able to drive it. If you use belts or chains in a critical component such as the weapon, consider the possibility of using a double pulley or double sprocket. During Touro's first match ever, at RoboGames 2006, its opponent was able to tear one of the drum V-belts. However, Touro's weapon continued to spin because of the redundancy from the second V-belt, allowing it to win the match by knockout.

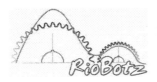

9.1.2. Prevent Common Failures

The 20 most common causes for a combat robot to lose a match, according to the website http://www.solarnavigator.net/robot_building_tips.htm, and our proposed solutions, are:

1. Battery connectors or other wires getting loose – always use good quality connectors such as Deans or Powerpole, always use ring terminals (not fork terminals), tighten each terminal connection using pressure washers (but never place them between the electric contacts, since their electric resistance may be high), and use liquid electrical tape or hot glue;

2. Motors, batteries or other components getting loose – avoid using nylon ties or clamps, even the metal ones, always verify any loose parts and tighten critical screws before each match, use threadlockers or spring locks;

3. Chains coming off from the sprockets – make sure that the sprockets are well aligned, avoid exposed sprockets (as in the special drive system for ice arenas from the robot pictured to the right); if possible, replace the chains with timing belts (such as in the weapon system of the same robot to the right), which can withstand larger misalignments;

4. Radio interference or signal loss – if using 75MHz or lower frequency radios, place the antenna outside the robot, without touching the metal surfaces, and use an amplified antenna such as the Deans Base-Loaded Whip (chapter 7); if using 2.4GHz or higher radio frequencies, or if the covers are not metallic, then the antenna can be left inside the robot;

5. Improperly charged or low capacity batteries – always use electronic chargers such as Triton, check the battery voltage before each match, and calculate and test the robot's consumption under real conditions; before an event it is a good idea to apply several discharge-charge cycles to the batteries;

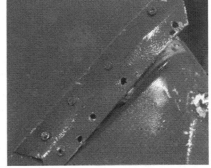

6. Smoking speed controllers – always match the maximum acceptable current in the controller with the motor specs; if a wheel or a spinning weapon gets stuck during a match, turn it off to avoid stalling the motor;

7. Rupture of rivets, screws, nuts – never use rivets (seen in the picture to the right), always use hardened steel screws and nuts, class 8.8 or 10.9 for hex screws and 12.9 for Allen, and with appropriate diameters as discussed in chapter 4;

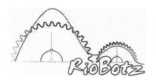

8. Low clearance robots getting stuck in the arena – from hobbyweights to super heavyweights, make sure that you have a ground clearance of at least 1/4", preferably 3/8" or more; don't forget to consider the wear and tear of the wheels, and use flat head screws on the robot's bottom cover to make sure they won't get stuck on the arena floor;

9. Low power wheel and weapon motors – use motors with enough power to guarantee that your robot's drive system acceleration is high enough for your strategy, preferably taking less than 2 seconds to reach top speed; if you have a spinning weapon, make sure that it can reach damaging speeds in less than 4 seconds;

10. Burning fuses – do not use fuses in combat, otherwise you can lose a match only because of a brief current peak; use a current limiting circuit if needed, but not a fuse;

11. Shorting of wires and electric components – always isolate the terminals with electrical tape (either regular or liquid, as pictured to the right), heat shrink and/or hot glue; always protect any electronic component that can be shorted out if metal debris enters the robot;

12. Overheating motors – avoid overvolting too much the motor; several motors can take up to twice their nominal voltage, but it might be necessary to use a current limiting circuit; in a few cases it is possible to mount fan blades onto the rear end of the motor shaft to improve cooling; avoid stalling the motors during combat;

13. Broken gears – all gearboxes need to be well designed and built, with well aligned and precisely spaced gears; the gear thickness and teeth dimensions must be proportional to the torque it carries; therefore, to optimize weight, use heavier duty gears in the last stage and lighter ones in the first; always use hardened steel gears instead of mild steel or cast iron ones;

14. Internal combustion engines that die or won't start – use an automatic ignition system, controlled by a separate radio channel (see chapter 5);

15. Shaft mounted components getting loose – never use set screws or pins, either in shaft couplings (as pictured to the right) or in other shaft mounted components such as pulleys or sprockets; always use keys and keyways (or keyless bushings such as Trantorque) to transmit torque, and very tight shaft collars to avoid axial displacement;

16. Broken or bent shafts – never use shafts made out of mild steel (also pictured to the right) or aluminum, always use hardened steel or titanium grade 5; make sure that the shaft diameter is large enough to keep the stresses below the material yield strength;

17. Wheels getting stuck in the robot's bent structure or armor (also pictured to the right) – leave a significant clearance between the wheels and any armor or structural part of the robot that could get bent;

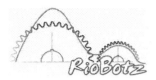

18. Flat tires – use solid wheels such as Colsons or, if using pneumatic wheels, make sure that they're filled with polyurethane foam, such as the NPC Flat Proof wheel pictured to the right; a few pneumatic kart wheels, aimed for rental karts, are so sturdy that they could be used in a robot without having to be inflated or filled with foam;

19. Robot failure due to arena hazards – this only applies to arenas with hazards, such as saws coming out of the floor or large sledgehammers; against saws, make sure that your robot has a thick bottom plate or cover it with alumina tiles; against sledgehammers, use a shock mounted top cover;

20. Home-made speed controllers and electronics – building a reliable speed controller that can withstand hundreds of amps is not a simple task, do your research and thoroughly test your system if you plan to develop it by yourself; see chapter 7 for more information.

9.1.3. Lose Weight

Make sure that your robot is not over its weight limit. When designing it, estimate the weight of all the components, to avoid unpleasant surprises. CAD programs can provide very precise calculations if you feed them with the correct part weights and material densities.

And don't forget to include the weight of the screws. We forgot to include the screws when carefully designing and calculating Touro's weight back in 2006, just to find out after it was built that it was almost 6.5lb (almost 3kg) overweight. Just because of the screws. To lose weight, there are a few techniques, as described next.

Rearrange your robot's components

If you're still in the design phase, try to rearrange the robot's internal components to reduce the chassis dimensions. If your robot has several empty spaces in it, it won't be difficult to make it smaller.

Consider all possible component arrangements, but don't forget to leave enough space for the wiring. Try placing the batteries in different orientations.

If it's a 4 wheel-drive design with 4 motors, try using only 2 motors with a timing belt or chain transmission to drive all wheels. If your design does not depend too much on traction, such as with powerful spinners, try using only 2 wheels, with the robot's center of mass located close to the line that joins their centers.

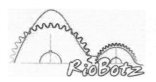 

Change the battery type

Switch SLA batteries to NiCd or NiMH. Most 24V SLA batteries have a capacity density of about 1.25A·h/kg, however this number does not consider the de-rating factor (see chapter 8) for a 3 minute run time. It does not consider the worst case scenario, where it will be fully discharged at the end of a 3 minute match. The de-rating factor in this case is about 0.28, which would result in a capacity density of only 0.28×1.25A·h/kg = 0.35A·h/kg.

The de-rating factor of nickel batteries in 3 minutes is much better, about 0.9, therefore a typical 24V NiCd pack would have 0.9×2.1A·h/kg $\cong$ 1.9A·h/kg, while a typical 24V NiMH pack would have an even better 0.9×2.9A·h/kg $\cong$ 2.6A·h/kg. In this way, without decreasing the robot's battery capacity, you can lose 80% of the battery weight when changing from SLA to NiCd. Changing from NiCd to NiMH will result in an additional weight loss of about 30%. But be careful with NiMH packs because, despite their greater capacity, they cannot supply the high current peaks that a NiCd pack with same capacity can, which makes a big difference especially for the weapon acceleration.

To lose weight even more, you could migrate to lithium batteries (see chapter 8), such as Li-Po, Li-Mn, or Li-Fe-PO$_4$ (A123 or K2), however with a higher cost.

Reduce shaft dimensions

To lose weight, try reducing the diameter of the robot's shafts. This can make a difference especially if they're made out of steel, which has a high density. This will also reduce the size and weight of other components such as bearings and their mounts. Check if the shaft length can be reduced. Using a lathe, drill an internal hole through the entire shaft, as long as it hadn't been tempered, to transform it into a cylinder. If the shaft has diameter D and the hole d, the weight will be decreased by a factor d^2/D^2, while the bending and torsion strengths (which usually are the most important in shafts) will decrease by only a factor of d^3/D^3. In this way, for instance, if a hole with diameter d = D/2 is drilled, the shaft weight will decrease in $(D/2)^2/D^2 = 0.25 = 25\%$, while the bending strength will only be $(D/2)^3/D^3 = 0.125 = 12.5\%$ lower.

Change the shaft material

An excellent technique to lose weight, although costly, is to switch all steel shafts to titanium grade 5 (Ti-6Al-4V), without increasing their diameter. We had to do this with Touro, its 1.5" diameter weapon shaft was originally made out of tempered 4340 steel, weighing 6lb (2.7kg). Since it had already been tempered, it would be very hard to drill it, as explained above, to lose weight. The solution was to replace the shaft with a Ti-6Al-4V one, which only weighed 3.5lb (1.6kg), resulting in a 2.5lb saving. And the shaft strength was not significantly lowered, because this titanium alloy has excellent mechanical properties. The cost was not too expensive, considering that it is a critical part of the robot: about US$150, in the US.

Avoid the temptation to switch steel to aluminum in shafts. If the shaft diameter is maintained, low and medium strength aluminum alloys will easily yield, and most aerospace alloys will possibly break due to their lower impact toughness. Aerospace aluminum could result in a lighter shaft with the same strength as a steel version, but the increased diameter needed by the resulting shaft would significantly increase the weight of its bearings, collars, and all other shaft mounts. So, as

extensively discussed in chapter 3, choose titanium and hardened steel shafts instead of aluminum or magnesium ones.

Change the material and dimensions of robot components

Wisely changing the material of a robot component is not a simple task. This was thoroughly discussed in chapter 3. The best material choice to reduce weight depends on the functionality of the component. For instance, if a robot's armor is shock-mounted to its structure, then most structural parts could have their stiffness maximized without worrying too much about impact toughness, while the armor should withstand impacts without worrying too much about its stiffness. In this case, very thick magnesium or aluminum alloys would be a good choice for a light structure, while thinner titanium Ti-6Al-4V would make a tough and light traditional armor.

But there are several other cases and options. See chapter 3 for a more detailed discussion on weight saving techniques based on changing both the material and dimensions.

Reduce the thickness of plates

If after optimizing the materials of the entire robot it is still too heavy, then it might be necessary to decrease the thicknesses of its plates.

The first idea that comes to mind is to drill holes in the plates, turning them into Swiss cheese. This should only be considered in an emergency, during the event. Holes are a bad choice, because they let debris enter the robot, which can short out the electronics, not to mention the higher vulnerability against hammerbots, spearbots or overhead thwackbots with thin weapon tips, which could reach internal parts, as well as against flamethrowers. Besides, circular holes have a stress concentration factor of about 3 under tension and 2 under bending. In other words, even a small hole will locally multiply the mechanical tensile stresses by 3 and bending stresses by 2 (the stress concentration factors of several geometries can be seen in the Appendix C). These higher stresses make it easier to initiate cracks at the borders of the hole.

In addition, you would need too many holes to significantly lose weight, as seen in the next example. Consider, for instance, a 0.5m × 0.5m cover plate made out of 1/4" (6.35mm) thick aluminum. Its mass is approximately $2800kg/m^3 \times 0.5m \times 0.5m \times 0.00635m = 4.45kg$ (9.8lb). Let's try to lower its mass in 25%, to 3.33kg (7.3lb), using a hole saw to drill several 1" (25.4mm) diameter holes. Each hole would only relieve $2800kg/m^3 \times \pi \times (0.0254m)^2/4 \times 0.00635m = 0.009kg$ (0.02lb). In other words, to lose $4.45 - 3.33 = 1.12kg$, you would need $1.12/0.009 \cong 124$ holes! In addition to the hours spent drilling 124 holes, the robot would suffer from the problems discussed above regarding debris, piercing opponents and flamethrowers.

A better solution is to mill the aluminum plate. In the previous example, we could decrease the plate thickness down to 3/16" (4.76mm) through milling, resulting in a 25% lighter plate. The bending stress of a plate depends on the square of its thickness, therefore it would be multiplied by $(6.35/4.76)^2 = 1.78$ in the thinner plate, which is lower than the factor 2 that would be obtained by drilling holes. And, since the tensile stress along the plate depends directly on its thickness, it would be multiplied by a factor $6.35/4.76 = 1.33$, much smaller than the tensile factor 3 of the holed version. Therefore, the milled plate would have a higher strength than the holed one.

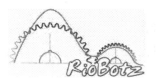

An even better solution is to selectively mill the plate. In other words, to reduce the thickness of the plate only in a few areas, leaving it with the original thickness at the most stressed areas. This was the procedure adopted in the 5/16" thick top cover plate of our middleweight Touro, to lose weight. We've selectively milled 1/8" deep pockets on its outer surface, as pictured to the right. The thickness was neither reduced near the screws (not to compromise strength) nor where the weapon motor is mounted (by

the RioBotz logo in the picture). The strip-shaped area between the pockets was also kept with its original thickness, acting as a rib to keep high the plate bending stiffness.

9.1.4. Travel Preparations

Once your robot is built and tested, making sure that it is not overweight and that it complies with all the event rules, then the next step is to make travel arrangements. Plan the trip well in advance, you'll get cheaper fares and hotel rates.

What to bring

Before the trip, make a list of tools. Avoid the temptation (which I have) of bringing your entire machine shop to the event. Choose wisely which tools you'll really need, among the ones listed in chapter 2. A few very useful items, but usually forgotten, are a portable vacuum cleaner (pictured below, to clean the robot interior in between matches, because small metal debris can cause shorts), a large fan (to cool down the batteries after each match, before charging them), a 220V/110V transformer if needed (pictured below, rated to at least 1kVA if using several power tools and chargers at the same time), heavy-duty electric extension cord, plug strip (pictured below), flashlight (for repairs inside the robot, preferably with a swivel head, as seen below) or headlight (for hands-free operation), telescopic mirror (to inspect the robot's interior without disassembling it, see picture), telescopic magnet (to pick up screws or nuts that fall inside the robot, see picture), and J.B.Weld and duct tape (for desperate emergency repairs in the robot). And don't forget the battery chargers and their power supply.

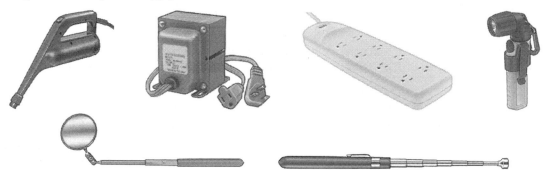

Have at least 2 sets of batteries, 3 or more if possible, and bring spare parts. It's a good idea to have robots that share parts, in this way you'll need to carry fewer spares. For instance, because both Touro's drivetrain and Touro Light's weapon system use Magmotors S28-150, it might be enough for both robots to only bring one spare. They also use the same front skids, battery packs, receivers, Victor speed controllers, TW C1 solenoids and MS-2 switch, not to mention several of their 8mm diameter screws. This helps a lot with spare part management and transport, in special if competing in overseas events.

Bring spare screws, in special if they're oddly sized or difficult to find. Remember that it will be more difficult to find metric screws to borrow in US events, and vice-versa, few Brazilian builders will have inch sized screws to lend. This also applies to tools that come in different systems of measurement, such as wrenches or sockets.

Traveling by plane

If you're traveling by plane, remember that most robot parts won't be allowed in the cabin, they'll need to be checked. Since your checked luggage will most likely be X-ray inspected and opened, it is a good idea to write down on each checked robot part what it is, such as "discharged dry cell battery pack" or "aluminum plate."

When traveling to RoboGames, we also include in every luggage a copy of Dave Calkins' invitation letter, explaining that the robots and parts are for competition purposes only. We also carry a picture of each robot (or its trading card, if it has one), in case the luggage is opened in our presence. In this way, it is easier to explain to the TSA officer why we're checking a sharp spinning bar or so many aluminum plates. If you're lucky, the TSA officer may even know your robot, as it once happened with Touro, making the inspection process very fast and friendly.

We've never had problems traveling by plane with NiCd packs, as long as they are discharged and placed into the checked luggage. Since no wet cell batteries are allowed in the plane, it's a good idea to write "discharged dry cell battery pack" on every pack.

Also, make sure that all the electric wires are very well organized and placed inside a different luggage than the one with the battery packs. Trust me, a luggage full of NiCd packs and random electric wire will draw a lot of unnecessary attention in the X-ray.

Apparently, lithium batteries such as LiPo or A123 can be carried with you inside the cabin, we've never had problems with them even when they were inspected. Otherwise, notebooks and their batteries would need to be forbidden as well. But they can be dangerous if shorted, so we always carry them partly discharged (not too much to avoid damaging them) and inside a fireproof LiPo sack such as LP-Guard (pictured to the right) or LipoSack.

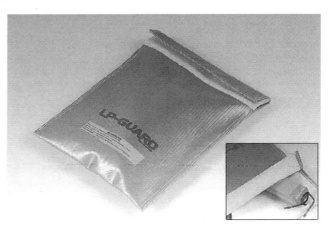

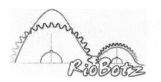

You can ship your robot fully assembled in a crate. However, for international flights, you might need to apply for a temporary export of your robot if it is shipped fully assembled in a large crate. This is usually expensive, and it involves a lot of bureaucracy. Shipping the robot by sea is also risky, because even if sending it well in advance it might arrive at the event after it is over.

The cheapest solution is to carry your robot in your checked luggage, not in crates. If your robot is a middleweight or from a heavier weight class, you will need to partly disassemble it if you want to split it into two or more pieces of luggage. Lightweights or lighter robots can be checked fully assembled if the weight limit allows and if they fit inside the luggage. Our lightweight Touro Light is checked inside a 10lb luggage, reaching exactly the 70lb allowance for each checked item in international flights originated in Brazil.

A very good investment is to buy a digital scale, bringing it with you to weigh all pieces of luggage before each flight, using up the entire weight allowance. The 150lb capacity Pelouze digital scale with remote display, pictured to the right, is a good option.

Note that fees are less expensive to check in an extra luggage than to have several overweight bags. For instance, international tickets bought in Brazil have a 70lb allowance per bag, allowing 2 checked bags, with a US$100 fee for a third piece of luggage. Checked items between 70 and 100lb are subject to a US$100 fee per item, and bags over 100lb are forbidden. Therefore, if you're carrying for instance 180lb worth of robot parts, it is better to pay US$100 for an extra 70lb bag (which will allow you to carry 60lb worth of parts in each of the three bags, as long as each empty bag weighs up to 10lb), instead of paying US$200 for two overweight bags with 100lb each (90lb worth of parts in each plus the own weight of the luggage).

Get to the airport well in advance, in special because of the odd and heavy luggage you'll be carrying. For international flights, register the robot parts at the customs office from your airport of origin before leaving the country, it will simplify your reentry with them. You can do this in the same day of your departure, before checking your bags, but you need to arrive early. You only need to register foreign parts, but it is also a good idea to register custom-made parts such as the robot itself or large disassembled parts of its chassis. Parts might need to have a serial number to be registered. If they don't have it, check with the customs officers if they accept serial number plaques issued from a University, for instance. Most manual and electric tools don't need to be registered, they're considered as tools for professional use. But it is always a good idea to register any expensive part of tool.

Finally, if you're carrying a lot of weight, then rent a car or van at your destination. Choose to pick it up and also to drop it off at the airport. It's less expensive and more practical than riding a taxi carrying heavy luggage full of robots all over town.

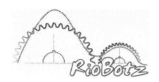

9.2. During the Event

Finally, the great day(s) has arrived. How will the event be? We will describe the typical procedures based on our personal experience at RoboGames, in the US.

9.2.1. Getting Started

After getting your badge and the ones from your other team members, you will be assigned a table in the pits, where you'll place all your tools and robots. Unless you are competing with a single featherweight, you'll probably have to manage well the pit space to store all the robots and tools, as pictured to the right. Try to place the robots, all important and frequently used tools, radios, batteries and chargers on the table.

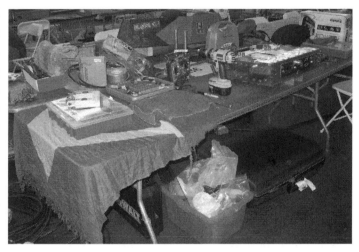

Place all electronic equipment (such as soldering iron, electronic board support), radios, batteries and chargers close together on one end of the table, and the robots closer to the other end. In this way, there's less chance of dropping some heavy component on a delicate electronic board. Make sure you have plug strips on both ends of the table, one for the delicate electronic equipment, and the other to be used in power tools while working on the robots. The remaining items, such as large or infrequently used tools, should be placed under the table in an organized way, easily accessible.

Do not place any items on the neighboring tables, used by other teams, even during a frantic pitstop, unless they allow you to do that. I'm not a good example of this, I'm sorry if RioBotz ever invaded your table...

Arrive early. Try to pass safety inspection on the first try, and as early as possible. "This will give you time to relax and socialize with the rest of the competitors," as pointed out by Mr. Tentacle in his webpage http://architeuthis-dux.org/tips.asp.

Organize the team members so that each one of them has a defined function. Label your tools. Make sure that everybody knows where each tool is stored, either on top or under the table, and always return this tool to its place after using it. This can make a difference during a quick pitstop.

Have a notepad to write down all the ideas that you have during the event. Ideas will come either from what you've learned talking with other builders, or from what your robot learned while struggling in the arena. This information will be very useful later. Several important upgrades in our robots came from crumpled pieces of paper covered in grease and pizza sauce written during the event. Also, don't forget to tape and to photograph the entire event, several ideas will come up while reviewing the pictures and videos.

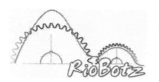

During the event, it is important to keep in mind that rivalry should stay inside the arena. Unlike these famous builders on the right, do not tease your opponents! Unless if it's playfully, of course.

Talk with other builders, show them your robots, exchange information, lend tools. This sport is still relatively small, it is fundamental to help other teams and to learn from them, to improve the level of the competition, attracting spectators and sponsors.

Don't be afraid to show the interior

of your robot to other builders, even if you'll face them in the next match. There aren't many secrets in this sport that haven't been revealed, in special if you search through the great number of websites, posts, build reports, tutorials and books on the subject. If you don't show your robot to other builders, you'll probably waste the chance of learning from their comments about your robot or from exchanging information by looking at theirs.

Let other people take pictures or tape your robot. This is good for your sponsors, in special if their logo shows up in the pictures or videos. Let them take pictures from your robot's interior. Even if another builder discovers a small weak point in your robot, he/she won't be able to explore it in the middle of a fight, there's not enough precision in combots to deliver a surgical strike. If your robot has a serious weak point, any experienced builder will figure it out even if you try to hide your bot. So, let them look at it, thoroughly if they ask to. In the next chapter, all RioBotz robots are exposed in details, including their interior components, through pictures and CAD drawings.

Walk along the pits to check the robots from the other teams, as pictured below. Unless the

other builders are too busy repairing their robot, try to talk with them. But always ask for their permission to take pictures from their robots, in special if you want to touch some part of the bots. If you need to borrow a tool, these teams will most certainly be much more helpful if you have been polite with them.

A nice picture from the RoboGames pits is shown in the next page.

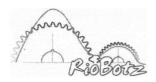

9.2.2. Waiting for Your Fight

Pay attention to the schedule of your next fight, not to get caught by surprise. Even if your fight will be much later that day, have your robot ready and checked. If you check your robot well in advance, you'll have more time to fix any eventual problems. In addition, if your opponent is also ready beforehand, you both can ask the event organizers for an earlier fight. Win or lose, this will leave more time after the match for you and your opponent to fix your robots.

About 30 to 40 minutes before the scheduled time of your fight, charge the robot's batteries one last time, to compensate for any self-discharge, which can be significant in nickel batteries. After this brief charging period, check the battery voltage with a voltmeter and close the robot.

If you're using wheels with polyurethane treads, such as Colsons, it is a good idea to clean their treads using WD-40. Just spray a little bit on the tread, all around the wheel, and wipe it off with a

dry cloth or paper towel. Even though WD-40 is a lubricant, it will start to react with the polyurethane tread surface, making it very sticky and improving the robot's traction. The downside is that the arena dirt will also tend to stick to the treads, meaning that you'll need to clean the wheels before every match. But it is worth it. Another great suggestion to improve traction is to engrave grooves on the polyurethane treads. In addition to the use of WD-40, we manually carved in our hobbyweights the Z-shaped grooves seen on the right, improving wheel traction a lot, in special in dirty arenas.

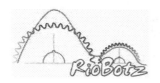

It is also a good idea to mark the robot's bolt heads with, for instance, a Sharpie. Then it will be easy to know if one of them got loose and needs to be retightened. After applying threadlockers and tightening each bolt, you'll just need to draw a short straight line starting on the bolt head, and extend it onto the robot structure. Before the match, the very existence of the markings will help you make sure that all bolts have been tightened and have threadlockers. And, after the match, it

will be easy to spot any loose bolts just by checking the alignment between the markings on the bolt head and on the robot structure, such as in the middle screw from the picture to the right. We've developed this technique after riding too many roller coasters and observing their similar bolt head markings.

If your robot does not use a spread spectrum radio system such as a 2.4GHz one, then you have to pick up the appropriate transmitter clip featuring the channel you're using, as pictured to the right. For instance, for a 75MHz radio system, you'll need to pick up a clip corresponding to one of the channels between 61 and 90 (see Appendix D). It is forbidden to turn on any radio without the clip, to avoid accidents that could happen if another robot uses the same channel as you. More recently, several competitions have required the use of radio systems

featuring some sort of binding, such as the 2.4GHz ones, eliminating the need for radio clips.

If the event staff allows, make a quick drivetrain test at the pits, but with your robot well secured and with its wheels lifted off the ground. Do not test the weapon, and use some weapon restraint at all times. The restraint should only be removed inside the arena, after you've been told to do so.

Go to the queue with your robot as soon as you're called. Lightweights or heavier robots should be carried on a dolly (as pictured to the right) or pushcart, to avoid accidents such as dropping them on the floor. Once at the queue, you will be standing beside your opponent (as seen in the picture). Exchange conversation, show

your robot. Do not be afraid of answering any questions about your robot. At this point it won't make any difference, it is just a way to talk and relax.

If your next opponent asks for the match to be postponed, and the event staff allows it, then don't hesitate to agree. You came all the way here to fight, not to win by WO. The spectators, pictured to the right, came here to see exciting combats. They might boo you and your robot if you don't agree to grant a brief postponement.

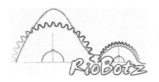 

9.2.3. Before Your Fight

A typical arena in the US has two doors, one for the robots from the next match to enter, and another one for them to leave, as pictured below. Next to the arena there is usually a table with a computer that allows you to check in real time the fight brackets and schedule. Next to the arena there is also the judges' table. If you're the robot driver, enter the arena from its entry door when your called, carrying your robot with a dolly or pushcart.

After entering the arena, you will take your robot to its starting position, as pictured to the right, which is determined by the event staff. Wait beside your robot. When requested, turn on the robot and remove the weapon restraint (if any).

Get outside the arena and position yourself in the areas reserved for each driver. After the arena is locked, you can touch very briefly the radio control just to see if the robot is responding. When you're ready, press the button "Ready / Surrender" from your driver area, pictured to the right. A few seconds after both drivers have pressed their buttons (which seem to last eternally), a series of lights will be turned on, until the green one is lit, starting the fight.

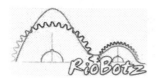 

9.2.4. During Your Fight

The matches usually last up to 3 minutes, except for insect classes in the small arena, where the matches are restricted to 2 minutes. Check the specific rules of your competition. The complete set of RFL rules can be found in www.botleague.net/rules.asp. A few of them are described next.

If a robot does not move for 5 seconds after the opponent has ceased attacking, a 10 second countdown will be issued, at the end of which it will lose by KO if it doesn't show any controlled translational movements.

Pinning or lifting your opponent is allowed, however it is limited to 15 seconds (10 seconds for antweights or lighter). After releasing the opponent, you must move far enough away to let it escape from that pinning position.

If a robot gets stuck on the arena through its own action, not due to some direct action of the opponent, then, depending on the event rules, it may (or may not) be granted one free release per match. If the combatant becomes stuck again during the same match, no intervention will take place: it will have 10 seconds to free itself not to lose the match by KO.

Arenas usually have a Death Zone. The first robot to contact the floor on the Death Zone is declared dead, regardless of which robot initiated the entry.

To surrender during a match, just press again the button "Ready / Surrender." Sometimes it is wise to surrender if your robot has suffered enough damage to make it impossible to win the match, in special if you're still on the winners' bracket in a double elimination competition. This will prevent further damage and allow you to rebuild the robot in time for the next match. But always think twice before throwing in the towel, not to regret it. Even if your robot is barely moving, there's a chance that the opponent robot suddenly dies for some reason, in special in very violent matches. But if your expensive electronics and batteries are hanging out of your robot, and your opponent seems to be in good shape, then don't hesitate to surrender.

At the end of a match, when requested to, put back your robot's weapon restraints (if any) and turn it off. Greet your opponent, independently of the result. Remember that your opponent was just trying to win, he/she didn't have anything to do with the judges' decision.

9.2.5. Deciding Who Won

Don't argue with the judges, even if you don't agree with their decision. Their decision is final. Sometimes, from the point of view of the loser combatant, a decision on a very close match might seem unfair. This does not mean that it was a wrong decision, it could just be a matter of point of view, since there always is a subjective aspect in the judgment. The proof of that is the very existence of split decisions.

This is why the judges follow very specific and objective guidelines, which are summarized below, extracted from the website www.robogames.net/rules/combat.php. It is important that all combatants are familiarized with these guidelines, so they can better understand the reasons behind the judges' decisions.

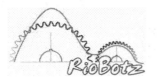

An odd number of judges, usually three, decide the winner of the matches where no robot is defeated during its 3 (or 2) minutes. There is also one Judge Foreman, who ensures that all judges are conforming to the guidelines.

In a judges' decision, the points awarded to the combatants by the panel of judges are totaled and the robot with the majority of points is declared the winner. Points are awarded by each judge in two categories: aggression, worth 5 points, and damage, worth 6 points. All 11 points must be awarded by each judge, who determines how many points to award each combatant. Therefore, a 3-judge panel will award a total of 33 points, which must be equal to the sum of the scores of both robots. Therefore, the closest possible win in this case would be by a score of 17-16, which can only happen in a split decision by the judges.

Aggression

Aggression is based on the relative amount of time each robot spends attacking the other, in a controlled way. Attacks do not have to be successful to count for aggression points, but the attacking robot must move towards the opponent, not just wait for it to drive into the attacker weapon.

The distribution of the 5 aggression points from each judge between the robots is of three types:

- a 5-0 (or 0-5) score, if one of the robots never attempts to attack the other, while the other consistently attacks;
- a 4-1 (or 1-4) score, if there's significant dominance of attacks by one robot, with the other only attempting to attack a few times during the match;
- a 3-2 (or 2-3) score, if both robots consistently attack each other, or if both robots only attack each other for part of the match. If both robots spend most of the match avoiding each other, then the judges will decide which one made more attempts to attack, awarding it 3 points and 2 to the other robot. Note that a robot that attacks a full-body spinner, intentionally driving towards it, is automatically considered the aggressor in the attack.

Note that there can be no ties in aggression, since its number of points is odd. Judges must decide which robot is more aggressive than the other.

Damage

Damage points are awarded to the robot that can make the opponent lose functionality in some way. Damage does not have to be visually striking, it has to do with functionality, with incapacitating the opponent. For instance, titanium will send off bright sparks when hit, but most of the time it will be undamaged. Also, a gash in an armor plate may be very visible, but it only minimally reduces the armor's functionality.

But a bent armor or wedge that prevents the robot from resting squarely on the floor, reducing the effectiveness of the drivetrain, counts as damage. A small bend in a lifting arm or spinner weapon may dramatically affect its functionality by preventing it from having its full range of motion, so it is also considered as damage. A wobbly wheel is also a sign of damage, probably indicating a bent shaft, compromising drivetrain performance.

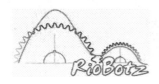

There are 6 levels of damage:

- trivial: being flipped over causing no loss of mobility or loss of weapon functionality (such as in an invertible drumbot that is able to spin its drum in both directions), direct impacts that do not leave a visible dent or scratch, sparks resulting from strike of opponent's weapon, or being lifted in the air with no damage and no lasting loss of traction.

- cosmetic: visible scratches to armor (as pictured to the right), non-penetrating cut or dent or slight bending of armor or exposed frame, removal of non-structural or non-functional cosmetic pieces (dolls, foliage, foam, or ablative armor), or damage to wheel, spinning blade, or other exposed moving part not resulting in loss of functionality or mobility.

- minor: being flipped over causing some loss of mobility or control or making it impossible to use a weapon (such as in an invertible drumbot with a drum that can only spin in one direction, because while inverted it would not be able to launch opponents), intermittent smoke not associated with noticeable power drop, penetrating dent or small hole (as pictured to the right), slightly warped frame not resulting in loss of mobility or weapon function, or removal of most or all of a wheel or weapon part without loss of functionality or mobility.

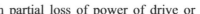

- significant: continuous smoke or smoke associated with partial loss of power of drive or weapons, damage or removal of wheels resulting in impaired mobility, damage to rotary weapon resulting in loss of weapon speed or severe vibration, damage to arm, hammer, or other moving part resulting in partial loss of weapon functionality, visibly bent or warped frame, and torn, ripped, or badly warped armor or large hole punched in armor (as pictured to the right).

- major: smoke with visible fire, armor section completely removed exposing interior components, warped frame causing partial loss of mobility or complete loss of functionality of the weapon system, internal components broken free from mounts and resting or dragging on the arena floor, significant leak of hydraulic fluid or pneumatic gases, or removal of wheels, spinning blade, saw, hammer, or lifting arm, or other major component resulting in total loss of weapon functionality or mobility.

- massive: armor shell completely torn off frame, major subassemblies torn free from frame (as pictured to the right), total loss of power, or loss of structural integrity such as major frame or armor sections dragging or resting on the floor.

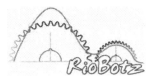 

If your robot is in good shape at the end of a close match, it is a good idea to demonstrate operability of the robot's drivetrain and/or weapon for a few seconds immediately after the end of the match, without touching the opponent. In this way, the judges will ascertain that your robot is still functional, and not sluggish or dead.

Scoring of damage points is based on relative grading of each robot's damage, as described below:

- a 6-0 (or 0-6) score is awarded when one robot suffers nothing more than trivial damage, and the other is at least significantly damaged, or one robot has suffered major or massive damage and the other is no more than cosmetically damaged.
- a 5-1 (or 1-5) score, if one robot suffers at least minor damage and the other suffers major or worse damage, or one robot has suffered cosmetic damage and the other has suffered at least significant damage.
- a 4-2 (or 2-4) score, if both robots have suffered nearly the same level of damage but one is slightly more damaged than the other.
- a 3-3 score, if both robots have suffered the same level of damage, or neither robot has even cosmetically damaged the other.

Damage that is self-inflicted by the robot's own systems, and not directly or indirectly caused by contact with the other robot or an active arena hazard, will not be counted against that robot for scoring purposes. In addition, any pre-existing damage in a robot before the match should not be counted against it.

9.2.6. After Your Fight

After your fight, immediately take your robot back to your pit and service it, even if it doesn't look damaged. Your first priority is to take care of the (probably hot) batteries.

Take care of your batteries

Immediately open the battery compartment. Carefully touch the battery packs to check their temperature. For NiCd packs, it's normal if they're warm or even fairly hot. But if they're too hot even to be briefly touched, this means that their temperature is much higher than 140°F (60°C), which is a cause of concern: the battery life might be significantly shortened. If one pack is much colder than the other(s), this might be an indication that its circuit was open during the match, either due to a broken solder inside the pack or to a connector malfunction, which can be easily checked with a multimeter. After checking their temperatures, immediately place the battery packs over a large fan to be cooled down (see chapter 8). Only start recharging them after they get cold.

Access damage

After taking care of the batteries, inspect the entire robot for any structural damage. Look for any large debris that might have entered the robot, such as metallic parts that could shorten your electronics or pieces of rubber or foam tire treads that could get stuck in the clearances between your wheels and structure.

Turn the wheels by hand to feel if there's any problem with the drivetrain. A stuck wheel could be either due to debris in the transmission (either foreign debris or from the own robot, such as a broken gear tooth inside a gearbox), or due to bent armor/structure interfering with the wheel. If the wheel gets stuck only in a few positions, this might be an indication of a broken gear tooth or a bent shaft. If the wheel gets stuck once at every turn, then the problem might be in its shaft or in the last stage of the reduction. If it gets stuck once every few turns, the problem might be in the previous stages. On the other hand, if the wheel easily turns by hand without any mechanical resistance from the motor, then you might be facing a broken key or shaft coupling, a stripped gear, a loose pinion, a ruptured belt, or a derailed chain from its sprocket, depending on your transmission system.

After checking the drive system, look for damage in the weapon, focusing on the most stressed parts, such as on the teeth of a drum, or the center section of a spinning bar. Check the condition of all belts and chains, from both the weapon and drive systems, and change them if necessary. If you previously marked the robot's bolt heads with a Sharpie, as explained before, it will be easy to spot if any of them got loose and needs to be retightened.

Remove damaged screws

There are several ways to deal with a screw with a stripped or broken off head. On socket head cap screws with hexes that have been stripped out, you can take a slightly larger Allen wrench and grind it just enough for it to be hammered into the stripped recess, and then unscrew them.

For screws with broken off heads, try to unscrew the remaining stub with a vise-grip. If this doesn't work, then use a Dremel (as seen in the left picture) to cut two parallel chamfers in such a way that an open-

end wrench could do the job (as seen in the right picture above).

If the stub is entirely embedded into the robot, then you can use a screw extractor, such as the ones pictured below. This tool drills a pilot hole into the screw stub, and then a left-handed thread takes care of unscrewing it. If the damaged screw is a high strength one, you'll need special screw extractors, which can drill even class 12.9 hardened steel bolts.

If you don't have screw extractors, another option is to use a Dremel to cut a channel in the middle of the embedded stub, and then use a large flathead screwdriver to unscrew it. This method is not as good as the previous ones, but it may work if the stub is not too bent.

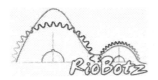

If everything else fails, then another tip is to weld a nut onto the stub (as pictured to the right), and then use an open-end wrench that matches the nut. If the stub is entirely embedded into the robot, it might be easier to first weld a washer to the stub, and then weld the nut onto the washer.

If the broken stub is deeply embedded into the robot, then another option is to weld a long and thin steel strip (as pictured to the right) and use a vise grip to unscrew it.

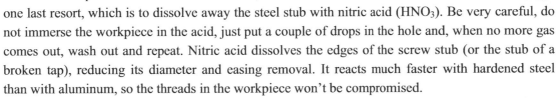

If the workpiece is made out of aluminum, then there's one last resort, which is to dissolve away the steel stub with nitric acid (HNO_3). Be very careful, do not immerse the workpiece in the acid, just put a couple of drops in the hole and, when no more gas comes out, wash out and repeat. Nitric acid dissolves the edges of the screw stub (or the stub of a broken tap), reducing its diameter and easing removal. It reacts much faster with hardened steel than with aluminum, so the threads in the workpiece won't be compromised.

But, if the workpiece has a high value, it might be worth looking for a bolt disintegrator device. In this technique, the workpiece is immersed in water or oil, while the bolt (or tap) is electrically eroded. You will probably have to look for a machine shop that offers this service.

Socialize

If your opponent from the previous match is not busy working on his/her robot, go to his/her pit after the match to check the damages, to talk about the match, to take pictures, and to invite him/her to see the damages caused in your robot.

It is very common to give your opponent unusable parts from your robot that were destroyed by him/her during the match. These are called "trophies," they are memories to keep from the combats. It is an honor to receive them, and giving them away is another way to be polite and to establish friendships with other builders.

The picture to the right shows a few of our most cherished trophies, which we had the honor to receive (or scavenge in the arena, in a few cases) along the years, since RioBotz was born, in 2003.

Our "trophy box" has now more than 50 pounds worth of good memories.

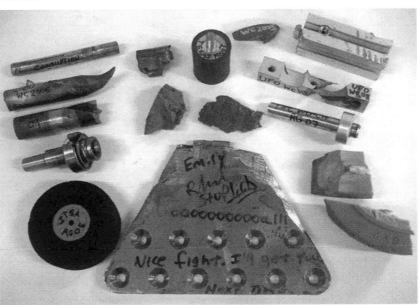

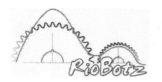

9.2.7. Between Fights

Between fights, do not perform any dirty jobs on your pit table, such as grinding large parts sending sparks everywhere. Most events have a designated area for this. Otherwise, find an isolated area to use such tools. If it is a very small job, then ask for other team members to form a protection barrier to avoid sending sparks to neighboring tables. Several builders are already stressed trying to get their bots ready for combat, so it is wise to avoid conflicts.

Never test your robot's weapon in the pits. Keep the safety restraints on the weapon at all times. It is usually OK to test the robot drivetrain if its wheels are lifted off the ground, but always check with the event staff.

A very useful accessory is a 2-way radio (pictured to the right), which can be used by 2 or more team members, especially to communicate with the driver. This gives more freedom for your teammates, allowing them to wander around the pits between fights, until their presence is required to fix a robot or to drive it. Use a headset (earphone and microphone) for a hands-free experience. It is important for the radios to have a vibrating alert, because loud noises and music from the pits and arena might make it difficult to listen to sound alerts from the incoming calls.

Even if you have been eliminated, try to attend the event until its end. In this way you won't miss the show, you'll watch the fights and championship matches from a privileged position from the pits, and you'll be giving prestige to your peer builders that are still competing. At the end of the event, you will have learned more than you could imagine. And you'll have made many friends and met great builders. After all, it's not every day that I get to meet legends such as Matt and Wendy Maxham, as pictured to the right.

Attending a combot event is a wonderful experience. And competing is even better, it is not easy to describe with words. You have to experience it yourself. Get ready for a major adrenaline rush.

9.3. After the Event

After each event, get together all your personal notes, and organize them while the information and memories are still fresh in your mind. They will be very useful to improve your robot and its future versions.

9.3.1. Battery Care

It is a good idea to store your batteries adequately, especially if you won't go to an event within the next months. If you'll be practicing driving your robot regularly, which is a good idea to improve your skills, choose perhaps 1 or at most 2 sets of batteries to be used on a daily basis, and store all others to save them for the next event.

SLA batteries should be stored at full charge, keeping their terminals very well isolated to avoid shorts. Recharge them at least every 6 months, even if you don't use them, due to self-discharge. You don't need to keep them in a refrigerator, as long as they're stored below 80°F (27°F).

Nickel batteries such as NiCd and NiMH, on the other hand, should be stored fully discharged, as discussed in chapter 8. But never below 0.9V per cell. It is a good idea to discharge them using an electronic charger such as Triton, see chapter 8. Then, place the batteries in a refrigerator at 5°C (41°F), not a freezer. This is so important that we have a dedicated refrigerator just for that, as pictured to right. But always store the batteries inside a sealed plastic bag such as ziploc, to protect them from humidity. In this way, the batteries can last up to 20 years, but you'll need to completely charge and discharge them at least every 6 months for that. When you remove the batteries from the refrigerator, wait for them to get to room temperature before charging. Never freeze the batteries.

Lithium batteries should also be stored in a refrigerator, inside a sealed plastic bag. But, instead of fully discharged, which could make them permanently unusable, they should be stored at about 40 to 60% of their charge level. Storing the battery at 100% charge level applies unnecessary stress and can cause internal corrosion. Recharge them back to 40 to 60% at least once per year, due to battery self-discharge.

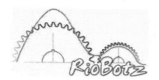 

9.3.2. Inspect Your Robot

After taking care of the batteries, disassemble your robot to access damage, several problems are not easy to spot in a fully assembled robot. Switch the screws that are in bad shape, either bent or with stripped heads. If you're having trouble removing a damaged screw, follow the screw removing techniques explained before.

Verify the condition of the belts or chains, this is a good time to change them if needed. Clean up very well your robot, acetone is very good to clean metallic parts (but don't use it on Lexan).

Then, visually inspect critical components, such as shafts, looking for cracks. Several times, visual inspection is not enough to spot a crack, because cracks usually have their mouth closed when the part is not loaded, only leaving a very subtle trace on its surface. One efficient way to detect a crack is to use a low-cost technique called dye penetrant inspection (DPI). DPI is based upon capillary action, where a low surface tension fluid penetrates into surface-breaking cracks.

To perform DPI, you only need two spray cans, one with the penetrant dye and the other with a white developer, found for instance at McMaster-Carr under Dye-Penetrant Detection Kit. The inspection steps are described next, applied to the tempered 4340 steel weapon shaft from our middleweight horizontal bar spinner Ciclone.

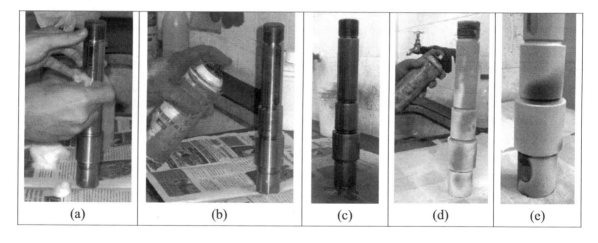

(a)   (b)   (c)   (d)   (e)

(a) pre-cleaning: the test surface is cleaned to remove any dirt, paint, oil or grease, using for instance acetone; do not leave fingerprints;

(b) application of penetrant: the penetrant is applied to the surface, in general as a spray; the penetrant dye can be colored, usually red as pictured above, or fluorescent, to be later inspected under ultraviolet light; always use gloves, because the dye will penetrate all the way under your fingernails if you're not careful (unless you want to save money on nail polish);

(c) waiting period: wait from 5 to 30 minutes for the penetrant to soak into any cracks or flaws; very small flaws may require a longer waiting time;

(d) application of developer: completely remove the penetrant from the surface using a dry lint-free cloth (do not use acetone in this step, it could remove as well the penetrant absorbed by the

cracks), and then apply the appropriate white developer to the entire surface, until the surface looks like it's frozen (as pictured above), forming a semi-transparent, even coating;

(e) inspection: wait for 10 minutes for the blotting action to occur, where the developer will bring any trapped penetrant to the surface, exposing cracks or flaws through the form of thin red (or fluorescent) lines under white (or ultraviolet) light; do not wait too long to inspect, because the thin lines may "bleed out" and make it difficult to evaluate the size of the crack, if any; inspect very carefully near geometry changes such as notches, where it is more likely to find a crack; beware with false positives, because very small harmless scratches (generated either during manufacturing or combat) can result in very thin lines - it is up to the inspector to distinguish between cracks and scratches, depending on the thickness of the developed lines; fortunately, Ciclone's shaft was free of cracks in the above inspection.

Finally, completely assemble your robot as soon as you finish servicing and inspecting it. With your robot fully assembled, it will be impossible to misplace any of its components. Misplaced components will most likely be lost forever if you only look for them several months later, near the date of the next event.

9.3.3. Wrap Up

Update your homepage as soon as possible. During or immediately after an event is the best time to do that, in special if you want to increase the number of hits in your webpage to please your sponsors. Most builders and enthusiasts that didn't attend the event will certainly be searching for photos and videos during it, and mostly everyone will be looking for them right after the event ends. Make sure you post announcements of your updates, for instance, on the RFL Forum in the appropriate topic related to the event.

If you don't have a webpage, make one. It is important that your team has visibility to be able to get sponsors. Nowadays it is very easy to design and upload a webpage. A basic one will take you less than an hour to prepare.

Check the current ranking of your robots at both www.botrank.com and www.buildersdb.com websites, and keep in mind the dates of the next events.

Now relax, and review the pictures and videos from the event. Win or lose, celebrate with your teammates and other builders. Cheers!

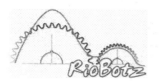

Chapter

10

RioBotz Build Reports

This chapter presents the build reports of all combat robots from RioBotz, including the entire Touro family. It also talks about the origins of our team, since our first combot Lacrainha.

10.1. Lacrainha

RioBotz was created in January 2003, when six undergraduate students from my University, PUC-Rio, asked me to be their advisor to help them build a combat robot. It didn't take long for me to get hooked. We then started to build our first combot, the overhead thwack robot Lacrainha ("Little Centipede" in Portuguese, pictured to the right), a **middleweight** (120lb) with structure completely made out of scrap metal. Note in the picture the SLA battery holder, made out of scrap perforated cable trays used in our University.

Its cylindrical shape was due to our very low budget: we were able to get several scrap aluminum disks used in the structure of the pigs built in our University. The pigs (pictured to the right) are used in the internal inspection and cleaning of oil pipelines, with the same cylindrical shape of our robot Lacrainha.

Bosch donated us a pair of GPB motors, and we bought two surplus worm gearboxes, made out of cast iron. Two heavy 12V 17A·h AGM SLA batteries, used in electric bikes, powered the robot and also acted as counterweights to help it strike with its hammer. We developed the entire electronics, both controller and power boards, using relays to provide a simple bang-bang control (no speed control at all). The radio was borrowed from the Aerodesign team from our University, and our first robot was

born. Lacrainha never saw combat, because it was soon replaced by its bigger brother Lacraia.

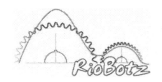

10.2. Lacraia

Still in 2003 we were able to get the support from our first sponsor, EPTCA Medical Devices. We then built an improved version of Lacrainha, the **middleweight** Lacraia (meaning "Centipede" in Portuguese, pictured to the right).

In spite of its better appearance, the robot was still very primitive: its 6063-T5 aluminum armor had a thickness of only 1mm, its electronics used bang-bang control with relays and a single MOSFET per motor, and SLA batteries powered the GPBs with a heavy and inefficient worm gearbox. Nevertheless, it was a competitive robot in the Brazilian competitions at that time. It was one of the only invertible robots. It achieved the 6th place during the III ENECA Brazilian national championship. The steel ball used in the hammer was later replaced by a sharp S1 tool steel piece.

Lacraia is now bolted to the ceiling of our lab (as pictured below), bearing a medieval axe.

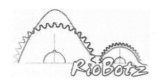

10.3. Anubis

In 2004 we designed our third **middleweight** (120lb), Anubis, another overhead thwackbot. With almost 10 times the power of Lacraia, Anubis is one of the fastest robots that we ever built. Its structure is made out of two 7050 aeronautical aluminum plates, 6061-T6 aluminum extrusions, and Lexan covers.

The tip of its weapon is made out of tempered S1 tool steel, designed to pierce the opponents. The shape of this tip reminded a lot the head of the Egyptian god Anubis (see its logo pictured right) - this is how it got its name. The robot also works as a rammer, since it has two tempered S1 steel plates that act both as counterweight and armor. The two NPC wheels, filled with polyurethane foam, are powered by two NPC T74 motors.

We specially developed a controller board for the two OSMC speed controllers that power the NPC T74 motors. Anubis was the first Brazilian combat robot that used NiCd batteries, it was powered by two 24V 3.6Ah NiCd Battlepacks, providing the necessary high currents and torques to accelerate the weapon and perforate armors.

Anubis only fought once, it won a rumble match

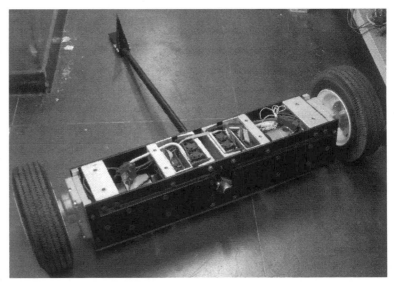

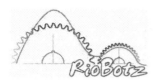

against 3 other robots (hey, that's a 100% win record!). However, the gearbox of one of the NPC T74 motors broke towards the end of the match. Today, we're converting Anubis into a Segway-type personal transporter.

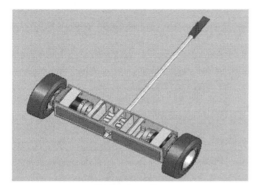
CAD drawing

7050 aluminum plates and 6061-T6 extrusions

Anubis (front) is much smaller than Lacraia

after the black electrostatic powder coating

compact electronics and NiCd batteries

practicing against a monitor

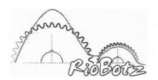

piercing an opponent in a rumble match	piercing once again

Still in 2004, we decided to create a robot to replace Anubis in case it broke in combat. We didn't want to build another thwackbot, we wanted to explore different possibilities. So we decided to follow the dark path known as Spin To Win.

Many people think of spinners (horizontal, vertical or drums) as the robots from the "dark side of the Force." This is because they are the quick and easy path to victory and destruction (and therefore follow the Sith philosophy).

But this is not entirely true: yes, they can generate a lot of destruction, but they are probably one of the most difficult robot types to properly design. It is not at all an easy path. It is easy to spin a heavy bar or disk using a high power motor, but it is very hard to design a robust structure and weapon system that can cause huge damages to the opponent without harming itself. The spinner is always challenging Newton's third law. Besides, vertical spinners and drumbots have directional weapons, and therefore they need as much a skillful driver as any wedge, hammer, lifter, launcher. Even a horizontal spinner involves some strategy, as it needs to maneuver around the opponents to hit their weak points, or to run away from the adversaries while its weapon is still spinning up.

The result from this dark path was our first spinner, Ciclone. It is the little guy standing beside Anubis in the picture. It was meant to be just a spare robot in case Anubis broke, but it ended up so destructive that it was promoted to our main combat.

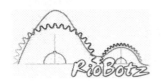

10.4. Ciclone

Our 120lb **middleweight** Ciclone ("Cyclone" in Portuguese, pictured to the right) is, essentially, an Etek motor surrounded by a robot. This motor, as already discussed in chapter 5, is extremely powerful. It is an excellent choice for heavyweights and super-heavyweights, but it is too heavy for a middleweight. We had a lot of trouble to fit an Etek into a middleweight without sacrificing the resistance of the structure. The complete weapon system, including the 5160 tempered steel bar, torque limiter, shaft collars, Etek motor with its mount, weapon shaft with mounted bearings, timing pulleys, and belt, added up to almost 50% of the robot's weight. It is not respecting at all the 30-30-25-15 rule, which would suggest that only 30% of the robot weight should be used in the weapon (see chapter 2).

To compensate for that, we had to sacrifice a little the remaining 30-25-15 from the rule. We used only two 18V RS-775 motors with DeWalt gearboxes set at high torque to drive the two wheels, which resulted in only 15% of the robot weight in the drive system, instead of the 30% from

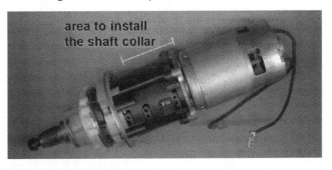
area to install the shaft collar

the rule. This caused Ciclone to be a little slow for US standards, however it was not too bad for the Brazilian competitions back in 2004. The two 18V gearmotors, powered by a specially developed control board and two OSMC speed controllers, were attached to the robot using shaft collars in the region pictured to the right.

Due to its weight budget, Ciclone only used two NiCd packs: one 24V Battlepack to power the Etek (weapon system) and one 18V pack from our DeWalt cordless drill for the drive system. The batteries accounted for only 7% of the robot's weight, instead of 15% from the 30-30-25-15 rule. Thus, the robot only had left about 100% − 50% − 15% − 7% = 28% of its total weight for the structure and armor, a reasonable value that is compatible with the rule.

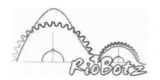 

The motors and gearbox for the drive system were obtained from disassembling two 18V DeWalt cordless drills. We used not only the motors (number 1 in the picture to the right) and planetary gearboxes (number 2), but also the batteries (number 5) and chargers (number 6). It is an excellent cost-benefit to disassemble cordless drills.

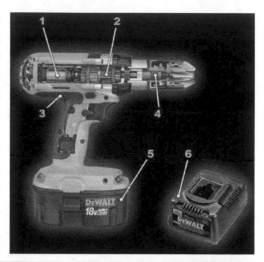

The structure of Ciclone was all made out of 4" high 1/4" thick aluminum extrusions. Unfortunately, we were only able to find 6063-T5 extrusions with those dimensions (the 6063-T5 is a very low strength aluminum alloy).

To compensate for that, the plates used to hold the mounted bearings from the weapon system (in the center of the picture to the right) were all made out of 7050 aeronautical aluminum, which has a much higher strength than the 6063-T5. The covers for the electronics and batteries were made out of Lexan.

The spinning weapon is a 5160 steel flat bar used as a leaf spring in the suspension system of trucks. The bar was bent using a servo-hydraulic machine from PUC-Rio's Fatigue laboratory, and later tempered. We used the torque limiter DSF/EX 2.90 (see chapter 5) to connect the bar to the 1.5" 4340 steel weapon shaft. This torque limiter acts as a clutch, to allow slippage during the impacts. The shaft is powered by a pair of timing pulleys, using an 8M size timing belt.

Subsequent improvements included a front armor made out of titanium grade 5, internal wheels, and a Hella key to turn on/off the robot. The cast iron mountings of the weapon shaft bearings survived combat, but not without a few cracks. So we later press fitted the bearings directly into the aluminum plates of the robot structure, and never had another problem. Cast iron is very brittle and heavy, it is not a good option in combat robot designs (except for the fact that cast iron bearing mounts are very easy to attach to a robot).

Ciclone became the Brazilian champion in 2004 and 2005. The 2005 event was particularly interesting, it was held on an ice arena, we had to develop special wheels to guarantee traction (see chapter 2). In its third competition, Ciclone was flipped over by a wedge. This was when we realized that we needed an invertible robot. Suddenly, drumbots started to sound like a good idea...

The pictures that follow show a detailed anatomy of Ciclone, as well as photos taken during its building, tests and combats.

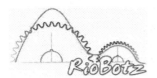

CICLONE

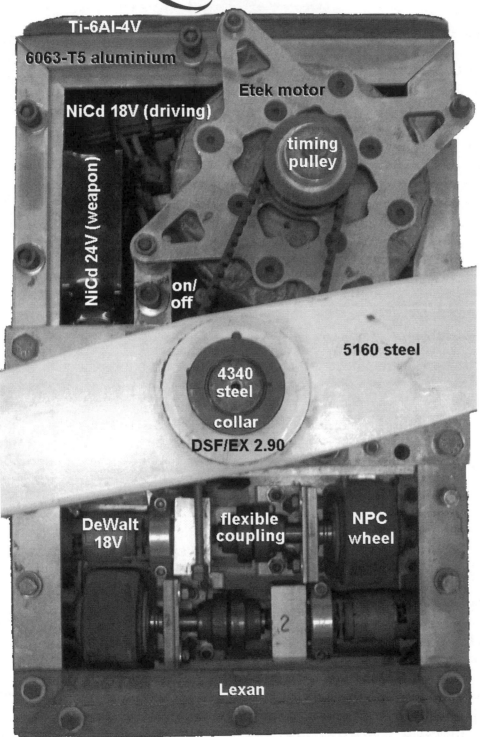

Ti-6Al-4V

6063-T5 aluminium

Etek motor

NiCd 18V (driving)

NiCd 24V (weapon)

timing pulley

on/off

5160 steel

4340 steel collar

DSF/EX 2.90

DeWalt 18V

flexible coupling

NPC wheel

Lexan

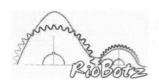

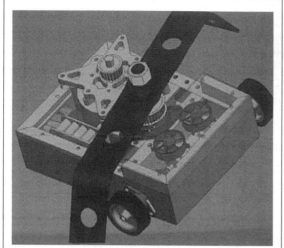

CAD drawing of the 2004 version

Ciclone and Anubis being built

nice black electrostatic powder coating

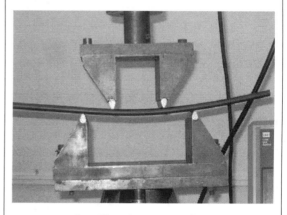

bending the weapon bar

assembling the OSMC speed controller

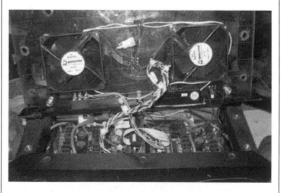

control board, OSMCs, and fans

Ciclone, 2004 version

modifications in 2005 to drive on ice

moving the wheels inside the robot

Ciclone, late 2005 version

you don't have to be Einstein to drive on ice

Ciclone's very first opponent

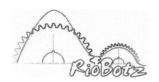

challenging a concrete block

smashing a monitor

spinner vs. spinner in the 2004 finals

2004 semifinal against a pneumatic flipper

flipping Vingador in the 2005 finals, on ice

getting a piece of Donatello

10.5. Titan

In 2005 we designed the **middleweight** Titan (pictured to the right), a horizontal spinner that incorporated a series of improvements over Ciclone.

Our first concern in the design was the 30-30-25-15 rule, which had been severely violated by Ciclone. This time, we designed the entire weapon system keeping in mind to use in it less than 25% of the robot's weight. We used two Magmotors S28-150 to drive a single 90° conical gear that powered the weapon shaft up to 3000 RPM. The spinning bar, a tempered 5160 steel leaf spring, was attached to the weapon shaft using a large Belleville washer and a threaded shaft collar. In the 2006 version, we added a Ti-6Al-4V titanium wedge to make it effective against wedges or very low robots. The total weight of the weapon

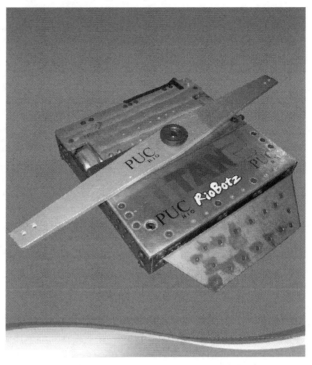

system, if we consider the wedge as part of it, reached about 30% of the robot's total weight, as recommended by the rule.

The robot's drive system used two Magmotors S28-150 with TWM3M gearboxes and 4" diameter Colson wheels. The drive system used only 15% of the robot's weight, half of the recommended value, but even so its speed was already much higher than Ciclone's. Traction was also better, because in Titan the two wheels were closer to the robot's center of mass.

We used two OSMCs to control the drive system, which were later replaced by Victors to gain some space. The weapon motors were powered by a TW-C1 solenoid, with a Hella key as on/off switch. The three 24V NiCd battery packs that Titan uses in parallel, together with the entire electronics, resulted in less than 15% of the robot's weight, well in agreement with the 30-30-25-15 rule.

Titan's integrated structure/armor was entirely made out of 8mm (5/16") thick Ti-6Al-4V titanium, with a 2mm thick titanium bottom cover. The top covers were made out of 8mm thick 7050 aeronautical aluminum. The top plate where the weapon shaft is attached to was made out of 304 stainless steel. Angle extrusions, also made out of 304 stainless steel, were used inside the robot to join the titanium walls. These heavy steel reinforcements were only possible thanks to the 15% weight savings from the drive system. Therefore, the structure/armor ended up using respectable 40% of the robot's weight, well above the 25% from the rule. This sturdy structure was important to help the robot survive its own reaction forces from the inflicted blows.

The pictures below show a detailed anatomy of Titan, as well as building and testing photos.

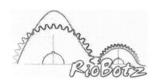

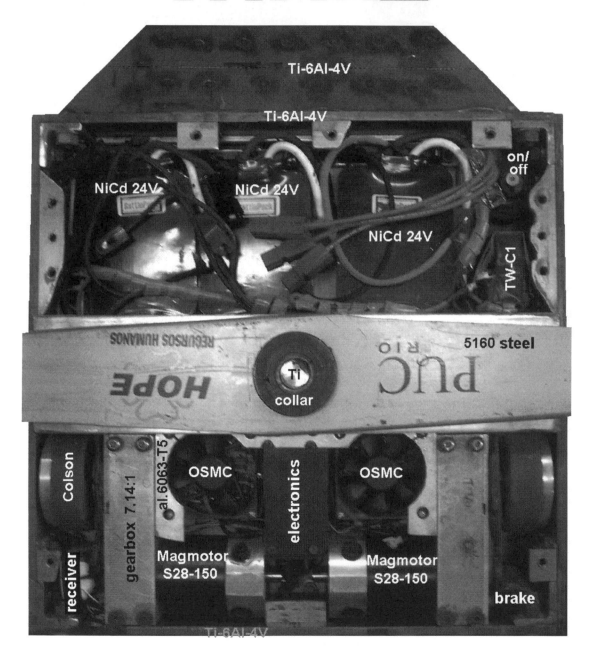

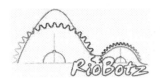

original CAD design, with 4 exposed wheels: it was soon changed to 2 internal wheels

titanium walls and bottom, and steel angle extrusions: the original NPC-02446 drive motors were soon changed to short Magmotors

milling the top covers, to reduce weight

Titan with the milled aluminum top covers

5160 steel bars for the weapon, before drilling and tempering

the center of gravity of each bar was found balancing them on the tip of a center punch

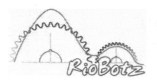

the weapon bars originally included a pair of S7 steel claws, later removed

Michelangelo, Leonardo and Rafael: 30kg (66lb) sparring ninja turtle-bots

weapon test: Titan vs. Michelangelo, at dawn, in the parking lot of PUC-Rio

poor Michelangelo after a whole night of beating: "shell shock"

Titan vs. the combat arena

318

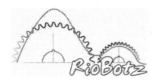

10.6. Touro

The design of our **middleweight** Touro ("Bull" in Portuguese, pictured to the right) focused on the idea of a low profile reversible robot with a kinetic energy weapon. Being reversible was a very important factor, because Touro would debut at RoboGames 2006, against several US robots that could easily flip their opponents. We went for the drumbot design, inspired by Falcon's compact size and motor choices, Tekka Maki's sloped front plates beside the drum, and Angry Asp's anti-wedge skids.

We started machining the drum, made out of a 1" thick ST-52 steel cylinder (similar to a 1025 steel, with 0.25% carbon). In 2007 the ST-52 was switched to 304 stainless, which has a much higher impact toughness. We've bolted to the drum two tempered S7 steel bars to work as teeth to catch the opponents.

We machined a double pulley to be fitted to the drum, allowing it to be powered by a pair of A-size V-belts. As discussed in chapter 5, V-belts work as a clutch, sliding during the impact. The drum was mounted to a 1.5" diameter solid shaft made out of tempered 4340 steel, which was later replaced with titanium grade 5 to save weight. The drum was powered by a Magmotor S28-400, the longer and more powerful version of the S28-150, at 24V. In 2006, the motor and drum pulleys had the same diameter, resulting in a drum top speed of about 4,900RPM. In 2007, the diameter of the drum pulley was reduced, increasing the weapon speed to 6,000RPM.

The entire weapon system resulted in almost 35% of the robot's weight. That value was a little over the 30% from the 30-30-25-15 rule, but this is not too bad for drumbots, because due to its small radius the drum needs to be heavy to generate a significant moment of inertia. Besides, the drum can also be considered as part of the armor, since sturdy drumbots can also do a great job as rammers. Several drumbots from various weight classes use about 20% of their weight in the drum, and up to 15% to power them (including shaft, pulleys, belts, bearings, motor and its mounts), adding up to 35% as in the case of Touro.

Touro's drive system is similar to Titan's, it used 2 Magmotors S28-150 with TWM3M gearboxes. We used 6" diameter Colson wheels, instead of Titan's 4" wheels, increasing Touro's top speed in 50%. The drive system ended up using about 15% of the robot's weight, well below the 30% value from the rule. Note that most robots spend something closer to 30% of their weight, rather than 15%, in their drive system. We were only able to reach 15% because we only used 2 wheels, powered by motors and gearboxes with very high power to weight ratios. Any rammer, wedge, thwackbot or overhead thwackbot, which depend a lot more on a robust and powerful drive system, as well as any robot with 4 (or more) active wheels, will need to get closer to the 30% value

to be efficient. Therefore, the 15% value would probably be a lower limit for the drive system weight, which could be enough only for robots with very powerful weapons.

A MS-2 switch (more compact than the Hella key) was used to turn the robot on or off. The weapon motor was controlled by a TW-C1 solenoid. To keep Touro compact, we used Victors instead of OSMCs for the drive system. We developed a small electronic control board specifically to power the TW-C1 and to act as BEC (Battery Elimination Circuit, see chapter 7) for the receiver.

Note in the following pages that we used a braided mesh (in light red, in the center) to organize and to protect the wiring, avoiding shorts due to friction with metal parts from the structure. The entire robot was powered by two 24V NiCd Battlepacks connected in parallel. The entire battery and electronic system added up to about 10% of Touro's weight.

Touro's integrated structure/ armor (see CAD to the right) used 3/4" thick 7050 aluminum walls, covered by a layer of Kevlar and another of titanium Ti-6Al-4V. The wall sections that hold the drum were 1" thick. The top and bottom covers were made out of 1/4" thick 7075-T6 aluminum. Pockets had to be selectively milled in all walls and covers to relieve weight, see chapter 9. A few internal mounts that required high stiffness, but not high strength, were made out of 6063-T5 aluminum extrusions, which were easier to find than 6061-T6 (all aluminum alloys have roughly the same stiffness and density, but very different strengths).

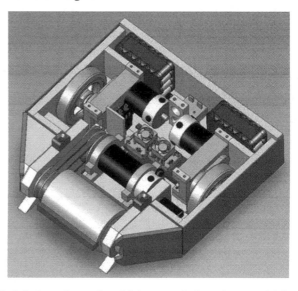

Almost all the screws made use of threaded holes along the thickness of the plates, which simplifies a lot the assembly task, without the need for nuts. Touro has 423 screws, but only 10 nuts (which are used in the Victors and MS-2 switch contacts).

Similarly to Titan, Touro's integrated structure/armor ended up with respectable 40% of the robot's weight, way above the 25% value from the rule. To be able to reach this 40% value, without compromising too much the drive system, weapon and batteries, is not an easy task. These 40% caught the attention of several US builders when they first saw Touro in 2006. A few builders asked us back then if Touro was a lightweight, judging from its size, and a couple asked if it was a heavyweight, looking at the 1" thick walls near the drum. These 40% also help to explain how Touro survived the violent fights against The Mortician in 2006 and Prof. Chaos in 2008.

In summary, a good drumbot might follow a small variation of the weight rule, which would be 20-35-30-15: 20% of the robot's weight in the drive system (a little more than Touro), 35% in the weapon, 30% in the structure and armor (or a little more, but always between 25% and 40%), and 15% in the batteries and electronics (a little more than Touro's 10% to be able to use more batteries). As for other types of robots, certainly other more specific rules can be proposed, however the original 30-30-25-15 rule is always a good starting point.

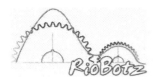 

To make it easier for a new team member to get used with (and service) our robots, we've generated exploded assembly views of most of them using Solidworks, as pictured below for Touro.

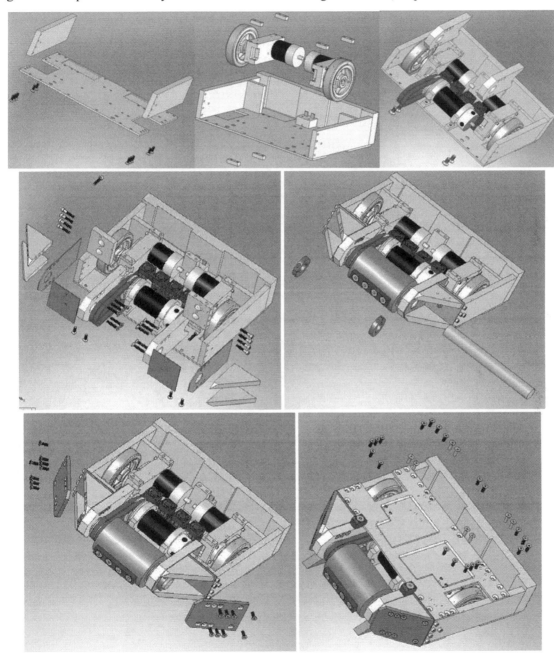

Touro got bronze, gold and silver medals at, respectively, RoboGames 2006, 2007 and 2008. It had also won 5 Brazilian championships until 2008, including the 2006, 2007 and 2008 editions of the RoboCore Winter Challenge. The pictures in the next pages show a detailed anatomy of Touro, as well as several action shots in combat.

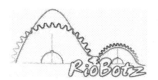

steel drums, just turned in the lathe

mechanical structure ready, now let's wire it

just born in our San Francisco hotel

repairs during a pitstop

our overwhelmed driver learning that Touro had won the RoboGames 2007 gold medal

Touro had to beat Pipe Wench and Sub Zero to get gold in 2007

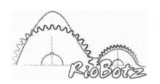

testing the welds from Wiz

destroying the Destroyer

sending The Mortician to the graveyard

telling Pirinah 2 that "size matters not"

making Stewie look like an UFO

playing around with Ice Cube

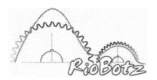

beautiful titanium sparks against Sub Zero

Pipe Wench righting itself

breaking Terminal Velocity's bar

flipping Dolly

Vingador's retirement

Orion 3 getting airborne

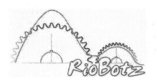 

making Emily fly

chewing Pirinah 3's blue tires

damaging Argus' flamethrower

making TSA Inspected pop a wheelie

getting some air time from Prof. Chaos…

…and giving some too

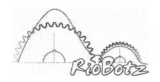

10.7. Mini-Touro

Touro has a father. And it only weighs 1.5kg (3.3lb). That was the (non-official) weight class in an internal combat robot competition at PUC-Rio University, organized by RioBotz for 30 freshman students that were enrolled in the subject "Introduction to Engineering." Each one of the 8 teams (pictured to the right, with their robots) developed during the term a 1.5kg radio

controlled combot using aluminum extrusions, Lexan, toy components, and scrap metal. The competition gathered several students around a small arena in the campus of the PUC-Rio University in November 2005.

RioBotz also developed then a 1.5kg combot, to entertain the audience during the intervals between the fights. So, at the end of 2005, the very first member of the Touro family was born. Named Tourinho ("Little Bull" in Portuguese), this almost-beetleweight was made out of a single 6063-T5 aluminum rectangular extrusion, with a Lexan top cover (pictured below). The radio-control, electronics, NiCd battery and wheels were all adapted from toys. The drum was a scrap piece of pipe with 6 flat head allen screws. The weapon shaft was simply a long 8mm diameter hex screw. The same name Tourinho was later used in our hobbyweight developed in 2006.

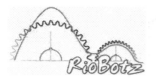

The motors used in Tourinho were quite unique. They were coreless (or ironless) DC motors from Faulhaber, meaning that their rotor did not have an iron core. The structural integrity of the rotor only depended on its windings. In this way, the rotor was hollow, allowing the permanent magnets to be mounted within the windings (as pictured to the right). Without the

iron core, the motor inductance was extremely low, increasing the life of the brushes and commuter, and the rotor inertia can get so low that a few very small models could reach accelerations of up to 1 million rad/s^2 (with an unloaded shaft, of course). The energy loss was very low, almost as low as in brushless motors. Their disadvantages were the tendency to overheat, because of the absence of the iron core to improve the heat exchange, and their high cost.

These motors had been used in a robotic rover project, until their embedded encoders were damaged. It would be more expensive to have them repaired then to buy new motor-encoder systems, so RioBotz basically got them for free. We used 2 of them to drive Tourinho's two wheels, and a third one to spin the drum.

The robot was a crowd pleaser, and with the lessons learned we were able to design Touro, using the scale factor principles described in chapter 2. Certainly it was much cheaper and faster to build our first drumbot with only 1.5kg, learn from it, and only then face the costs and challenges of creating a middleweight version. We've learned several things from building the 1.5kg version. For instance, the use of a single front ground support under Tourinho's drum seemed like a good idea to guarantee that both wheels would always touch the ground. But this made the robot tilt diagonally whenever it hit an opponent, so Touro was later designed with 2 front ground supports. Tourinho also had traction problems with the wheels so far behind, which helped us place Touro's two wheels close to its center of mass. Tourinho certainly saved a lot of redesign time for Touro.

During the building of Touro, we also decided to generate an improved version of its 1.5kg father. Instead of a 6063-T5 aluminum extrusion, we milled a unibody (pictured to the right, see

chapter 2) from a solid 7050 aluminum block. The Lexan cover was replaced by black garolite. We machined a new drum, and replaced the toy NiCd packs with two 11V LiPo batteries connected in series, to generate 22V. From the original Tourinho, we only took the Faulhaber motors, to power the drum and the 1.75" diameter DuBro rubber wheels. This is how our 3lb **beetleweight** Mini-Touro was born, in 2006.

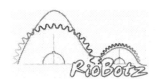 

But the Faulhaber motors we had did not include gearboxes. This meant that the drive system top speed was too high (even though its acceleration was not too bad), and that the drum was not as fast as it could get. So in 2007 we replaced them with B-Series 16:1 gearmotors for the drive system, and with a HXT 2835 2700kv (2700RPM/V) inrunner brushless motor, with a Phoenix 25 speed controller, to power the drum. A 3M-size timing belt was attached to the brushless motor through a timing pulley. This belt fitted inside a smooth groove in the robot's drum (pictured to the right), which allows it to slide during impacts.

The batteries, which were originally connected in series to deliver the required 22V for the Faulhaber motors, were then wired in parallel to provide 11V, in order not to burn the new lower voltage motors, and 1,450mAh. The resulting top speed of drum, considering the speed reduction from the belt, was about 10,000RPM.

Mini-Touro was able to get the gold medal at RoboGames 2006, still in its Faulhaber version. The final match was against the powerful spinner Itsa (pictured to the right).

Mini-Touro faced Itsa again two years later. After a tough final match against the undercutter One Fierce Weed Wacker, Mini-Touro was able to get another gold medal at RoboGames 2008.

Mini-Touro, so full of itself, was later spotted subjugating super-heavyweight Ziggy.

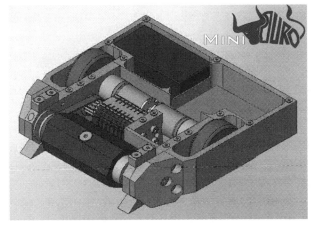

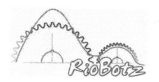

10.8. Tourinho

The RoboCore Winter Challenge featured in 2006, for the first time in Brazil, a **hobbyweight** (12lb) competition. So we decided to create a bigger brother to Mini-Touro. Tourinho (pictured to the right) was born. It is a hobbyweight drumbot, with walls made out of 1/8" thick 2" high 6061-T6 aluminum extrusions, a 2mm (about 5/64") thick 2024-T3 aluminum bottom, and a 12mm (almost 1/2") diameter titanium weapon shaft.

Tourinho originally had a Lexan top cover and an 8" wide ST-52 steel drum spinning at 5,700RPM. The Lexan top cover was later replaced with a 2024-T3 aluminum sheet, to avoid cracking around the countersunk holes. The ST-52 steel was replaced with 6351-T6 aluminum to increase the drum diameter to 2", which was then overvolted to reach an 11,000RPM top speed.

Two surplus Buehler gearmotors had been used in 2006 to drive two 3" diameter Colson wheels, controlled by a Scorpion XL board.

These drive motors were replaced in 2007 with 16:1 36mm planetary gearboxes from Banebots, powered by RS-540 motors, as pictured to the right.

In 2008, the not-so-reliable RS-540 was replaced with Integy Matrix Pro Lathe motors, adapted to the same Banebots gearboxes.

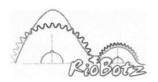

The weapon motor was an 18V DeWalt, powered by a 40A automotive relay. The 16.8V NiCd pack was obtained from removing the top cell from a DeWalt cordless drill battery, reducing its height to less then 2" to fit inside the robot.

In 2007, the NiCd battery pack was changed to two 2,100mAh 3S LiPo batteries (shown in red in the previous picture), connected in series to increase the weapon motor voltage to 22.2V.

In 2008, the relay was switched to a Victor speed controller. The Victor can not only control the speed of the weapon motor, but it also allows the drum to reverse its spin direction if the robot gets flipped over.

Also in 2008, the 2,100mAh LiPo batteries were replaced with new 2,200mAh LiPo with higher discharge rate (in blue in the CAD rendering below). The new batteries were repositioned inside the robot to allow the chassis to be sized down to 11" wide × 9.15" length. With the saved weight, it was possible to increase the thickness of the side walls to 1" in the region where the weapon shaft is supported, as shown below in a Solidworks rendering.

Similarly to Touro, Tourinho also has an assembly guide featuring several exploded views, which is summarized below.

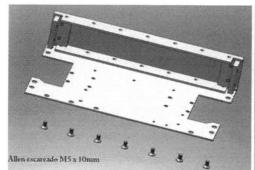

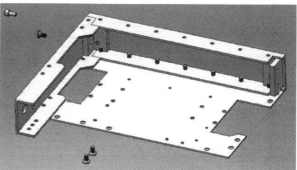

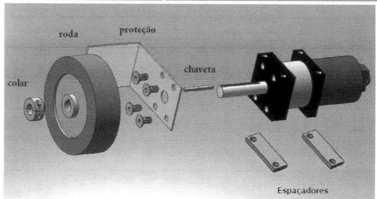

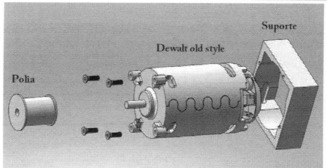

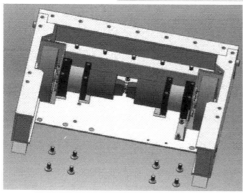

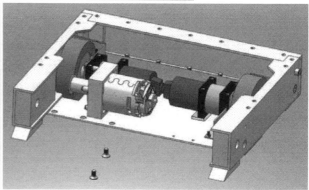

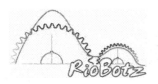

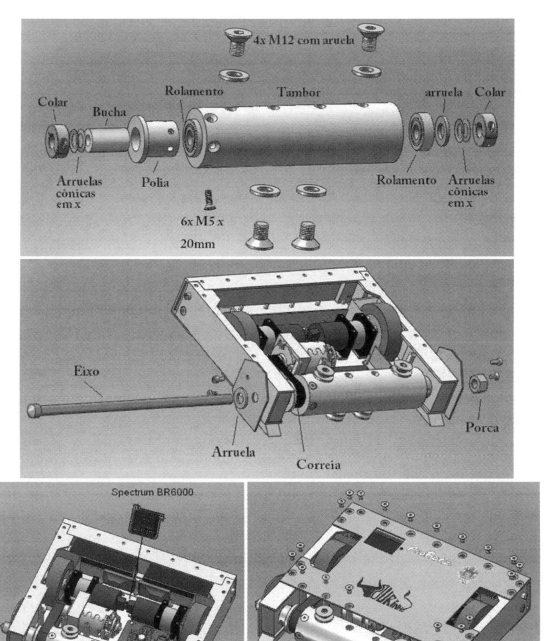

Tourinho became the champion of the 2006 and 2007 editions of the RoboCore Winter Challenge. More pictures from Tourinho are shown next.

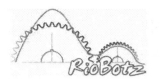

flipping the vertical spinner Agressor

hitting Lasca Bit from Team Proteus

drum vs. drum, against Xpow

launching the shell spinner Butcher

challenging a powerful featherweight

finishing Catatau

LTFD – Little Tourinho's Flipper Drum

in mid-air, trying to self-righten

untangling wet noodles

getting my daughter hooked since she was 3 y.o.

active RioBotz robots back in 2006: middleweights Titan, Touro and Ciclone in the back, and Tourinho (hobby), Mini-Touro (beetle) and Puminha (hobby) in the front

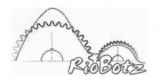

10.9. Puminha

As soon as we finished building Tourinho, we decided to build another **hobbyweight** (12lb) robot to compete at the RoboCore Winter Challenge 2006. We then built, in about a week, our first wedge, called Puminha ("Little Puma" in Portuguese, pictured to the right).

Originally, four surplus Pittman gearmotors (pictured to the right, partially covered by nylon mounts) were attached to 3" diameter Colson wheels, controlled by a single Scorpion XL board and powered by a 16.8V NiCd battery pack. The speed controller was later switched to a pair of Victors.

The side walls were made out of 1/4" thick 6061-T6 aluminum extrusions, and 1/8" thick for the front and rear. Lexan was used in both top and bottom covers, which was later replaced by 2024-T3 aluminum. The wedge was originally a 2mm thick titanium grade 5 plate, borrowed from Touro's armor, attached to the front wall using two stainless steel door hinges.

After a few broken gears from the Pittman gearmotors, we decided in 2007 to switch them to four 16:1 36mm planetary gearboxes from Banebots, powered by RS-550 motors, a little faster than the RS-540 previously used in Tourinho.

In 2008, we switched the RS-550 to even better motors, the Integy Matrix Pro Lathe, using the same Banebots gearbox, as pictured to the right. In addition, the gearboxes were modified following Nick Martin's recommendations, described in the March 2008 edition of Servo Magazine, to avoid any broken last stage pin.

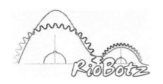

The new drive motors made Puminha become even faster and with more traction. In fact, during the first Brazilian multi-bot match ever, in 2008, Puminha was able to carry both its hobbyweight fellow Tourinho and its featherweight opponent Hulk all over the arena, as pictured to the right (Puminha is hidden under both robots).

By 2008, Puminha had already won two Brazilian championships organized by RoboCore: the 2007 ENECA and the 2008 Winter Challenge. The picture to the right shows Puminha launching the shell spinner Butcher all across the arena during the 2007 final match.

As seen below, the 2mm thick wedge was in a very bad shape by the end of the 2007 season. In 2008, the wedge was upgraded to a 1/4" thick titanium grade 5 plate, attached using heavier duty door hinges. In 2008, the battery was also upgraded to a 4S (14.8V) LiPo Polyquest with 4,500mAh, making Puminha so fast that only little Anakin can drive it.

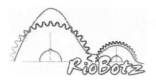

Puminha also features step-by-step assembly instructions, aimed to help new team members. The main steps are shown below, in an old-school style using photographs, instead of using exploded-view Solidworks images such as in Touro's and Tourinho's assembly instructions. Note in the pictures the 2008 version of the wedge, with thickness increased from 2mm to 1/4".

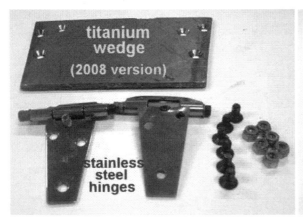

titanium wedge (2008 version)

stainless steel hinges

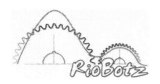

10.10. Touro Light

After Touro's bronze medal at RoboGames 2006, we decided to build another robot to compete in one of the upper weight classes. We decided to test once again our scale factor theories to see if we could create an effective **lightweight** (60lb) based on Touro's design. Touro Light ended up so similar to Touro that many people sometimes get confused about which is which (even ourselves).

Since Touro is 2 times heavier than Touro Light, their scale factor should be the cube root of 2, which is 1.26. The actual external dimensions of Touro's chassis, without the drums, wheels, and front ground supports, are 20.3" (width) × 19.25" (length) × 4.5" (height). Considering the 1.26 scale factor, these values would translate to 16.1" × 15.3" × 3.57", very close to Touro Light chassis' actual dimensions 15.6" × 15.9" × 3.50".

Touro Light used two 36:1 42mm Banebots gearboxes, powered by 14.4V Mabuchi RS-775 motors, to drive its two 5" diameter Colson wheels. Even with the speed controllers trimmed to a 20V limit, the overvolting of the 14.4V RS-775 motors caused them to overheat. This overheating forced us to replace them almost every 2 matches during RoboGames 2007.

In 2008, the RS-775 were upgraded to 18V DeWalt motors, adapted to the same Banebots gearbox, as seen in the PAD (Powerpoint Aided Design) drawing on the right. However, the more powerful DeWalt motors ended up causing the planetary gear pins from the last stage of the gearbox to break at almost every match during RoboGames 2008. This problem should be solved in the future by making modifications to the gearbox, increasing in about 1mm the diameter of the last stage pins, as recommended by Nick Martin from Team Overkill.

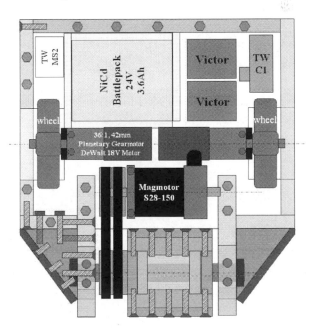

The robot's drive system ended up so light that it was possible to beef up the structure. Touro Light's integrated

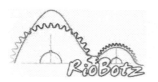

structure/armor had basically the same thicknesses as its bigger brother Touro. It used 3/4" thick 7050 aluminum walls, and the top and bottom covers were made out of 1/4" thick 7075-T6 aluminum. The wall sections that held the drum were 1" thick. Similarly to Touro, pockets had to be selectively milled in the walls and covers to relieve weight.

The drum was made out of a 1" thick 410 stainless steel cylinder, with two tempered S7 steel bar teeth. We machined a double pulley to be fitted to the drum, allowing it to be powered by a pair of 3L-size V-belts. The drum was mounted to a 1"

diameter titanium grade 5 shaft. It was powered by a Magmotor S28-150 to reach about 6000RPM.

To keep down Touro Light's development cost, we tried to use in its design several spare parts from Touro. They both used identical front ground supports, two Victors to control the drive motors, a TW-C1 solenoid to power the weapon, a MS-2 switch, and the same 24V NiCd Battlepacks (though Touro Light only used one pack instead of two). In addition, both used Magmotors S28-150, Touro Light for the weapon and Touro for the drive system. These shared components helped to keep low the number of spare parts needed in a competition, saving us a lot of excess baggage fees when traveling to overseas events.

Touro Light ended up getting the gold medal at RoboGames 2007, together with its big brother Touro (pictured to the right), both undefeated in their weight classes.

The photos below show Touro Light in action in 2007 and 2008.

cooling down Texas Heat

flipping Crocbot

ripping off Conniption's drivetrain

pounding Herr Gepoünden

facing K2 in the RoboGames 2007 final

launching the Rocket

10.11. Micro-Touro

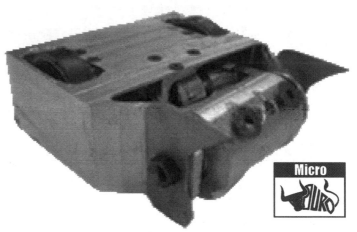

In early 2007 we decided to build Micro-Touro (pictured to the right), the first Brazilian **antweight** (1lb) combat. We tried to incorporate in its design several features from our beetleweight Mini-Touro, such as a unibody chassis, milled from a single 7050 aluminum block, brushless weapon motor, and LiPo batteries. However, we were not so careful with the scale factor between beetles and ants, which is 1.44 (cubic root of 3). Micro-Touro's chassis ended up with basically the same height as Mini-Touro's, instead of 1/1.44 = 69.4% of that value. Such big height makes it easier to get hit by opponents and to get flipped over. A drum with a large diameter was necessary to match the robot's height, which forced us to use aluminum in it (6351-T6 alloy), instead of steel, in order to save weight. The drum's teeth were class 12.9 flat head allen screws. The weapon shaft was simply a long 6mm diameter hex screw. Two titanium strips were used as front ground supports, working as well as armor. The top cover was made out of black garolite.

Two Sanyo 50:1 micro-geared motors were used to drive two 1.5" diameter rubber foam Lite Flite wheels, controlled by two Banebots BB-3-9 speed controllers and a micro-receiver. One 700mAh 3S (11.1V) LiPo battery was used to power the entire robot.

The weapon used a LittleScreamers "DeNovo" micro outrunner brushless motor (pictured to the right), with 1,250kv (RPM/V), capable of sustaining 11A, powered by a hexTronik PRO 10A speed controller. A 2M-size timing belt was attached to the brushless motor through a timing pulley. This belt fitted inside a smooth groove on the side of drum, which allowed it to slide during impacts.

Micro-Touro didn't do well in the 2007 RoboGames. The 1.5" diameter wheels ended up leaving a very low ground clearance. This caused Micro-Touro to frequently get stuck due to the floor deflections of the small combat arena. This will be taken care of in the future using a shorter chassis, which will also help lowering the robot's center of mass, making it more difficult to get flipped over.

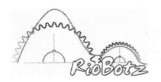

10.12. Touro Jr.

As the Brazilian hobbyweight robots were getting increasingly powerful, we've realized that Tourinho's 1/8" thick side walls started to look very thin. Keeping that in mind, we designed in mid-2007 our third **hobbyweight** (12lb), named Touro Jr. To be able to increase the thickness of the structure/armor, we faced the challenge to reduce the overall size of the robot. The result was a drumbot chassis with very short length (about 8") and height (1.75"), and respectable 5/8" thick 7050 aluminum walls. Nevertheless, pockets had to be selectively milled in the walls to relieve weight. The robot ended up a little wide, about 12.5", to be able to fit the drivetrain gearmotors.

Two 16:1 36mm planetary gearboxes from Banebots, powered by RS-540 motors, were used to drive two 3" diameter Colson wheels. Two Banebots BB-12-45 speed controllers were used in the drive system.

The 6351-T6 aluminum drum used flat head allen screws as teeth. It was powered, through a pair of 2L V-belts, by a Feigao 540-06XL 2779kv (RPM/V) inrunner brushless motor. The brushless speed controller was a hexTronik PRO 120A, powered by a 4S (14.8V) 4,500mAh Polyquest LiPo battery. Considering the V-belt speed reduction, this 2,779RPM/V brushless motor, in theory, would be able to spin the drum up to 30,000RPM. The actual top speed was certainly lower due to bearing friction and air resistance, but it was still so high that our tachometer was not

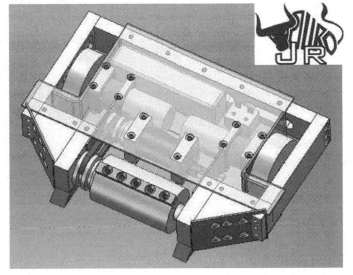

able to measure it. This very high speed makes it difficult to launch the opponents. The drum teeth, instead of biting into the other robots, end up just grinding them. During the weapon tests, it was not easy to bite into aluminum blocks. But when it did bit, the impact was so high that the windings from the Feigao brushless motor detached from the can and broke off the speed controller contacts. We're currently looking for a replacement motor for the weapon, one with a lower RPM/V value.

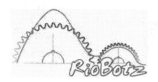

10.13. Touro Feather

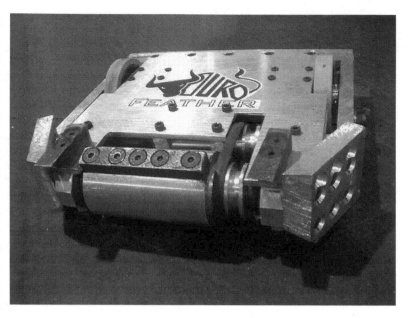

We started to build Touro Feather (pictured to the right) in early 2008, as soon as it was announced that the RoboCore Winter Challenge would debut a **featherweight** (30lb) class. Touro Feather was basically a longer and heavier version of our hobbyweight Touro Jr. Its structure/armor was made out of 5/8" thick 7050 aluminum, with 3/16" thick 2024-T3 top and bottom covers. Two 3/4" thick 7050 aluminum plates were mounted diagonally in the front to work as ablative armor.

The 6351-T6 aluminum drum used tempered S7 steel bars as teeth, the same ones from Touro Light. It was spun by a KB45-08L 2300kv (RPM/V) inrunner brushless motor using a pair of 3L V-belts. The brushless speed controller was a hexTronik PRO 120A, powered by a 4S (14.8V) 4,500mAh Polyquest LiPo battery. Considering the V-belt speed reduction, the theoretical top speed

of the drum would be a little under 30,000RPM. Even including the bearing friction and air resistance, the actual top speed was still so high that the drum teeth ended up grinding the opponents instead of biting them. We had to slow down the drum to less than half of its top speed to be able to bite and launch the other robots. Even at this lower speed, there was enough energy to launch the wedge Titanick a few feet up in the air during RoboGames 2008, as seen on the right.

Another problem we had with the high RPM/V weapon motor was regarding its low starting torque. Brushless motors inherently have a lower starting torque than DC motors, a problem that is exacerbated when their RPM/V is high. Since the 3L V-belts were relatively stiff and their pulleys had small radii, sometimes the drum would simply not start spinning due to such lack of starting torque. We hope to solve this problem in the future by switching the 3L V-belts to the thinner and more compliant 2L type.

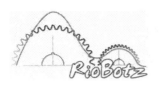

Touro Feather used two modified 12:1 42mm Banebots gearboxes with RS-775 motors for its drive system, controlled by two Victors, see the CAD drawing to the right. These RS-775 motors from Mabuchi were not the common 14.4V version. They were an 18V model, used in very old DeWalt cordless drills, borrowed from the drivetrain of our retired middleweight Ciclone. A MS-2 switch was used to turn the robot on or off.

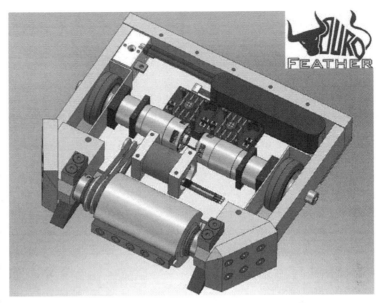

Touro Feather used two 4" diameter orange wheels from Banebots. The color denotes the wheel hardness: green for 30 Shore A (too soft), orange for 40 Shore A (soft) and blue for 50 Shore A (a little soft). Their low cost and finished keyed bore make them an attractive option. We've tested the three types. On a clean floor, the green type provided the best traction, followed by the orange and then the blue.

However, there are a few catches if using them in combat. Because both green and orange types were very soft, the dirt on the combat arena floor would stick very easily to them, compromising traction. They would also get worn out very quickly during aggressive driving tests. In all three types, the polypropylene cores were thinner than Colson wheel cores, which could lead to a low resistance to direct hits.

Finally, a critical feature was the poor bonding of their rubber tread to the polypropylene core. A shallow cut from the blade of the vertical spinner Hulk, at the RoboCore Winter Challenge 2008 final, was enough to rip off the entire orange rubber tread from the black core. The remaining rigid

black core (pictured to the right), besides barely touching the ground, was not able to provide any traction. Fortunately, we could count on Daniel's ability to drive with only one wheel to win the match. Colson wheels under similar conditions would only break locally (as pictured to the right) and still be functional, as we've experienced several times against other spinners. Until those issues are corrected, we'll be using Colson wheels in Touro Feather, not

only due to their better resistance, but also due to the higher 60 Shore A hardness, better suited for dirty combat robot arenas.

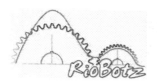

Despite these drivetrain and weapon problems, Touro Feather was able to get fourth place in its debut at RoboGames 2008. One month later, it became the champion of the RoboCore Winter Challenge 2008. The figures below show the exploded assembly views of Touro Feather.

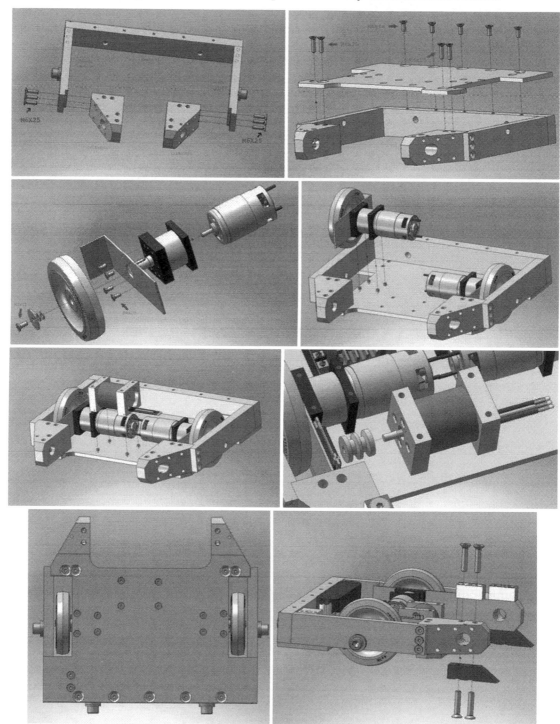

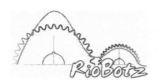

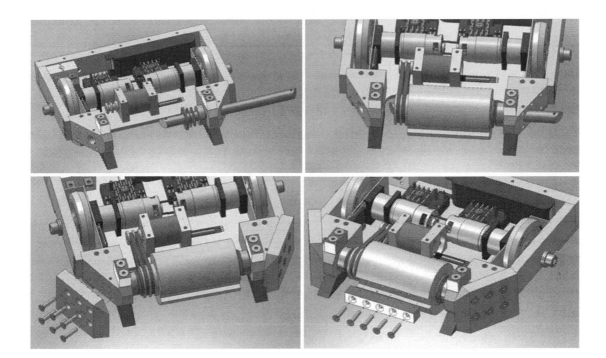

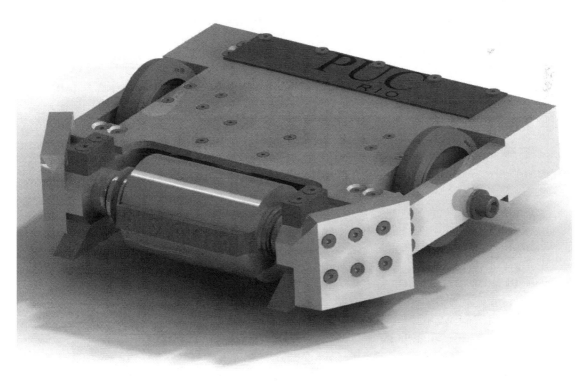

10.14. Pocket

A couple of months before RoboGames 2008, we decided to build our first **fairyweight** (150 grams). In its original design, Pocket Touro was supposed to be a drumbot. Its design was extremely compact to avoid going over the 150 gram weight limit. But this made it impossible to find commercial DC motor speed controllers that could fit inside the robot. Increasing the chassis would surely make it go over the weight limit. We then aimed to develop our own speed controller, small enough to fit inside our vaporbot.

However, the building of Touro Feather was demanding most of our spare time in the University. We finally realized we wouldn't be able to develop the speed controller in time for RoboGames 2008. So, one week before the competition, we completely changed our design, from drumbot to wedge. Since the robot would not be a drumbot anymore, we dropped the "Touro" from its name, calling it simply Pocket.

The fairyweight wedge Pocket (pictured to the right) was built from two carbon fiber (CFRP) rectangles, joined together in a V-shape using four triangular pieces of balsa wood and some Gorilla glue. A strip of CFRP was used as the robot's rear wall, glued as well to the wood. Two triangular pieces of 1/16" thick titanium grade 5 sheet were glued to the outer wooden triangles to work as side armor, as pictured to the right.

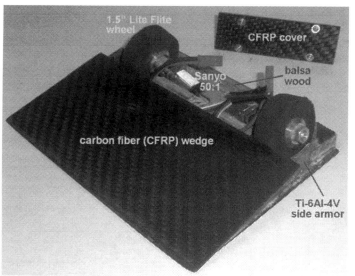

Pocket used two Sanyo 50:1 micro-geared motors to drive two 1.5" diameter rubber foam Lite Flite wheels. The motors were controlled by two Banebots BB-3-9 speed controllers, powered by a 250mAh 2S (7.4V) LiPo battery and connected to a 75MHz nano-receiver. The drive system was held in place with the aid of a CFRP cover, pictured above.

Pocket didn't do well at RoboGames 2008. It had a hard time getting under other fairyweights, because its carbon fiber wedge was not sharp enough. Anyway, it was a great learning experience. Building such a light robot is an extremely challenging task. Every part must be carefully planned and weighed, even the wires and the amount of solder used in the electronics.

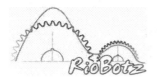

Conclusions

I hope this tutorial is useful for the entire combat robot community, as well as for other builders of competitive robots in general. A great thing about such competitions is that they promote hands-on learning and exchange of information among the teams, bringing together people with completely different backgrounds. This is the spirit behind this tutorial. We would appreciate receiving any corrections, suggestions and contributions, through the "RioBotz Combot Tutorial" RFL Forum post, so that the text can evolve into next versions. Every contribution will be acknowledged. This is just version 2.0, we'll always try to go deeper into the most interesting subjects and keep the content updated with recent technology advances.

Try not. Do, or do not. There is no try.

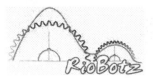

FAQ - Frequently Asked Questions

The questions below were taken from the Forum at www.robocore.net

Design Fundamentals

I would like to know the best way to build a very basic combat robot made out of Lego.
You'll probably be able to build at most a hobbyweight or a lighter robot. You would need to bond the pieces with professional (24 hour) epoxy, otherwise the robot would fall apart after the first impact. Even so, the plastic pieces would not resist the tool steel and titanium weapons that most of the teams use nowadays, you would need some metallic armor. As for the electronics, you would need to adapt the RCX or NXT control module to a radio-control system. You'll probably need more powerful motors than the ones from Lego, for that it would be necessary to add power electronics to amplify the outputs of the RCX/NXT. The VEX system, used in the FIRST competition, is similar to Lego, but its parts are made out of metal and joined by screws, and it already comes with radio and receiver. But, for instance, its wheels and gears are made out of plastic, which would be a weak point.

How can I make a tank tread?
There are basically two types of tank treads: the ones used in toys (made out of rubber) and the ones from war tanks (made out of metal). They are completely different. Don't use the toy ones, they fall off easily, and the rubber tread would stretch out too much or even rupture when applying high torques. A possible solution is to use timing belt pulleys as driving wheels and double-sided timing belts (pictured to the right) as the tread. Several timing belts have steel wire reinforcement, which would prevent the stretching issue. The treads used in war tanks are made out of several articulated steel parts, requiring a lot of machining knowledge to build.
They are expensive and, even so, a powerful spinner could easily knock them off. The best bet is to use wheels. If you want more traction, use 6 wheels, with all-wheel drive, using a system of pulleys or sprockets, or even use 6 motors for redundancy.

What are the advantages of using a rubber tire as armor?
A rubber tire creates a good protection against blunt spinners, working as a damper, but not against very sharp blades. There are not many advantages against other types of robots. Wedges usually get underneath tire-robots very easily. Vertical spinners and drumbots have a greater advantage, because their weapons will grip better onto the tire to fling it high. If you intend to use the tire as both armor and structure, then install some metallic protection layer between the tire and the robot's interior, to help shield it against perforating weapons.

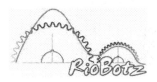

What is the maximum weight of a weapon so that the robot does not get too sluggish?

Follow the 30-30-25-15 rule discussed in chapter 2. The total weight of the weapon, including its motor and accessories, should not be much above nor below 30% of the robot's weight. This is not an exact rule because it depends on the robot type (for instance, 35% might be a good value for drumbots), but certainly 10% of the weight would be too little, and more than 50% would be too much.

Is there any limitation regarding the number of weapons that a robot can have?

No, but it is probably better to use only one. Unless the weapons work together on the opponent, at the same time, as it was discussed in chapter 2. Or if the secondary weapon is a wedge.

I wanted to know if there's any middleweight with dimensions 2m × 2m × 2m (6.6ft × 6.6ft × 6.6ft, the size limit for middleweights in Brazilian events) or anything close to that.

No middleweight would get even close to such size, it would be very fragile. A hollow cube made out of aluminum with those dimensions and wall thickness of only 1mm would have 67kg (148lb)! And this would happen without any parts inside it. The robots need to be compact so that their structure and armor can be thick.

Which is the best software to draw robots?

The most used in the US in combat robotics are probably Solidworks and Rhyno3D. The 3D modeling capabilities of those programs are better and easier to use than in Autocad.

Motors and Transmissions

I'd like to know how other teams build such fast robots with such small motors. Which motor do you recommend to power my robot's drive system?

Watch out for the main indicators of motor performance: the ratio between the maximum power and the motor weight is one of them, and the ratio between I_{stall} and I_{no_load} is another. Compare these values with the ones from the motors listed in chapter 5. Depending on the motor, it is possible to double the input voltage, multiplying the power by 4. But there's a chance of overheating, so limit such overvolting to tests that take no longer than 3 minutes, and check if the speed controller can take so much current (especially if the motor stalls during a match). The S28 series Magmotors are an exception, they're already optimized for 24V, so use them at most at 36V, not 48V, unless you have a current-limiting system. Test a lot, and always keep spare motors.

Does anybody know a mechanical trick to increase the power output of a motor?

There is no mechanical trick, the first law of thermodynamics, which deals with the conservation of energy, doesn't allow that. What you can do is to align or lubricate your transmission system to reduce friction, reducing power losses. But there is no way to mechanically increase its power unless you provide more power, for instance, by increasing the input voltage of the motor.

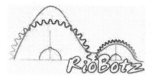 

Are the Bosch motors bi-directional? Their datasheet says they only work in one sense.

All permanent magnet DC motors work in both senses without problems, it is enough to invert the input connections. However, these motors usually have advanced timing, where the permanent magnets of the stator are rotated with respect to the brushes, turning faster in one sense than in the other (see chapter 5). In a few cases, the speed difference is so large that the manufacturer recommends that it is used only in the faster direction. You can reverse them, but they will be slower.

What do you need to power up a very high speed spinner bar from a middleweight?

You need a powerful motor, preferably using NiCd batteries instead of NiMH. SLA batteries could be a good option for heavier classes. Lithium batteries such as A123 would be the best choice, however the cost is higher. The stored energy depends on the bar, but typical speeds for middleweight bar spinners can go up to 3,000RPM, such as in our robot Titan.

Which motor can I use to drive a hobbyweight?

The gearmotors from Pittman and Buehler are good and inexpensive choices, if bought second-hand. Power drill motors are also a great choice, especially for the weapon system. Disassemble an 18V drill and you'll get a motor, gearbox, battery and charger for a relatively low cost. Stock motors such as the 540 and 550 series are also a great choice, but you'll need to gear them down.

What does it mean to have a servomotor with 4 kg×cm torque?

This servo is able to bear, for instance, the weight of a 4kg mass hanging 1cm away from the output shaft. Or a 2kg with a leverage arm of 2cm.

I want to know how to use a step motor.

A step motor is a brushless electric motor that can divide a full rotation into a large number of steps. The motor's position can be controlled precisely, with the resolution of one angular step, without the need for position sensors. Hence, they are a good option for open-loop position control. But they are not the best option for speed control, which is what you usually need in combat. Their torque and power are relatively low if compared to a DC motor of same weight. Most electronic systems that power step motors do not tolerate the high currents required in combat. Almost all the electric motors used in combat are either brushed or brushless permanent magnet DC motors.

Can anybody send me a scheme/drawing of how to build a lifting arm for a hobbyweight, using an electric motor instead of a pneumatic cylinder?

To build an electric lifter, you can use a rack and pinion system coupled to an electric motor to convert rotary motion into linear movement. A few gearmotors have an endless screw that drives a cursor, translating it forward and back. You can also use a 4-bar mechanism, such as the one from BioHazard or Ziggy, see chapters 5 and 6. This mechanism could be powered by either linear or rotary electric motors.

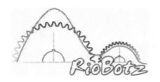 

Electronics

With the relays I'm using, my motors can only move forward.
Use an H-bridge. It can be implemented, for each motor, using 4 relays with single contacts or with 2 relays with double contacts. The scheme is in chapter 7. However, this "bang-bang" control is not the best option. Try to implement PWM (chapter 7), it will be easier to drive the robot and it will make the motors and electronics last longer.

How can you control a 12V motor with a 5V receiver?
You need to use a speed controller, see chapter 7.

Which speed controller is the best?
Chapter 7 describes OSMC and Victors, they are probably one of the best options available. For smaller robots (hobbyweights or under), the Scorpion XL and HX are also good options. Building your own battle-proof speed controller is very challenging, but you'll learn a lot in the process.

Does the Victor controller brake the motor?
Victor has a jumper to choose between coast (not braking) or brake. In the brake mode, Victor shorts the motor leads, turning them into generators, which will then dissipate energy from its internal resistance in the form of heat.

Does hot glue affect the RC boards?
No, hot glue guns are a great option to protect your electronics against shorts and to avoid loose contacts and screws in the electric system. Avoid using hot glue on components that need to dissipate heat, such as FETs.

Batteries

Which is the perfect and cheap combination for high power batteries? What type of batteries should I use?
Especially for middleweights, perhaps the best solution is to use NiCd batteries. The cheap solution is to use SLA batteries (AGM or gel). You can get high currents from SLA's, but for them to last the entire match you will end up adding a lot of weight to the robot. Good quality SLA's might be a good option for the heaviest robots such as super-heavyweights, especially due to cost issues. Unfortunately there is no powerful, cheap and light battery, you have to choose two of them: powerful and cheap (high capacity SLA's), light and cheap (low capacity SLA's), or powerful and light (NiCd, NiMH or lithium).

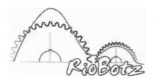 

In a few technical datasheets it is written "discharge rate: up to 3C." What does the "C" stand for?

Such 3C means that the battery tolerates, without problems, a discharge current in A equal to 3 times its measured capacity (C, then the name 3C) measured in A·h, as explained in chapter 8. If your battery has 3.6A·h, then it tolerates a discharge current of 3 × 3.6 = 10.8A. This is the same as saying that it can be discharged in 1/3 of an hour (because 10.8A × 1/3 h = 3.6A·h). But combot matches don't last 20 minutes (1/3 hour), they usually last 3 or 2 minutes, therefore the discharge rates for use in combat should be at least 8C, a value that you can find in high discharge NiCd and in most lithium batteries. This means that the 8C battery could be discharged without significantly warming up in 1/8 hour = 7.5 minutes, which in practice allows you to fully discharge it in 3 minutes without overheating. The ideal discharge rate would be at least 20C, to be able to discharge the entire pack in 1/20 hours = 3 minutes.

Can I assemble my own battery pack?

Yes, see the chapter 8. But take care not to overheat the batteries when soldering them together, it is necessary to solder them quickly and with localized heat.

Can I use a 12V car battery with a 45A·h capacity in my middleweight? Is it true that lead-acid batteries are not allowed?

Considering that the de-rating factor (see chapter 8) of SLA batteries is about 0.28 for 3 minutes, your actual capacity would be 0.28 × 45 = 12.6A·h. To completely discharge such battery during a 3 minute match (1/20 hours), your motors would need to draw 12.6A·h / (1/20)h = 252A continuously. This is a lot of continuous current for a middleweight. Use a battery with lower capacity, which will save you a lot of weight. The car lead-acid batteries, which can spill the electrolyte if upside down, are forbidden in combat. They need to have an immobilized electrolyte, such as the gel or AGM types.

Could anybody write a tutorial on how to build a combat robot?

Here it is! I hope this tutorial has helped. Several other tutorials can be found on the internet as well, along with great build reports and other FAQ lists. There's even a Combat Robot Wiki (http://combots.net/wiki). There are also several great forums for further research, such as:

- RFL Forum (http://forums.delphiforums.com/therfl);
- Antweight Forum (http://forums.delphiforums.com/antweights);
- BattleBots Forum (http://forums.delphiforums.com/BattleBot_Tech);
- RoboWars Australia Forum (www.robowars.org/forum); and
- RoboCore Forum ("Forum" link at www.robocore.net; most topics are in Portuguese).

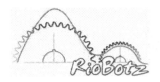

Bibliography

All the books below are recommended readings if you plan to build a combat robot. A lot of the information in this tutorial was learned from them.

	**[1] Art of Electronics, The.** Horowitz, P., Hill, W., 1125 pages, Cambridge University Press, 1989. This is one of the most complete books on electronics, a must-read if you plan to develop your own electronic system.
	**[2] BattleBots, The Official Guide.** Clarkson, M., 272 pages, McGraw-Hill, 2002. This guide presents in an informative way the most famous robots from the BattleBots league, as well as their builders, in an almanac style. The pictures are very interesting. There are a few very basic building tips as well.
	**[3] Build Your Own Combat Robot.** Miles, P., Carroll, T., 416 pages, McGraw-Hill, 2002. Excellent guide about building combots, with great chapters such as the ones on batteries and another about the main tips to design each type of robot. It also deals with special topics such as autonomous and sumo robots.
	**[4] Building Bots: Designing and Building Warrior Robots.** Gurstelle, W., 256 pages, Chicago Review Press, 2002. A good book for beginners, being able to teach even builders with little engineering background. It has good sections on materials, radio systems, and internal combustion engines. The book presents several basic physics equations to exemplify the concepts, making it very instructive. It also includes basic tips to organize an event.
	**[5] Combat Robots Complete: Everything You Need to Build, Compete, and Win.** Hannold, C., 311 pages, McGraw-Hill / TAB Electronics, 2002. With 22 chapters, this comprehensive book is able to approach several subjects in robot design. Naturally, since it covers so many subjects, a few of them are not presented with a very high depth. It includes excellent appendices, and it teaches step-by-step how to build specific robots from three different weight classes: antweight, featherweight, and heavyweight.

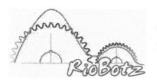

	**[6] Combat Robot Weapons. Hannold, C.**, 288 pages, McGraw-Hill / TAB Electronics, 2003. It's a follow up of the book Combat Robots Complete, it focuses on the main types of weapons and how to design them. It also presents great strategies related to each weapon type. Since it focuses on weapons, this book is able to go much deeper than the previous one from the same author.
	**[7] Electric Motors and their Controls: An Introduction.** Kenjo, T., 192 pages, Oxford University Press, 1991. Excellent introductory book on electric motors, it presents in a simple and didactic way the principles of operation of nearly all the existing types, and how to control them.
	**[8] Fatigue Under Service Loads.** Castro, J.T.P., Meggiolaro, M.A., 1200 pages, in Portuguese, to be published in 2009. It deals in depth with mechanical behavior of materials, presenting the main materials used in engineering, their properties, and how to select them depending on the application. It also deals with failure mechanisms such as fatigue, fracture, yield, plastic collapse, creep and corrosion, among others. Several sections from chapters 3 and 4 from this tutorial, as well as the appendices A, B and C, were adapted from this book.
	**[9] Gearheads: The Turbulent Rise of Robotic Sports.** Stone, B., 304 pages, Simon & Schuster, 2003. A great informative book about the history behind robot combat, from its beginnings.
	**[10] Kickin' Bot: An Illustrated Guide to Building Combat Robots.** Imahara, G., 528 pages, Wiley, 2003. An excellent combot building guide, it is one of the best combat robot books ever written. It thoroughly approaches several subjects from robot conception to combot events. It has great sections about building tools and how to efficiently use them. The pneumatics appendix is also a must-read.
	**[11] Robot Wars: Technical Manual.** Baker, A., 144 pages, Boxtree, 1998. It is a very good almanac with a lot of information on the Robot Wars league. Despite its title, it is not exactly a technical manual about robot building. The pictures are very interesting, and they reflect the success that robot combat has achieved in the United Kingdom.

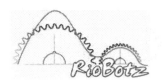 

Appendix A – Conversion among Brinell, Vickers and Rockwell A, B and C hardnesses

HB 3ton	HV	HRA 60kg	HRB 100kg	HRC 150kg	HB 3ton	HV	HRA 60kg	HRB 100kg	HRC 150kg
100	105	-	-	-	311	327	66.9	-	33.1
105	110	-	-	-	321	337	67.5	-	34.3
111	116	-	65.7	-	331	347	68.1	-	35.5
116	121	-	67.6	-	341	358	68.7	-	36.6
121	127	-	69.8	-	352	370	69.3	-	37.9
126	132	-	72	-	363	382	70	-	39.1
131	138	-	74	-	375	394	70.6	-	40.4
137	144	-	76.4	-	388	408	71.4	-	41.8
143	150	-	78.7	-	401	422	72	-	43.1
149	157	-	80.8	-	415	436	72.8	-	44.5
156	164	-	82.9	-	429	451	73.4	-	45.7
163	171	-	85	-	444	467	74.2	-	47.1
167	175	-	86	-	461	485	74.9	-	48.5
170	179	-	86.8	-	477	502	75.6	-	49.6
174	183	-	87.8	-	495	521	76.3	-	51
179	188	-	89	-	514	541	76.9	-	52.1
183	192	-	90	-	534	562	77.8	-	53.5
187	196	-	90.7	-	555	584	78.4	-	54.7
192	202	-	91.9	-	578	608	79.1	-	56
197	207	-	92.8	-	601	632	79.8	-	57.3
201	211	-	93.8	15	630	670	80.6	-	58.8
207	217	-	94.6	16	638	680	80.8	-	59.2
212	223	-	95.5	17	647	690	81.1	-	59.7
217	228	-	96.4	18	656	700	81.3	-	60.1
223	234	-	97.3	20	670	720	81.8	-	61
229	241	60.8	98.2	20.5	684	740	82.2	-	61.8
235	247	61.4	99	21.7	698	760	82.6	-	62.5
241	253	61.8	100	22.8	710	780	83	-	63.3
248	261	62.5	-	24.2	722	800	83.4	-	64
255	268	63	-	25.4	733	820	83.8	-	64.7
262	275	63.6	-	26.6	745	840	84.1	-	65.3
269	283	64.1	-	27.6	757	860	84.4	-	65.9
277	291	64.6	-	28.8	767	880	84.7	-	66.4
285	300	65.3	-	29.9	779	900	85	-	67
293	308	65.7	-	30.9	790	920	85.3	-	67.5
302	317	66.3	-	32.1	800	940	85.6	-	68

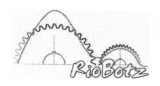 

Appendix B – Material Data

Relative prices per weight of several materials, with respect to 1020 carbon steel.

cement	0.10-0.15	aluminum alloys	3.0-11.0
charcoal	0.15-0.20	natural rubber	3.1-3.2
burning oil	0.45-0.50	regular glass	3.2-3.3
gray cast iron	0.55-0.60	copper alloys	4.0-7.7
reinforced concrete	0.60-0.70	glass fiber (GFRP)	5.2-7.2
soft woods	0.90-1.6	polycarbonate	5.5-5.7
laminated 1020 steel	1.00	stainless steels	6.0-13.0
structural A36 steel	1.08	tool steels	6.0-31.0
press.vessel A515 steel	1.3	nylon	7.1-7.3
low alloy steels	1.0-3.0	acrylic (PMMA)	11.5-12.0
PVC	1.7-1.8	titanium alloys	22.0-130
zinc	2.0-3.8	copper-nickel alloys	27.0-35.0
UHMW	2.1-2.3	nickel superalloys	50+
alumina (Al_2O_3)	2.4-3.9	carbon fiber (CFRP)	200+

Prices in US$/kg for several metals (data from 1998, possibly outdated).

ASTM A36 (plate)	0.50-0.90	Nickel 200	19-25
ASTM A36 (bar)	1.15	Inconel 625	20-29
SAE 1020 (plate)	0.50-1.45	Monel 400	15-17
SAE 1040 (plate)	0.75-1.30	Haynes 25	85-104
SAE 4140 (bar)	1.75-1.95	Invar	17-20
SAE 4140H (bar)	2.85-3.05	Super Invar	22-33
SAE 4340 (bar)	2.45-3.30	Kovar	30-40
304 stainless (plate)	2.15-3.50	C11000 copper	4-7
316 stainless	3.00-6.20	C17200 (Be-Cu copper)	25-47
440A stainless (plate)	4.40-5.00	C26000, C36000 (plate)	3.20-4.85
17-7PH stainless (plate)	6.85-10.00	C71500 (Cu-Ni, 30%)	8.50-9.50
gray cast iron	1.20-3.30	C93200 (bar)	4.50-12.20
malleable cast iron	1.45-5.00	AZ31B (plate, extrusion)	8.80-11.00
Al 1100 (plate)	7.25-10.00	AZ91D (cast)	3.80
Al 2024 T3 (plate)	8.80-11.00	lead	1.20-2.70
Al 2024 T351 (bar)	11.35	solder 60Sn-40Pb	5.50-7.50
Al 6061 T6 (plate)	4.40-6.20	tin	6.85-8.85
Al 6061 T651 (bar)	6.10	zirconium 702 (plate)	44-49
Al 7075 T6 (plate)	9.00-9.70	tungsten (pure)	77-135
Al 356 (cast)	4.40-11.65	molybdenum (pure)	85-115
Ti ASTM grade 1 (pure)	28-65	silver	170-210
Ti 5Al 2.5V	90-130	tantalum (pure)	390-440
Ti 6Al 4V	55-130	gold	9,500-10,250
zinc (pure)	1.20-2.45	platinum	11,400-14,400

Typical values of the impact strength of structural materials.

material	$G_{IC}(kJ/m^2)$	$K_{IC}(MPa\sqrt{m})$
pure ductile metals	100-1000	100-450
ductile low carbon steels	100-300	140-250
high strength steels	10-150	45-175
titanium alloys	25-115	55-115
aluminum alloys	6-35	20-50
glass fiber (GFRP)	10-100	20-60
carbon fiber (CFRP)	5-30	32-45
wood, $\perp$ to fibers	8-20	11-13
polypropylene (PP)	8	3
polyethylene (PE)	6-7	1-2
reinforced concrete	0.2-4	10-15
cast irons	0.2-3	6-20
wood, // to fibers	0.5-2	0.5-1
acrylic (PMMA)	0.3-0.4	0.9-1.4
granites	~0.1	1-3
Si_3N_4	0.1	4-5
cement	0.03	0.2
glass	0.01	0.7-0.8
ice	0.003	0.2

Values of E/ρ, $E^{1/2}/\rho$, $E^{1/3}/\rho$, S/ρ, $S^{2/3}/\rho$ and $S^{1/2}/\rho$ of a few materials, where E is the Young Modulus (in GPa), S_u is the rupture strength (in MPa), and ρ is the relative density.

material	E/ρ	$E^{1/2}/\rho$	$E^{1/3}/\rho$	material	S_u/ρ	$S_u^{2/3}/\rho$	$S_u^{1/2}/\rho$
steels	26	1.8	0.8	1020 steel	56	7	2.7
				304 stainless	77	9	3.1
				4340 steel	184	16	4.8
				5160 steel	196	17	5.0
				S7 steel	251	20	5.7
				18Ni(350)	305	23	6.1
aluminum alloys	26	3.0	1.5	Al 2024 T3	174	22	7.9
				Al 6061 T6	115	17	6.5
				Al 7075 T6	196	24	8.4
titanium alloys	25	2.3	1.0	Ti-6Al-4V	224	22	7.1
magnesium alloys	25	3.7	2.0	AZ31B-H24	143	23	9.0
				ZK60A-T5	169	25	9.6
beryllium alloys	164	9.4	3.6	Be S-200	415	45	15
polycarbonate (PC)	2	1.3	1.1	PC	54	13	6.7
Delrin	2	1.3	1.0	Delrin	54	13	6.2
UHMW	0.7	0.9	0.9	UHMW	43	13	6.8

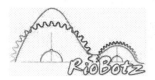

Appendix C – Stress Concentration Factor Graphs

K_t is the ratio between the notch root stress σ_{max} and the nominal stress σ_n (or between the notch root shear stress τ_{max} and the nominal shear stress τ_n). The nominal stresses σ_n are defined in each graph. The notch root stresses are therefore $\sigma_{max} = K_t \cdot \sigma_n$, which can be used for design against yield, fatigue, etc.

(i) holed plates subject to a traction force **P** or bending moment **M**:

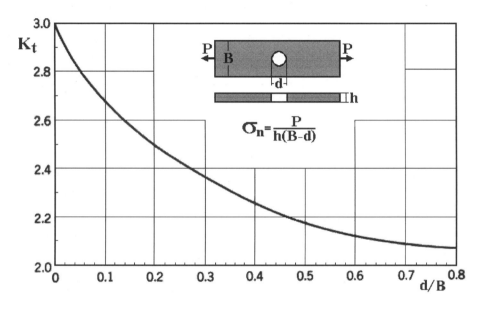

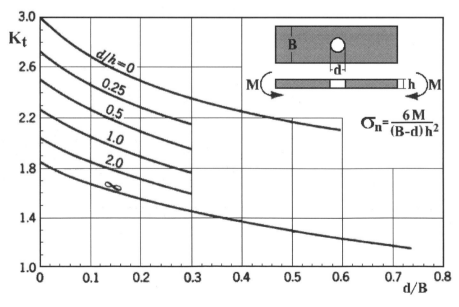

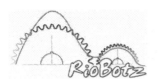

(ii) notched plates subject to a traction force **P** or bending moment **M**:

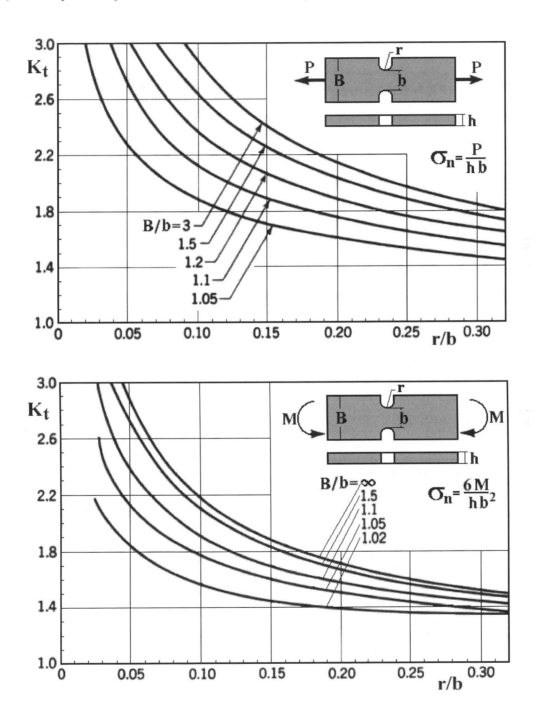

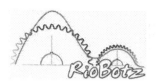

(iii) filleted plates subject to a traction force **P** or bending moment **M**:

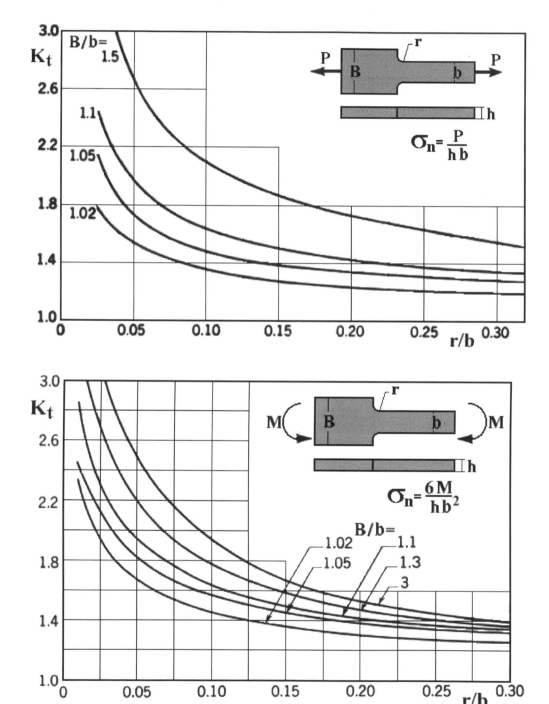

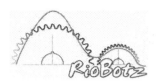

(iv) notched shafts subject to a traction force **P**, bending moment **M**, or torque **T**:

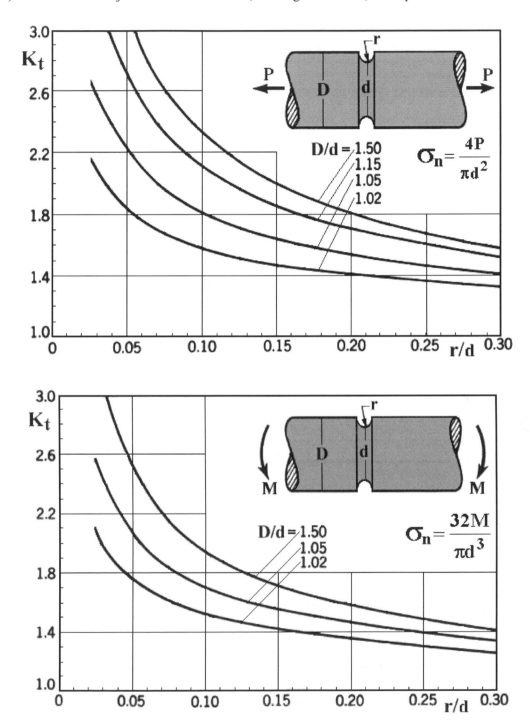

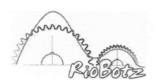

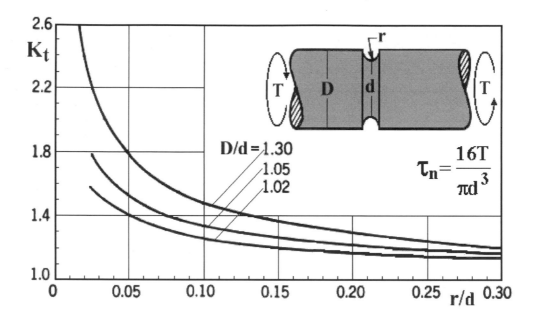

(v) filleted shafts (shoulders) subject to a traction force **P**, bending moment **M**, or torque **T**:

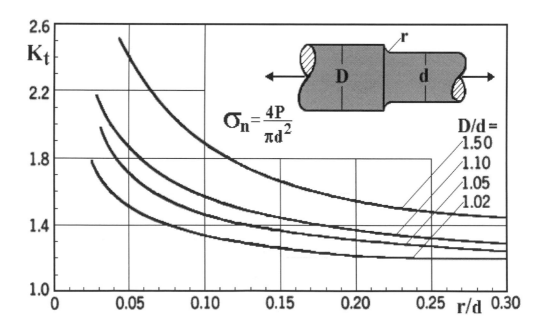

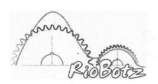

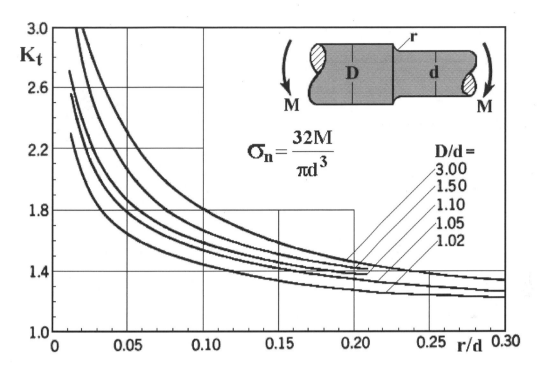

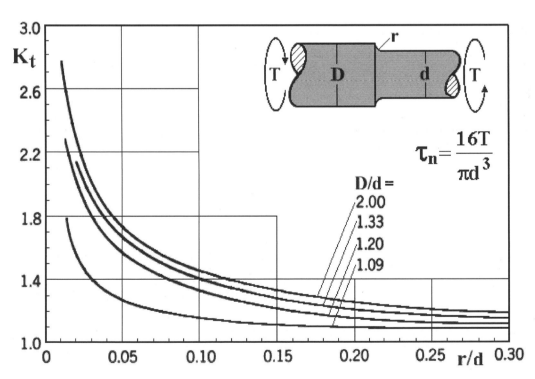

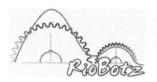

(vi) holed shafts subject to a traction force **P**, bending moment **M**, or torque **T**:

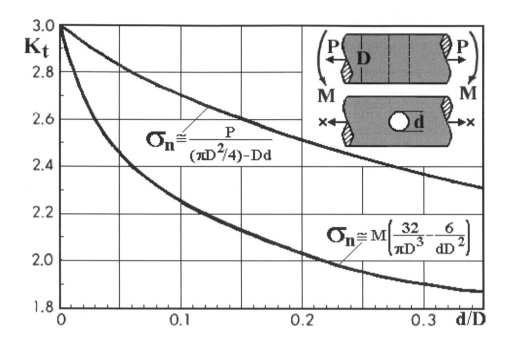

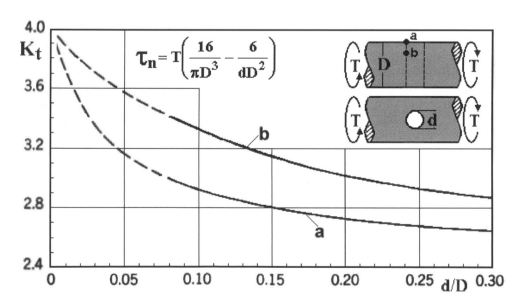

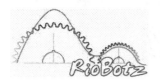 

Appendix D – Radio Control Channels and Frequencies

27 MHz band (air / cars / boats)	72 MHz band (continued)	75 MHz band (cars / boats)
26.995 MHz - Channel 1. Brown		
27.045 MHz - Channel 2. Red		
27.095 MHz - Channel 3. Orange	72.310 MHz - Channel 26	75.410 MHz - Channel 61
27.145 MHz - Channel 4. Yellow	72.330 MHz - Channel 27	75.430 MHz - Channel 62
27.195 MHz - Channel 5. Green	72.350 MHz - Channel 28	75.450 MHz - Channel 63
27.255 MHz - Channel 6 - Blue	72.370 MHz - Channel 29	75.470 MHz - Channel 64
	72.390 MHz - Channel 30	75.490 MHz - Channel 65
	72.410 MHz - Channel 31	75.510 MHz - Channel 66
50 MHz band	72.430 MHz - Channel 32	75.530 MHz - Channel 67
(air / cars / boats)	72.450 MHz - Channel 33	75.550 MHz - Channel 68
50.800 MHz - Canal RC00	72.470 MHz - Channel 34	75.570 MHz - Channel 69
50.820 MHz - Canal RC01	72.490 MHz - Channel 35	75.590 MHz - Channel 70
50.840 MHz - Canal RC02	72.510 MHz - Channel 36	75.610 MHz - Channel 71
50.860 MHz - Canal RC03	72.530 MHz - Channel 37	75.630 MHz - Channel 72
50.880 MHz - Canal RC04	72.550 MHz - Channel 38	75.650 MHz - Channel 73
50.900 MHz - Canal RC05	72.570 MHz - Channel 39	75.670 MHz - Channel 74
50.920 MHz - Canal RC06	72.590 MHz - Channel 40	75.690 MHz - Channel 75
50.940 MHz - Canal RC07	72.610 MHz - Channel 41	75.710 MHz - Channel 76
50.960 MHz - Canal RC08	72.630 MHz - Channel 42	75.730 MHz - Channel 77
50.980 MHz - Canal RC09	72.650 MHz - Channel 43	75.750 MHz - Channel 78
	72.670 MHz - Channel 44	75.770 MHz - Channel 79
	72.690 MHz - Channel 45	75.790 MHz - Channel 80
72 MHz band	72.710 MHz - Channel 46	75.810 MHz - Channel 81
(air only)	72.730 MHz - Channel 47	75.830 MHz - Channel 82
72.010 MHz - Channel 11	72.750 MHz - Channel 48	75.850 MHz - Channel 83
72.030 MHz - Channel 12	72.770 MHz - Channel 49	75.870 MHz - Channel 84
72.050 MHz - Channel 13	72.790 MHz - Channel 50	75.890 MHz - Channel 85
72.070 MHz - Channel 14	72.810 MHz - Channel 51	75.910 MHz - Channel 86
72.090 MHz - Channel 15	72.830 MHz - Channel 52	75.930 MHz - Channel 87
72.110 MHz - Channel 16	72.850 MHz - Channel 53	75.950 MHz - Channel 88
72.130 MHz - Channel 17	72.870 MHz - Channel 54	75.970 MHz - Channel 89
72.150 MHz - Channel 18	72.890 MHz - Channel 55	75.990 MHz - Channel 90
72.170 MHz - Channel 19	72.910 MHz - Channel 56	
72.190 MHz - Channel 20	72.930 MHz - Channel 57	
72.210 MHz - Channel 21	72.950 MHz - Channel 58	**700MHz band**
72.230 MHz - Channel 22	72.970 MHz - Channel 59	
72.250 MHz - Channel 23	72.990 MHz - Channel 60	
72.270 MHz - Channel 24		**2.4GHz band**
72.290 MHz - Channel 25		

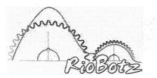 

Appendix E – Judge Dave's Guide

This appendix reproduces a very famous guide written by Judge Dave Calkins. He has judged over 2000 robot combat matches between 2000 and 2009, in three continents and around 100 events. He has organized 30+ events, and built more robots than he can name. The rules below come from all that experience. If you want to win at the sport of robot combat, you need to read these rules and live by them.

Rule # 1. Read Carlo's Law and Live By It.

Carlo Bertocchini, builder of Biohazard (three times Heavy-weight Champion), wrote an article for Battlebots on-line, and he put it better than I (Dave) can, so here it is (abbreviated for space - reprinted with permission of Battlebots, Inc.):

"Finish your robot before you come to the competition!"

This seems too obvious to even mention, let alone to place at the very top of a list of secrets to success. Besides, so what if you just have a little wiring to do, or that one last gear to mount? It's 3:00 AM you have been working for 36 hours straight and it is almost time to load the robot in the car and get to the competition. You can do that last bit of wiring in the pits, right? Well, the fact is, if you are in this situation, you have probably already insured a loss in the BattleBox.

"Moe" got in late with just a few "minor" adjustments left to do. He spent his whole day trying to work on the robot while at the same time getting through all the required procedures. He was somehow able to convince the inspectors that his robot was safe and able to move under its own power.

Now it is the first day of competition. Moe is still working on his robot after having slept just two hours last night under the pit canopy. Moe found that the minor adjustments took longer than he expected, and he found a few more changes that just had to be made.

Now Moe is called to battle. He drags his toolbox to the battle queue and continues to wrench on his robot. Time to fight. He sets up his robot and steps out of the arena. The box is locked. The blue driver is ready. "Red driver are you ready?" "Uh, I guess."

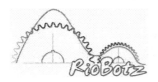

Three thousand people watch anxiously from the stands as the starting lights count down to green.. Three thousand people watch with disappointment and ill-disguised hatred as Moe walks in to the box to collect his robot, which never got off the red square.

What I am suggesting here is not easy. It takes good planning, discipline, and lots of free time to get the job done. But here is one simple way to guarantee that your robot will be finished: If it looks like time is running short, rather than drive or fly hundreds of miles just to work on your robot in a tent, why don't you just leave your robot home! Come and see the show, have the time of your life, learn a few things, and set your sights on doing well and enjoying the next competition."

Think the above is silly? Happens one in twenty times during the prelims. I don't even get annoyed anymore. I just use the extra time to grab another soda. Don't make me say "I told you so." Your bot should be finished two months before the event. If it's not, just go to the event without your bot and watch and learn.

Rule # 2. Practice Driving.

Sounds obvious, I know. So do all the other rules. But less that 60% of contestants obey this rule. So many competitors spend countless hours making tiny little changes to their robot to make it "perfect" that they don't spend any time driving it. I am not making it up when I say that I've met at least 100 competitors who have had less than an hour of driving practice before they step into the arena.

Listen guys, do you think Barry Bonds woke up one day, walked onto the field and became the slugger he is? Thousands of hours in the batting cage. Dale Earnhardt didn't just hop into the driver's seat and start winning at NASCAR. Pay very close attention to this next sentence, because if you want to compete in RoboGames, the ComBots Cup, Battlebots, or any other competition, it's the most important thing I can say to you: The single greatest common denominator to winning is driving ability. Get that?

The first time I saw Jason Bardis (two-time BattleBots champion) and Dr. Inferno Jr. I thought it was a joke. His robot was an old plastic toy with a couple of motors attached. No one thought he'd make it past the first round. But let me tell you, most of Jason's competitor's never got off more than a single blow against him. He drove circles around them. He deftly avoided arena hazards. He struck and dodged -- like the finest boxer. He won time and time again. Watching Gary Gin or Matt Maxham (both multi-event champions) drive Original Sin or Sewer Snake is like poetry in motion. They are one with their remote.

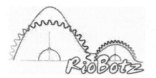 

Spend one hundred hours practicing driving before you ever get to the event. Robot not done yet? Fine. Go spend $20 on a cheap R/C car and drive until your robot's ready. Switch to your bot as soon as the drivetrain is finished, even if the weapon isn't done and the armor isn't on. Spend two full hours each day driving – go find some empty lot, parking garage, or cul-de-sac. Now chase that $20 R/C car around with your bot (let the kid next door drive the car, he's probably a better driver than you anyway.) Make sure you can catch it. Corner it. Out-maneuver it. Dominate it.

Got that down pat? Good. Now disconnect a motor on your robot. Learn to drive with any given motor disabled. Tape a shim to the underside of a corner and see how you can drive. All of these things will happen in the arena, and you can either learn now, or learn then. Your choice.

This is also the time to find out if you're going to burn out a speed controller. These are the weak links of most robots. Make sure that your battery-pack/speed-controller/motor system can handle heavy loads, and if you're over-volting, make sure you don't burn out your speed controller. You will, but it's better to do it now than at an event.

Rule # 3. Be Able to Self-Righten.

It is not a question of if your robot will be flipped over, it is only a question of when your robot will be flipped over. I have seen competitors, their eyes filled with tears as they take their magnificently engineered robot out of the arena after a loss, saying "I was so sure we wouldn't get flipped." Hoo boy…

Guys slide into second in baseball. Wrestlers get body slammed. Quarterbacks get dog piled. Skiers dump skis along a quarter mile path. What makes you so sure you won't get flipped? I've seen at least fifty matches where Robot A was utterly dominating Robot B and would have won by a landslide if it was a judge's decision. And then, by bad luck, an arena hazard, or just a big collision, SPONK! Robot A is upside down. Five-four-three-two-one Knockout! Your robot design must be able to either self-righten (flipper, actuating arm, whatever) or operate upside down.

If you can't self-righten, you'll never make it to the finals. Count on that. If there is any position in which your robot is a helpless kitten, count on it ending up that way at some point during the competition. And don't count on the other guys' freeing you. They're there to win, not help you recover from your short-sightedness.

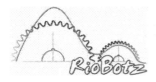 

Rule # 4. Simulate Getting Attacked.

OK, so you've finished your bot with a few months to spare. This is the piece of advice that you are just not going to want to take. The one that's hardest of all to comply with. The one that will make your kids cry.

I want you to go to the hardware store.

Buy the biggest sledgehammer you can find (the really big kind that make you strain when you lift them.)

Now raise it above your robot.

And beat the living hell out of it.

Harder.

Harder.

I SAID HARDER!

Awe, did it bweak? Issums widdle wobot in a big pile uv parts??? Well, I just saved you the indignity of having that happen while 3000 people watched. If your robot cannot survive a good bashing with a sledgehammer and free-fall from 12 feet, it will not last in the arena. Use good 6061 or 7075 Aluminum, Steel, or (preferably) Titanium. My personal philosophy is that Lexan may be bulletproof, but it isn't bot-proof. Make sure you have a good enough infrastructure to support your outer shell.

Ensure that all components are securely mounted. They're going to get knocked around. I long ago lost count of the number of battery packs that I have seen knocked out of a robot and gone flying across the arena because the robot didn't have wrap-around armor. You lose your batteries, you lose the match. It's that simple.

Drop your bot off the roof of your garage. It's a good simulation of what's going to happen when a bot like Toro flips it, or a spin-bot like Morpheus whacks it one. If it can't run after that fall, you need to re-work the guts so that it can withstand that kind of hit.

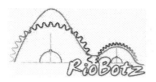 

Even if you're the best driver in the world, you're still going to take lots of knocks (including on the bottom of your bot, so have undercarriage armor as well.) You must be able to survive those hits, and your first match is the wrong time to find out where your weak spots are.

Rule # 5. Have A Weapon System.

Better yet, have two. This is robot combat. You don't play baseball without a bat, you don't go to war without a gun, and you don't become a pro-wrestler without having at least two frontal lobotomies. If you want to beat the daylights out of the other robot, bring a weapon!

Wedges can be very effective, but it's extremely rare for a wedge with no other weaponry to make it to the finals. Watch lots of matches (everything you can from TV, or better still, buy a week pass for an event) and take lots of notes. See what weapons work, and which don't. Think about why things work. Two weapon systems that look identical may operate completely differently, with greatly different results.

Better yet, come up with a new and unique weapon's system. Something that hasn't been tried before. Every time I go to a competition, somebody has brought along a new robot which garnishes lots of ooh's and ah's from the masses – and more than a few "Why didn't I think of that."

OK, now you've thought up a great new weapon system. You've built a prototype, figured out the kinks, and gone on to build the full system. That brings us to:

Rule # 6. Simulate Attacking.

I swear some people show up to a competition having only ever tested their robots on kittens. Sure, it may give your garage a nice new primer coat of kitty juice, but that doesn't mean it will even scratch the paint on another robot.

I walk the pits before competitions and between matches to see who's doing what and how this year's robots are sizing up. During BattleBots 5.0, I saw this very well designed super heavy weight with a horizontal spinning mass (that's our technical term for a big spinning hunk of metal). Except the metal bar had not a single ding on it. You can give something a nice coat of paint, but you can't hide the dings. No scratches. Nothing. On further inspection, I noticed that the bar (which

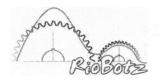 

probably weighed 40 pounds) was held to the rotating shaft with a half ounce pin. An aluminum, 1/32 inch cotter pin. The kind your six-year old niece can bend with her pinky.

"You guys test this against anything?"

"Of course not, it could hurt someone!"

First time that metal bar hit another robot, the pin sheared, the bar went flying and they're as defenseless as my editor is when I miss a deadline after he's given me my advance. [They lost.] If they had spent five minutes in their garage or at some junkyard testing their weapon against a solid object, they would have realized the cotter pin was a weak link and they could have fixed it.

I remember after BattleBots 3.0, Son of Whyachi had gone from a nobody to Heavy Weight Champion (guess which team tested their robot), and come Season 4.0, there were about ten SOW clones. They all looked just like SOW, but not one of them had replicated the parts of the weapon system that made it so effective – the one-foot diameter hardened steel direct drive system. Putting a couple hammers on some pipe doesn't do the trick.

There's a term for this – "Cargo Cult" – it comes from south pacific islanders who got used to planes coming in during world war two and bringing supplies. After the war, the planes stopped coming. So islanders fashioned headsets from coconuts, built runway towers, and made landing lights. But the planes never came. Just because something looks the same, doesn't mean it will work the same. Don't be a cargo cult competitor.

And while you're testing, make sure that you're able to actually push twice the amount of dead weight as the maximum in your weight class. This will be a good simulation of a bot pushing against you. If you can't push that much weight, you're probably going to lose. A great many matches come down to pushing matches (5th round, both bots' weapons systems out, half your armor gone, and a burned out speed controller), so you need to be sure you can win under these circumstances. It's also another time to find out if your speed controller can handle the load, or if it's going to give up its magic smoke.

Rule # 7. Go To A Competition, Watch As Many Matches As You Can, And Take Notes.

And if you're a contestant and you've lost, don't go home and sulk. Go sit in the stands and watch every damned match until the finals are done. I've seen too many crybaby first timers leave

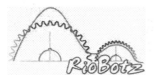

immediately after their first loss (Michael Jordan got cut from his High School basketball team – he did NOT go home and sulk.) You can learn more from other people's victories and mistakes than just your own, so sit back, relax, and enjoy the show.

And take notes. Your memory's not that good, trust me.

Rule # 8. Use Good Batteries, Have Spares, And Make Sure They'll Last Five Full Minutes.

When you start building bots and playing with them, you're going to learn one lesson the hard way (you won't learn it here, trust me.) Batteries get hot. REAL hot. And they take forever to recharge. At least in robot-combat time. So you should have easy access to replace your batteries between matches. Have at least two full sets. One on the charger, and one in the bot. As soon as a match is over, put your just-used batteries on the charger. Just before a match, take the fresher pair off and install them. No matter how good a driver you are, or how well built your robot is, if the batteries don't last the match, you're not gonna win.

Rule # 9. Don't Let The Judges Decide The Match For You.

I judge matches based on the full 3 minutes. The first minute is as important as the last. The fact that you kicked ass the last 20 seconds doesn't make up for the first 160 seconds when your competitor was mopping the floor with your rivets. You want to avoid narrow losses? Want to avoid a screaming match with the officials because they didn't share your belief that your completely out-of-control robot was actually using a strategy? Want to see me taking naps and drooling all over the horrible jacket they make me wear?

Simple. Go for a knockout. Don't let the match go to the last second. Design your robot and operate it so that you kill the other robot. So the referee counts him out. Keep your fate in your hands, don't put it in ours. In judging matches, we're painfully fair and unbiased. The problem is -- you're not unbiased. You want your bot to win and the other team's bot to lose. I don't care who wins. I'm not picking on you when you get the loss, nor have I been bribed to give the other guy the win. It's just that in a close match, we make the call. Both sides think they've won, but only one of them will be correct.

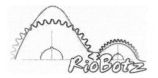 

Avoid hating me. I really am a nice guy. I don't want you to lose. But if the other guy did a better job, he's going to win. But my opinion doesn't have to matter. All you have to do is knock him out. Do not hesitate. Do not get him unstuck. Do not try to avoid extra damage. You are there to win. There's only one way to absolutely ensure that you win.

Go for the knock out.

Every single match.

Besides, I could use the sleep…

Rule # 10. Read The Damned Rule Book

I cannot count the number of builders who have spent hundreds of hours and thousands of dollars building their dream-come-true, and didn't spend one small hour reading the rule-book from cover to cover. You need to do this for every competition; they vary from game to game and change from season to season.

- Know exactly what the judging criteria are. Hint: Number of hits is not part of the judging criteria.

- Know what's allowed and what's not.

- Understand how to pass safety (if you don't pass safety, you don't compete.)

- Understand what can get you disqualified.

If you can't spend the hour reading the rules (don't think that you know them just because you've seen every episode on TV) you probably will never get to compete, much less win.